RADAR AT SEA

Radar at Sea

The Royal Navy in World War 2

Derek Howse

on behalf of the Naval Radar Trust

Foreword by Admiral of the Fleet Lord Lewin

First published 1993 by
MACMILLAN PRESS LTD
Houndmills, Basingstoke, Hampshire RG21 6XS
and London
Companies and representatives
throughout the world

ISBN 0–333–58449–X

A catalogue record for this book is available
from the British Library.

11 10 9 8 7 6 5 4 3
03 02 01 00 99 98 97 96 95

Printed in Great Britain by
Antony Rowe Ltd
Chippenham, Wiltshire

Contents

Illustrations

Foreword

For thousands of years after man made his first voyage astride a log the mariner depended upon his eyes and their range of vision for knowledge of his whereabouts and his safety. As the centuries went by, skills acquired by the three Ls – Lookout, Log and Leadline – identified the true professional seaman; even so the most experienced navigator could miss an important discovery just out of sight over the horizon or come to disaster on an unknown coast in fog, while in war whole fleets could pass in the night, each unaware of the other's proximity. The outcome of the Battle of Jutland, the last great battlefleet action, depended on visibility.

I belong to the diminishing band of Ancient Mariners who knew Life Before Radar – but only just. My very first sea experience was standing a middle watch trick as the starboard lookout in the training cruiser *Vindictive* on passage through the outer reaches of the Thames Estuary. Bewildered by the innumerable lights of every colour and intensity I was quickly convinced of the importance of acquiring the seaman's eye. We soon picked up the tricks of our trade, looking for the masts of the ship hidden by the curve of the earth, judging range by the distance of the sea horizon, keeping station at night by the size of the guide in the field of the binoculars. Then came radar . . .

I was lucky to serve in one of the earlier ships to be fitted with an Air Warning set – the battleship *Valiant*. The Norwegian campaign gave us our first action experience of this remarkable invention. From the Top Secret hut nestling at the foot of the mainmast which only the operators were allowed to enter came a stream of reports of detections made. Received by telephone, these ranges and bearings were solemnly entered in a W/T Operator's log book by the midshipman acting as Assistant Air Defence Officer, but our faith was severely shaken when the occasional stick of bombs fell close alongside without any warning. It was not until we learnt to plot the reports that they began to make sense. Bristling with anti-aircraft guns, *Valiant* was well able to defend herself, but the gunnery depended entirely on visual direction. Using the radar warning as an aid to early sighting of enemy aircraft was all-important.

Later *Valiant* and the new aircraft carrier *Illustrious* joined Admiral Cunningham's Mediterranean Fleet, bringing radar and modern fighter aircraft to a theatre where the Italian high-level bombers had been threatening ascendancy. The balance was transformed; for a time the Royal Navy enjoyed near air supremacy, until the arrival of the Luftwaffe proved again that in air warfare overwhelming numbers will always saturate defences, however sophisticated.

By then radar was making its mark on surface and anti-submarine warfare. The Battle of Matapan, the hunting of the *Bismarck*, depended upon radar. Radar, airborne and seaborne, made a critical contribution to the winning of the Battle of the Atlantic. Radar's application to gunnery made possible the many successful night actions in the later stages of the war.

In this invaluable book, Derek Howse – another of those who knew Life Before Radar – describes with graphic clarity how the imperative pressure of war spurred the scientists to make rapid progress. In only ten years, the advent of radar, giving the ability to see far beyond the horizon, transformed the professions of the aviator and the seaman; in war and peace life for them would never be the same again. The new generation quickly accepted radar as essential but were perhaps slow to appreciate its limitations. A spate of 'radar assisted' collisions at sea exposed the need for a proper understanding of its use and added a new aspect to many of the cases in the Admiralty Court. Today it is difficult to imagine how we managed without this remarkable invention, but we have, I hope, learnt that it is an aid, omnipresent but not infallible. The seaman's eye remains the last line of defence!

Lewin of Greenwich

ADMIRAL OF THE FLEET LORD LEWIN
OF GREENWICH, KG, GCB, LVO, DSC

Preface

There is no doubt that the development of radar during the short period from 1938 to 1945 had a greater impact on naval warfare than any other development since steam replaced sail. This book tells in non-technical language how the British Navy contributed to that development. Addressed to the general reader, it tells not only the technical story in simple terms, but also that of the operational use of shipborne radar at sea – for warning, for fire control, for fighter direction, for navigation – in the Battle of the Atlantic, in the Mediterranean, in the Normandy landings, in the Pacific, and all other theatres of war – and particularly about the people who designed and fitted the equipment, and those who used it at sea.

The idea of such a book was conceived at the end of 1985, the brain-child of Professor J.F. Coales, who had been intimately involved with naval radar before and during World War 2. In June 1985, half a century after the first historic experiments for the Air Ministry, the Institution of Electrical Engineers in London organized a seminar on 'Fifty Years of Radar' to which Coales, in collaboration with J.D. Rawlinson, contributed a paper dealing with the early stages of naval radar in Britain. The realization that so little had been published about the Navy's contribution – as opposed to that of the Air Force or Army – led him to put forward the idea of assembling a comprehensive collection of archives relating to the history of British naval radar from its inception in 1935 to the end of World War 2, not only for the historical record, but also in the hope that one day it would lead to a published account.

THE NAVAL RADAR TRUST

A start was made by contacting civilian and naval officers who had been involved, and by gathering archival material such as personal notebooks, recollections, photographs and so on. In December 1985 a working reunion of more than forty wartime colleagues was held in Cambridge. Since all concerned were at least in their sixties, many in their seventies or eighties, it seemed important to get on with the collection and digestion of data as soon as practicable.

From these beginnings, the project steadily gained momentum. An Administrative Committee was elected to manage the enterprise and this was subsequently formed into the Naval Radar Trust, with charitable status and the following membership:

- **Sir Hermann Bondi,** *KCB, FRS, then Master of Churchill College, Cambridge, former Chief Scientific Adviser, Ministry of Defence.*
- **Professor J. F. Coales,** *CBE, ScD, FEng, FRS, Emeritus Professor of Engineering, University of Cambridge.*
- **Basil Lythall,** *CB, formerly Chief Scientist, Royal Navy, Member of the Admiralty Board, and Deputy Controller of the Navy for Research and Development.*
- **D. Stewart Watson,** *CB, OBE, formerly Director of the Admiralty Surface Weapons Establishment, Deputy Chief Scientist, Navy, Director General of Establishments, Ministry of Defence, and last Chief of the Royal Naval Scientific Service.*

By December 1986, when the second reunion took place, Coales had contacted some 150 wartime colleagues. At that meeting, the Naval Radar Trust was set up and I was appointed to be the author of the book.

SOURCES

For primary material, the most important records seen were those preserved in the Public Record Office, though a significant amount of material was also found in the Naval Historical Branch of the Ministry of Defence, Churchill College at Cambridge, the Admiralty Research Establishment at Portsdown, the Royal Signals and Radar Establishment at Malvern, HMS *Collingwood* at Fareham, and HMS *Dryad* at Southwick.

In April 1987, I wrote to the editors of various nautical and technical journals, describing the project and asking for relevant reminicences, anecdotes and so on. The response was overwhelming – over 150 replies, from Admirals of the Fleet to Ordinary Seamen, from Fellows of the Royal Society to laboratory assistants, and from many other individuals – from Europe, America, Australasia. A great deal of valuable historical detail has resulted, particularly anecdotal material so necessary to flesh out a work of this kind. Together with the material collected by Coales since 1985, this forms a most valuable archive which will be deposited in the Archive Centre at Churchill College, Cambridge, where it will be professionally cared for, in company with many other naval papers of World War 2.

All in all, an enormous amount of information on the subject of wartime naval radar has come to light since 1985, despite wartime secrecy. In writing this book, the most difficult thing has been to decide what to leave out.

ACKNOWLEDGEMENTS

My first thanks must go to Professor John Coales, for all his encouragement, support and help in a thousand ways, as well as to his colleagues in the Naval Radar Trust, particularly Basil Lythall and Stewart Watson. These also accepted the chore of reading through and commenting upon early drafts, as did (in alphabetical order) Alec Cochrane, Commander Tony Fanning, Pat Hansford, Fred Kingsley, Alan Laws, Commander Frank Morgan, Harry Pout, Alex Rae, Peter Redgment, the late Alfred Ross, Jack Shayler, Captain Sir David Tibbits and Commander Bobby Woolrych. Of these, Cochrane, Kingsley and Pout must receive special mention in that they actually drafted (anonymously) some of the passages where they had expert knowledge, while Basil Lythall contributed the appendix on height determination and Alex Rae prepared the index. Was any book so thoroughly vetted before publication?

My special thanks must go also to David Brown and his most helpful staff at the Naval Historical Branch, and to him personally for contributing an appendix on naval airborne radar; Janet Dudley, senior librarian at the Royal Signals and Radar Establishment; John Briggs at the Admiralty Research Establishment, Portsdown; Lieutenant-Commander Bill Legg of HMS *Collingwood*; Eryl Davies of the Science Museum; and Lieutenant-Commanders Peter Selfe and Peter Lee of HMS *Dryad*. I must also mention Vice-Admiral Sir Arthur Hezlet, author of one of the only two relevant books on the subject, who made many helpful suggestions on how I should approach the subject, as well as giving help on submarine matters, as did Vice-Admirals Sir John Roxburgh and Sir Ian McIntosh; Commander Christopher Dreyer, David Cobb, Len Reynolds and Robin Board for help on Coastal Forces matters; Sir Bernard Lovell for a most fruitful correspondence on the interaction between air force and naval radar; the late Captain Andrew Yates, Captain Lord Mottistone, Captain John Naish, Commander Jack Emuss and John Lindop for help and suggestions; MacK Lynch and Jock Maynard for Canadian matters, and the Naval Officers' Association of Canada for permission to quote from *Salty Dips*; Fritz Trenkle for sharing his knowledge on German naval radar; the late Sir Clive Loehnis and Edward Thomas for their help on the wartime intelligence organization; Commander John Somerville, Ms Gaynor Ellis and Mrs Nancy Wilkins for permission to quote from the papers of Admiral of the Fleet Sir James Somerville, Captain R. M. Ellis and A. F. Wilkins respectively, deposited in the Archive Centre, Churchill College, Cambridge; and Mrs R.W. Fleming and George Willett for giving me access to the papers of their late father, Captain Basil Willett, who was so important in the development of early naval radar.

For illustrations, our particular thanks must go to the Imperial War Museum, the National Maritime Museum, and Times Newspapers Ltd for waiving reproduction charges. For help in picture research, my thanks go to Paul Kemp of the former museum, Bob Todd of the latter. Picture credits will be found on the appropriate caption.

The names of many other individuals who contributed are acknowledged in the text or in the endnotes. For the great number which could not be included, please accept my sincere thanks.

And finally, I must thank Admiral of the Fleet Lord Lewin, not only for doing me the honour of writing the Foreword but also for his encouragement and help from the very beginning of the project; and Admiral Sir Jeremy Black for contributing the Epilogue, which so aptly links the main story with the present day.

Sevenoaks, Kent DEREK HOWSE
February 1992

ACKNOWLEDGEMENTS BY THE NAVAL RADAR TRUST

The Naval Radar Trust acknowledges with gratitude the Ministry of Defence, Mr David Packard, and the Medlock Charitable Trust for major financial help, without which the research could not possibly have been carried out nor the archival material collected.

Other valuable contributions were received from the Royal Society, the Fellowship of Engineering, BICC plc, and GEC-Marconi Ltd, as well as many generous contributions from individuals, both naval and civilian, who were involved in the developments during World War 2.

Without their financial help and the support and industry of so many old colleagues combined with the enthusiasm and determination of the author, Derek Howse, the book could never have been completed.

Introduction

Radar is an electronic system that uses radio waves to detect objects that may be invisible to the naked eye because of distance, darkness or cloud. Radar can also determine the position of an object, its distance from the observing station – on land, at sea, or in the air – and, if the object is moving, its speed and direction of travel.

With pulse radar, with which we shall be principally concerned, very short pulses of radio-frequency energy are generated and transmitted, and their reflections or 'echoes' from solid objects are detected and displayed in such a manner that the location of the objects – generally bearing and distance from the observer – can be found. The essential components of such a system are:

- *a radio transmitter*, with its associated *pulse generator*;
- *an aerial array*, or *antenna*, to send the pulses in the direction needed, and to receive the echo returned;
- *a receiver*; and
- *an indicator* or *display* (generally a cathode ray tube), to show the results.

The word 'RADAR' is an acronym of '**RA**dio **D**etection **A**nd **R**anging', coined in the United States Navy in 1940 and adopted for general use by the Allied Powers in 1943. Previously, it had been known in Great Britain by the code name 'RDF', devised in 1935 to be deliberately misleading but often thought of as an abbreviation for 'Radio Direction Finding' or 'Range and Direction Finding'. Sometimes 'radio-location' was used in Britain and 'radio echo equipment' in the USA. The Germans originally called it by the cover name 'DeTe Gerät' (decimeter telegraphy equipment), changed at the start of the war to 'FuMO' (**F**unk**M**eß**O**rtungsgerät), 'radio position measurement equipment'). The French called it 'DEM' (*détection électromagnétique*), the Italians 'RDT' (*radio detector telemetro*).

For convenience in this book, the term 'radar' will be used throughout to describe the appropriate techniques even in the early days before the word came into general use. To conform to British practice in World War 2, the word 'aerial' will be used rather than the more modern (and originally American) 'antenna'.

A glossary of technical terms and abbreviations will be found on p. 314.

1
Setting the Scene

These same Wise Men, it is related, when about to engage in any Venture, do make use of that which they name the Devil Finding Mirror, the same being a Species of Speculum wherein are displayed such Demons of the Middle Air as may draw nigh to molest them on their Occasions. (Nikolas Ruyter, *Journal of Travels in Cathay and the Eastern Indies,* 1683)[1]

I don't know whether you have ever really received the thanks of the Fleet for the incredible work you and your party have done in providing us with R.D.F., the biggest thing that has happened to the Navy since we changed from sail to steam. No doubt one day merit will get its reward. (Commodore M.M. Denny, CB, Chief of Staff, Home Fleet, in a personal letter dated 2 April 1943 to Captain B.R. Willett, Captain Superintendent, ASE)

In March 1935, it was stated that 7-power binoculars were the only instrumental aids used by Britain's Royal Navy to detect aircraft; otherwise, the naked eye was relied upon. Not four years later, just after the Munich crisis, the battleship *Rodney* and the cruiser *Sheffield* had apparatus which could detect aircraft 60 miles away, apparatus not even on the drawing board four years before. And by the end of World War 2, hardly ten years after the binocular statement, all British naval vessels of any size had such apparatus, many with ten or more sets for different purposes, for gunnery as well as for warning, of both air and surface targets. This apparatus was, of course, radar – Radio Detection And Ranging.

The story of the Royal Navy's radar really starts in August 1935, when Vice-Admiral R.G.H. Henderson, Third Sea Lord and Controller (responsible for the Navy's ships and equipment), directed HM Signal School (the Navy's radio experts) 'to start work as soon as possible on naval application of the detection and location of aircraft by wireless methods. . .'.[2] However, before beginning that story, we must set the scene, first on the subject of radar itself, then on the Navy.

THE BEGINNINGS OF RADAR

Who invented radar? Did this happen in Britain, the United States, France, Germany, Italy, Japan? The answer is not simple. It was a development, rather than an invention, which occurred simultaneously in several countries after the technical prerequisites had themselves been developed – directional aerials, pulse transmitters, electronic displays (particularly the cathode ray tube), and the ability to generate ultra-short radio waves of sufficient power for the tiny fraction of the transmitted pulse which returned as an echo from the target to be perceived.

Though the first practical radar set was not operational until 1936, the idea of radio echo detection at sea and in the air had occurred to many people from the beginning of the twentieth century onwards. Indeed, as early as 1886, Heinrich Hertz, the discoverer of radio waves, had observed their reflection from metal, while in 1904, Christian Hülsmeyer of Düsseldorf was granted a British patent for his Telemobiloscope, a radar anti-collision device which he had successfully demonstrated in Cologne but which was never developed. Between the wars, two things in particular paved the way for radar proper – ionospheric research into the propagation of short radio waves, and the development of television – and the results of these developments were freely available internationally in the scientific press. From 1935, however, the threat of war and the realization that radar could be of great military importance drew a veil of secrecy over developments, so that work in this field in Britain, USA, Germany, Holland, France, Italy and Japan proceeded independently under the strictest conditions of secrecy. By 1940, Britain, Germany and the USA all had radar sets at sea.[3]

EARLY RADIO IN THE ROYAL NAVY

Thanks to the then Commander H. B. Jackson (Fellow of the Royal Society in 1901 and First Sea Lord in 1915 and 1916), the Royal Navy was very early in the field of wireless telegraphy. In 1895, a year before Marconi lodged his patent in the UK, Jackson had successfully signalled by radio from stem to stern of HMS *Defiance* in Plymouth harbour.

At that time, wireless telegraphy was regarded as a matter for electrical, rather than signal, experts, so responsibility for its operation and maintenance was given to the Torpedo Branch, which looked after electrical matters in the Navy. Consequently, the development of wireless for Fleet purposes was entrusted to the Torpedo School, HMS *Defiance* at Plymouth, where a Wireless Section was set up under Commander Jackson in 1896, moving to HMS *Vernon* at Portsmouth in 1901. In 1905, the first civilian wireless expert came to *Vernon* – A. W. Madge, a Naval Instructor. By 1911, the Wireless Section included three civilian scientists, eight draughtsmen, three naval officers and a Captain of Royal Marines who was responsible for instructing telegraphists at sea. The responsibilities of the Section thus included research and experiment, development for manufacture, and instruction. It was essentially a naval unit, with some civilian experimental staff.

As a result of experience during World War 1, responsibility for wireless in the Fleet was in 1917 transferred to HM Signal School in the Royal Naval Barracks at Portsmouth, and *Vernon's* Wireless Section transferred its allegiance at the same time, becoming the Experimental Department of HM Signal School, under the charge of the Wireless Commander, whose title was changed to Experimental Commander. In the period immediately following the Armistice, the organization of the Experimental Department was established formally by its then head, Commander (later Admiral of the Fleet Sir James) Somerville, along lines very similar to those which had existed for the previous 15 years and which were to continue broadly unchanged for the next 25 years.[4]

So it was that when radar arrived in 1935 it became the responsibility of the Navy's Signal Branch, research and development falling to the Experimental Department of HM Signal School at Portsmouth.

It is ironic that the Navy might well have started work on radar some years earlier. In 1928, L. S. Alder, one of the most brilliant radio engineers in Signal School, had proposed a scheme for the detection and location of objects by radio, which the provisional patent application shows to have presaged radar quite precisely. In the Admiralty, the Signal Department appear to have suggested Alder's proposals should not be proceeded with, passing the papers to the Director of Scientific Research where they were dealt with by a retired Commander in the Patent Section who, without consulting any scientific staff, concurred with the Signal Department and shelved the proposal. 'A distressing lapse for which we were responsible in DSR's department . . . It taught us a lesson, but too late,' wrote Sir Charles Wright (one of the scientists not consulted) after the war.[5]

Mention must also be made of another early British service proposal, equally abortive, for a pulsed radar system to detect ships for coast defence, submitted to the War Office by W. A. S. Butement and P. E. Pollard of the Army's Signals Experimental Establishment at Woolwich in 1931. They were grudgingly allowed to do experiments at Woolwich – but they were not to be done in working hours. These proved successful as far as they went but, when they offered to give a demonstration, Butement and Pollard were informed: 'There was no War Office requirement.'[6]

THE ANTI-AIRCRAFT PROBLEM

While radar was eventually to become an integral part of the Navy's fire-control systems, to start with it had to be grafted onto existing optical systems, which were the responsibility of the Director of Naval Ordnance (DNO) who, in the years immediately preceding World War 2, attempted under the very stringent financial conditions to respond to a 1926 report of the Imperial Defence Committee on Anti-aircraft Defence, a report which predicted that accurate bombing – of the kind necessary to attack ships, for example – was only possible if the aircraft flew straight and level at constant speed, likely heights and speeds being suggested. The Admiralty placed a contract with Vickers Ltd to develop the high angle control system (HACS) for capital ships, aircraft carriers and cruisers, and this system continued to evolve through several successive 'Marks' up to the outbreak of World War 2. For smaller ships (mostly with 4.7in guns), a simplified system known as the fuze-keeping clock (FKC) was developed by Elliott Brothers of Lewisham, to be used in association with the Fire Control Box, which was the responsibility not of DNO but of the Director of Torpedoes and Mining (DTM) because, as we have seen, the Navy's Torpedo Branch was then responsible for all electrical apparatus, including surface (or low angle) fire control systems. These weapon systems included the optical directors, which were hand-driven and provided range and directional data on the target needed by the fire-control equipment.

In due course, gunnery radar characteristics were to be strongly influenced by the nature of these prewar gunnery developments, constrained as they were by the combined effects of financial stringency and the many technical difficulties (even assuming straight-and-level attacks) of operating weapon systems from ships at sea.

1934: THREAT FROM THE AIR

During the summer of 1934, the suspected growth of Germany's air force caused very considerable disquiet in Britain. On 7 July, Winston Churchill made a speech urging Prime Minister MacDonald and the Government to double the strength of the RAF immediately – and then to redouble it. On 8 August, a letter by Professor F.A. Lindemann, Churchill's adviser on scientific matters, was published in *The Times*, calling for a concerted scientific effort in the field of air defence research. Back in 1932, Stanley Baldwin had said, apparently with Air Ministry agreement: 'The bomber will always get through!' The lamentable results of the summer Air Exercise of 1934 seemed to prove that this was so.

This disquieting state of affairs had induced A. P. Rowe, assistant to H. E. Wimperis, Director of Scientific Research at the Air Ministry, to make an informal survey of existing methods of air defence on the basis of some fifty-three files on the subject he read through. Rowe's conclusion was that, unless science evolved some new method of aiding air defence, we were likely to lose the next war if it started within the next 10 years.[7]

Wimperis, having consulted various scientific friends, sent a minute to the Air Council on 12 November 1934. After discussing various possibilities, he went on:

> Scientific surveys of what is possible in this, and other, means of defence at present untried are best made in association with two or three scientific men specially collected for the purpose: their findings may sometimes prove visionary, but one cannot afford to ignore even the remotest chance of success: and at the worst a report that at the moment 'defence was hopeless' would enable the Government to realise the situation and know that so far retaliation was the sole remedy – if such it can be called. I would submit, therefore, that the formation of such a body be now considered.[8]

He proposed that the chairman should be H. T. Tizard, a former Royal Flying Corps pilot; and that members should include Professor A. V. Hill, a physiologist who had worked on anti-aircraft problems in World War 1; Professor P. M. S. Blackett, a physicist who had served in the Royal Navy; and himself. He suggested the following terms of reference:

> To consider how far recent advances in scientific and technical knowledge can be used to strengthen the present methods of defence against hostile aircraft.

First to comment on this proposal was Air Marshal Hugh Dowding, Air Member for Research and Development. 'I agree generally,' he said. 'Possibly the terms of reference of the committee might be criticised as being so wide as to approach to infinity so far as the labours of the committee are concerned, but the presence of DSR [Wimperis] should suffice to confine its researches to what he considers to be practicable.'[9] All agreed, and approval to set up such a committee was swiftly given by the Air Minister, Lord Londonderry.

1935: THE TIZARD AND SWINTON COMMITTEES

The new Committee for Scientific Survey of Air Defence, known colloquially as the Tizard Committee, met for the first time at the Air

Ministry on 28 January 1935 – Tizard, Hill, Blackett, Wimperis, with Rowe as secretary – and was destined not only to transform Britain's air defences in a very short period, but to start the train of events which led to the development of British radar at sea, on the ground, and in the air.

Meanwhile, on 18 January 1935, Wimperis had written to R. A. Watson-Watt, Superintendent of the National Physical Laboratory's Radio Department at Slough, to ask his opinion on the possibility of using radio waves to damage aircraft or to incapacitate the crew – the so-called death-ray. No, said Watson-Watt in a memorandum considered at the Tizard Committee's first meeting, he and his colleague A. F. Wilkins thought the death ray was not a practical proposition, but that radio-detection as opposed to radio-destruction might well be possible and they would submit proposals if needed.[10] The first draft of a second memorandum, 'Detection and location of aircraft by radio methods',[11] describing what later came to be called radar, was in Rowe's hands by 4 February. This so convinced Wimperis that on 15 February he asked Dowding for £10,000 to investigate this new method of detection. Dowding was cautious, however, and said that the calculations must be confirmed by experiment.

On 26 February 1935, therefore, Wilkins – with Watson-Watt and Rowe acting as witnesses – carried out the historic 'Daventry experiment' when an aircraft, flying in the beam of the BBC's 50m short-wave station at Daventry, produced a detectable echo in a receiver when it was eight miles away. In the Air Ministry, Dowding was delighted and, according to Wimperis, promised him all the money he wanted within reason.[12] On 1 March, Watson-Watt had visited a remote site at Orfordness in Suffolk where trials in radio location could be carried out in secrecy, under the guise of ionospheric research. Treasury sanction for the expenditure of £12,300 in the first year was obtained on 13 April,[13] and on 13 May, a team of six under Watson-Watt moved from Slough to Orfordness to start the experiments which were to have such outstanding results in the development of radar in Britain.

While this was going on, there had been considerable activity on the political front – largely the result of plotting by Churchill, Austen Chamberlain and Lindemann – resulting in the setting up at Baldwin's initiative of a political committee to control the technical Tizard Committee. Meeting first on 11 April 1935, the new committee had the even more bureaucratically complicated name – the Air Defence Research Sub-committee of the Committee for Imperial Defence – but was generally known as the Swinton Committee as it was chaired by Sir Philip Cunliffe Lister, later Lord Swinton, who became Air Minister when Baldwin succeeded MacDonald as Prime Minister on 5 June. The Navy was represented by the Controller – Admiral Henderson – and Winston Churchill joined at the fourth meeting on 25 July at Baldwin's invitation.

Churchill and Henderson (a distinguished gunnery officer) were old friends from the former's time as First Sea Lord in World War 1, and the minutes of the Swinton Committee show them sitting together at all meetings. At Churchill's insistence, Lindemann joined the Tizard Committee at their eleventh meeting on 25 September 1935, an event which nearly caused the break-up of that committee.[14]

At their meeting on 27 May, the Swinton Committee resolved to give every encouragement to the Orfordness trials 'in conjunction with those [trials] required by the Navy'. And it gave the scientists something to aim at – a range of detection of 50 statute miles within 5 years.[15]

As for the Tizard Committee, in its first few meetings it reviewed the existing methods of detecting aircraft used by the three services – visual, acoustic, infra-red – and then looked at methods of dealing with enemy aircraft once detected. Having talked to the Air Force and the Army, their fourth meeting on 18 March 1935 was devoted to the Navy, represented by C. S. Wright, the Navy's Director of Scientific Research (DSR); A. B. Wood from the Admiralty Research Laboratory (ARL), who had done experiments on infra-red detection in 1926; G. Shearing, Chief Scientist of HM Signal School's Experimental Department; and Commander N. G. Garnons-Williams, a Gunnery Officer from the Naval Ordnance Department.

First, the Navy's visual methods were discussed – it was then that Garnons-Williams made the statement about 7-power binoculars which was cited at the start of this chapter – as were the acoustic and infra-red experiments carried out by the Navy during the 1920s and 1930s. Then Shearing told the Committee he thought the most promising method of detection would be 'that in which the aeroplane is used to reflect energy from some source,' citing Post Office experience when using waves of 5m, the reception of which were interfered with when aircraft were in the vicinity. Tizard did not mention the Daventry experiment or the Orfordness proposals explicitly, though the possibilities were hinted at. Shearing added that Signal School was doing work for communications purposes on ultrashortwaves, at present of the order of 50cm, using small valves which had been produced in America, but that a magnetron valve had been developed capable of working down to about 36cm.[16]

On returning to Signal School, Shearing wrote a report for the Experimental Commander. The committee, he said, appeared to place no reliance on acoustics or infra-red as a solution, and considered the only possible method was one which involved the use of a source of energy independent of the aircraft, such as ultra-high-frequency radio.[17] And both Wright and Shearing agreed wholeheartedly.

The setting up of the Tizard and Swinton Committees had such an influence on the development of all radar in Britain, naval radar included, that it has been treated in some detail here. However, subsequent activities by these committees will be mentioned only where they affect the naval story directly. Both continued to meet regularly until the outbreak of war.[18]

THE NAVY STARTS RADAR DEVELOPMENT

Watson-Watt and his team started work at Orfordness on 13 May 1935. The whole operation was shrouded in the greatest secrecy, but Wright, the Admiralty's Director of Scientific Research (DSR), was invited to witness Orfordness's first demonstration on 8 June,[19] and a month later managed to get permission to bring Shearing and one or two Signal School officers to Orfordness,[20] by which time aircraft had been detected out to 40 statute miles, though at that time only range could be measured, no direction or height. However, it was enough to give the naval visitors some idea of the potential for naval purposes of this new and secret RDF – already so named by Rowe and Watson-Watt.[21]

But this was not to be a one-way deal – Orfordness needed help from Signal School as much as the latter needed help from Orfordness. To explain this, we must go back to the period immediately following World War 1, when a small group was set up at Signal School to undertake radio valve research. An important factor influencing this decision was the Navy's interest in transmitting valves with envelopes made of silica instead of glass, a type of construction used because it was the only way of making high-power valves which were both small and robust. However, with the exception of small numbers for the other services, there was no demand for silica valves outside the Navy and they did not therefore interest the valve manufacturers. Signal School was therefore forced to undertake both research and manufacture. In 1932, the Air Ministry was having trouble obtaining certain valves and the Admiralty agreed that Air Ministry should put their valve problems to Signal School for solution.[22]

Therefore, when Wilkins and his colleagues (then still at Slough) started to design their first radar transmitter, they selected silica valves for their initial experiments. In April 1935 – certainly before Wright, perhaps even the Controller himself, was let into the radar secret – Wilkins and L. H. Bainbridge-Bell had visited Signal School with a cover story about new ionospheric research. They sought advice from H. G. Hughes (head of the valve section),[23] and picked the brains of C. E. Horton and J. F. Coales (of the direction-finding section) on directional aerials for short waves (below 20m wavelength).[24]

Shearing was formally let into the radar secret before his visit to Orfordness on 7 July, so Watson-Watt was able to discuss valves with him and to arrange a visit to Portsmouth. Three days later, Wimperis

1. Sir Charles Wright (1887–1975), Canadian-born Director of Scientific Research at the Admiralty, 1934–46, who contributed so much to the development of British naval radar, particularly in the earliest years. This photograph of him in his retirement shows the camera he used as one of the scientific staff in the *Terra Nova* during the British Antarctic Expedition, 1910–13, led until his death in 1912 by Captain R. F. Scott, RN.

(*Pat Wright*)

wrote to Wright at the Admiralty, asking that facilities be afforded to Watson-Watt to obtain any assistance he might require at Signal School. 'The Orfordness investigation,' he added, 'has been given the highest possible priority and is, of course, of a strictly secret character.'[25] Captain G. W. Hallifax, Director of the Signal Department (DSD) forwarded Wimperis's letter to Captain J. W. S. Dorling of Signal School (the first in the file to be stamped Most Secret), confirming the arrangements already made by Shearing for Watson-Watt to visit Portsmouth on the 15 July, adding: 'From what DSD has heard of the Orfordness demonstrations, this work appears highly promising and of the greatest importance not only to the Air Force but also to the Navy. In the latter case it seems possible that it may have applications other than for the detection of aircraft.' He said that Signal School should ask for additional staff if needed.[26]

In the event, E. G. Bowen deputized for Watson-Watt on the visit, and Signal School agreed to produce urgently the special silica valves needed by Orfordness. In his letter to DSD reporting this, Dorling thought that 'every effort should be made to explore the naval possibilities' but, to achieve this, asked for only one scientific officer and one assistant extra.[27]

By 13 August, all was agreed and DSD told Dorling:

> the Controller has decided that Signal School should start work as soon as possible on naval application of the detection and location of aircraft by wireless methods, work on the production of transmitting apparatus being begun at once.
> To enable this work to be carried out without interfering with other commitments, Treasury approval is being sought for an increase of Signal School staff of one Scientific Officer and one Assistant II.
> As regards receiving gear, permission is being sought for a Signal School Officer to visit Slough and Orfordness for a period of about six weeks.[28]

Treasury approval was received on 29 August, but the new staff did not arrive until December.[29]

On Wright's initiative, A. B. Wood from ARL was already on a two-month attachment at Orfordness. Now, R. A. Yeo, a scientific officer from Signal School, was relieved of his radio communications duties and arrived at Orfordness on 2 September 1935 for a five-week attachment.

* * *

HM Signal School – and the Royal Navy – was in the radar business – just!

2

1935–9: The Beginnings of the Navy's Radar

1935: TYPE 79 IS BORN

Yeo returned from Orfordness on 9 October and DSD lost no time in calling a meeting at the Admiralty – attended by the Director of the Tactical Division, DSR and his deputy, and the Captain and Experimental Commander of the Signal School – to hear Yeo's report and to decide what ought to be done next.

After Yeo had reported on progress at Orfordness, preliminary requirements for naval radar were formulated – very tentative because no one had any idea what might be possible:

	Aircraft	*Ships*
Warning of approach	60 miles	10 miles
Precise location	10 miles	5 miles

Note: When 'mile' is used in this book as a unit of distance at sea, the nautical mile rounded off to 2,000 yards should be assumed. See Glossary.

No order of priority was laid down explicitly, but a hand-written note by Yeo makes it clear that at that time the Naval Staff feeling was that ship detection was more important than aircraft detection.[1]

Orfordness was working on a wavelength of 25m, but the Navy would have to concentrate on considerably shorter wavelengths, first because long wavelengths mean very large aerials – practicable ashore but not in a ship – and secondly because the longer the wavelength, the easier it is for the enemy to detect the transmissions.

Wright felt strongly about this. Wilkins makes this point in his account of the former's visit to Orfordness in June 1935:

> My main recollection of the visit was that Wright regarded metre-wave RDF as an interim device for naval vessels because, by using it, they would be acting as beacons for the enemy. He said that microwaves were essential for the future and that work would have to be done to

produce powerful radiation at such wavelengths. It was undoubtedly this belief of Wright's which later led to the contract being placed with Randall at Birmingham which resulted in the invention of the cavity magnetron.[2]

To do the experimental work needed to achieve equipment operating at these shorter wavelengths – and suitable for fitting in a ship – the meeting decided the Signal School would need three additional scientific officers and three assistants (one of each was already sanctioned by the Treasury) and two fitters, with more staff later, and yet another Scientific Officer for liaison with Orfordness. They would also need a site and a hut with a clear view to seaward at, say, Eastney or Southsea Castle near Portsmouth, and additional laboratory and workshop space in the main Signal School complex at the Royal Naval Barracks in Portsmouth.[3]

It took more than two months for the Admiralty to obtain Treasury sanction for the very modest staff and accommodation proposals, and formal Admiralty approval was not given until 30 December 1935.[4] Meanwhile it had been decided that the RAF radar scientists should, from May 1936, move to Bawdsey Manor, some 8 miles down the Suffolk coast from Orfordness. Largely at Watson-Watt's instigation, the Air Ministry was pressing the Swinton Committee to recommend that all British experimental work on radar should be centred at Bawdsey. Signal School resisted strongly: conditions in a ship at sea, gunfire, topweight, interference with wireless and other apparatus, tactical considerations in a fleet – all of these could best be dealt with by the Signal School, where there already existed a comprehensive organization with civilian scientists, engineers and naval staff working together for the development, design, procurement, manufacture, testing and fitting of radio equipment tailored for naval use. Besides which, the Navy was at least as interested in the detection and location of surface craft as aircraft. Ship targets would be essential for experiments and Bawdsey was far from ideal for this. 'Considered purely from the point of view of obtaining solutions to the two sides of the Naval problem [air and surface], it appears that everything is to be gained by carrying out the experiments at Portsmouth, provided that the liaison with Orfordness already recommended is set up.'[5]

The Navy won the day. At the Swinton Committee's meeting on 12 December (with C. S. Wright in attendance) it was resolved that, whereas Bawdsey Manor must be the headquarters for research on location of aircraft, 'the Admiralty might prefer RDF location at sea to be investigated at the Signal School at Portsmouth.'[6] For the Army, a group drawn mainly from the Air Defence Experimental Establishment at Biggin Hill (with a few from Woolwich) joined the Air Force scientists at Bawdsey Research Station in 1936.

Would the Navy's interests have been better served if early naval radar had been developed in an inter-service civilian establishment dedicated to radar, rather than in a naval establishment already heavily committed in another field albeit closely related technically? The latter choice would certainly have been correct had it been backed up by adequate scientific resources with proper priority, technical support from Admiralty departments as needed, and able to establish a fruitful dialogue with naval operational authorities. In fact, it emerges that, for the first two years at least, this did not happen.

Even so, an inter-service solution would have been at least as imponderable. There would have been potential benefits from the much larger pool of scientific effort, but would the Admiralty have pressed strongly enough for adequate priority for naval applications? If so, would they have been successful? Would there have been adequate liaison with – and support from – people familiar with the marine environment and with the practicalities of ship-fitting? On balance, the decision taken was probably correct despite the initial failure to support it properly, but it was not to be until after the end of 1937, when radar was at last given a more reasonable proportion of experienced Signal School effort, that the potential of this method of working was demonstrated by the dramatic progress then achieved.

1936: DEVELOPMENT ON FOUR-METRE WAVELENGTH

For naval research on what came to be called type 79 – in 1936, radar was treated in the Navy as just another wireless set, and type 79 happened to be the next number in the wireless series – the year 1936 was one of consolidation but, largely through lack of resources, all too little achievement. A hut was erected near the shore facing Spithead at Eastney Fort East on land occupied by the Royal Marine Barracks, together with two tubular structures to carry aerials, to the disgust of the Colonel Commandant, Royal Marines, whose hitherto pleasant view they spoiled.[7] Installation of apparatus at this Signal School Extension began on 14 July.

As far as staff was concerned, radar had no priority within the Signal School, the only person transferred from communications work being Yeo, who had to work alone until joined in December 1935 by W. P. Anderson, the other new member of staff approved in August, E. M. Gollin, being drafted to do Yeo's old job in communications. (He was to work on radar from 1937.) By October 1936, there were still only four scientific officers (one of whom was permanently at Bawdsey) and one assistant working on radar, with two more assistants due to join in November.[8]

2. Eastney Fort East, c.1940. The Signal School Extension near Portsmouth to seaward of the Royal Marine Barracks, facing Spithead. Much of the Navy's early development and trials of radar took place here, using the lattice structures to support aerials at approximately the same height above sea level as in a ship.

(NRT)

Radar was then Most Secret, which meant that nothing could be discussed with anyone not on a named list, reported to include less than two hundred names nationwide. So Yeo and his small party started work at Eastney Fort East with only W. Ure, Head of the Transmitting Section and Yeo's immediate superior, knowing anything about it except for Captain Dorling, Experimental Commander F. J. Wylie, Application Commander the Hon. J. B. Bruce (at Signal School, naval 'application officers' were appointed to provide the link between the scientists and the Fleet), Chief Scientist G. Shearing, and presumably F. W. Brundrett, the 'secretary' (administrator), a scientist who for some years had been the power behind the throne in the Experimental Department but was soon to join Wright at the Admiralty. The rest of the Signal School got on with equipping the Fleet with the best possible communications and direction-finding equipment, quite unaware of what Yeo was doing.

Yeo chose a wavelength of 4m for his principal experiments in 1936 and 1937 on the grounds that an aerial for any longer wavelength would be too large to be accommodated in a ship. In the event, this proved an unfortunate choice because the design of silica envelope valves at that time incorporated very long lead-filled seals which would only operate efficiently down to a wavelength of about 7.5m.[9] He also did some work – which never achieved much – on 1.5m, a wavelength already being used by the Navy for an aircraft homing beacon and which was to be used for the RAF's first airborne radar (ASV and AI), for the RAF's CHL, for the Army's first Coast Defence radar (CD), and – presumably unbeknown to the Signal School at the time – for the US Navy's first air search radar XAF, which went to sea for trials in the battleship *New York* in 1938 and led to the wartime SK and SR radars on the same waveband.

Yeo and Anderson designed a receiver which they took to Bawdsey for trials on 16 March 1936 and by September an experimental transmitter and receiver on 4m had been built to a stage where it was possible to transmit without paralysing the receiver. With dipoles fixed to the Colonel Commandant's eyesores, echoes had been obtained from aircraft but not ships. Range could be measured, but not bearing or elevation. As we have seen, 1.5m had been investigated but there were technical difficulties with the transmitter and no trials had been possible. Work was in hand on centimetre waves for communications purposes which it was reckoned might be useful for detection in the future – as indeed it was to prove.

By September 1936, the Controller was getting restive, presumably irked at not being able to report spectacular Naval results at Swinton Committee meetings. He sent for DSD to ask what was happening – 'What arrangements have been provided or are contemplated for applying successful results for defence purposes?'[10] – in other words, once you have got it, how are you going to use it?

A special meeting was hastily called at the Signal School on 1 October, attended by DSD and DSR (but not by Yeo), where the radar team's achievements to date were discussed. DSD said he thought it was necessary to produce results before asking for unlimited resources. It was agreed that a second set should be made and fitted as soon as possible in the Signal School trials ship, the elderly coal-burning minesweeper *Saltburn*. Based on Portsmouth, *Saltburn* was shared with the Navigation School, which used her for cruises to give practical experience to budding specialist navigators. This sharing caused many difficulties, made no easier by the fact that the Navigation School's captain was not in the radar secret, so could not be told why the Signal School needed her so badly.

To satisfy the Controller's question on how it was proposed to use radar once it was fitted, it was agreed that a full-time application officer should be appointed specifically for radar, preferably to join before the *Saltburn* trials.[11]

The first sea trial of the Navy's first radar, by now known as Type 79X on a wavelength of 4m, took place on 15 December 1936 in *Saltburn*. Fixed receiving and transmitting dipoles were suspended 50ft apart on a cable slung between her masts, 75ft above sea level. With the ship lying stopped some 7 miles south-east of the Nab Tower, east of the Isle of Wight, an aircraft made runs in and out from the ship, contact being held out to 18 miles and regained at 15 miles. The aerial arrangement gave no indication of bearing, but surface echoes of what was believed to be the Nab Tower were detected at 5 miles, a light cruiser at 4 miles.[12]

1937: BRICKS WITHOUT STRAW

The other proposal at the October meeting came to fruition on 18 January 1937, when Commander the Hon. Henry M. A. Cecil, RN (retd.), joined HM Signal School as the first application officer specifically for radar. A very gifted member of a distinguished family – musically gifted as well as intellectually – he was a most welcome addition to the radar team, and he was to make many contributions over the next four years.

What would appear to have been Cecil's first impact on the Navy's radar scene was a succinct situation report sent to the Admiralty by Signal School, signed by Captain Dorling but obviously drafted by Cecil in consultation with Shearing and Yeo in February 1937. It opened thus:

> *Type 79X and centimetre wave development* – The field of experiment and development covered by the above heading is so large and indefinite at the moment and the staff available for the work is so limited that it appears essential to lay down a limited number of lines on which to concentrate in the immediate future.[13]

The most important points in this report were these:

- a priority list was proposed, putting detection of aircraft above that of surface craft;
- it was still intended to carry on with work on four separate wavelengths, even though the total staff working on naval radar included only four scientific officer grades and one of these was at Bawdsey (in October 1936, the total staff at Bawdsey had been no less than fifty);
- a proposal was made that high frequency work should be put out to a commercial firm;
- the 16cm ship and iceberg detection set in the French liner *Normandie* was mentioned,[14] and it was urged that work on type 79X should be kept secret.[15]

Thus it was that, throughout the first half of 1937, Yeo and his team of three at Eastney continued to work on type 79X on 4m wavelength, though they virtually stopped work on 1.5m and 60cm for lack of manpower. Early in the year, members of the DF section started work independently on 23cm wavelength at Southsea Castle (useful for both communications and radar), while A. B. Wood (he had joined Signal School from ARL), with one assistant at the RN Barracks, turned his attention to the generation of waves less than 23cm. There seems to have been very little contact between Eastney and the other two groups.

One of the many things still lacking in type 79 was the ability to measure the target's bearing. Signal School prepared preliminary designs for rotating aerials to make this possible and in mid-April asked the Admiralty to authorize Portsmouth Dockyard to prepare the necessary drawings and to fit the new aerials on the top of *Saltburn*'s masts in time for trials in July. In the event, largely owing to the ponderous bureaucracy of the Admiralty machine, fitting was not completed until a few hours before *Saltburn* sailed on 7 July for the Navigation School cruise to Guernsey and Harwich. Instead of having the hoped-for fortnight in harbour to try out the equipment with its new aerials, the tuning had had to be done at sea with no known fixed echoes during the couple of days before the arrival of the trials aircraft. Not surprisingly, the trials were very disappointing, aircraft being detected at a mere 7.5 miles, and the Eddystone at 7 miles. A demonstration to Bawdsey staff embarking at Harwich had been planned but this was cancelled.[16]

RAF PROGRESS

In contrast to the seemingly slow naval progress, the RAF team at Bawdsey had made great strides, thanks to government priority giving them what appeared to the naval team to be almost unlimited resources in money and men – and possibly thanks to greater resolve at the top. By 5 July 1937, the design of the Home Chain stations had reached a point where the Swinton Committee felt able to recommend the immediate provision of a chain of twenty RDF stations from St Catherine's Point to the Tees.[17] On 4 and 5 September, an Anson aircraft fitted with 1.5m ASV radar detected vessels of the Home Fleet on exercises at a range of five or six miles as well as *Courageous*'s fighters, before visual sighting.

> This was something of a landmark in the history of airborne radar. We had found the Fleet under conditions which had grounded Coastal Command, we had detected other aircraft for the first time with a self-contained airborne radar and, simply by returning home in one piece, had demonstrated some of its navigational capabilities.[18]

So wrote E. G. Bowen, one of the Anson's crew that day, to whom is due much of the credit for the design of that set, and of subsequent airborne radar.

THE NAVY'S LACK OF PROGRESS

With the war clouds gathering, the Navy's lack of progress was very worrying. Metric radar was in the experimental stage but, as things stood, it seemed unlikely that even the minimum requirements could be met within a reasonable time scale, while the centimetric work was even more experimental.

It so happened that this critical juncture in the development of the Navy's radar coincided with important changes in the naval team directing the project, all in accordance with the normal naval practice whereby an officer's tour of duty seldom lasted more than two or three years. At the Signal School, Captain A. J. L. Murray succeeded Captain Dorling in command on 25 August 1937, and Commander Basil R. Willett was to succeed Commander Wylie as Experimental Commander on 15 November. At the Admiralty, Captain Philip F. Glover succeeded Captain Makeig-Jones as DSD on 6 September.

A few days after he had taken over, Murray wrote to DSD about the *Saltburn* trials fiasco, grumbling about the long time taken to get Admiralty approvals. Type 79X was quoted as 'high priority' at Signal School. Could not Admiralty departments give it a similar priority? And *carte blanche* was needed for work on *Saltburn*.[19] He called a meeting at the Signal School for 8 September to discuss the whole matter.

DSD's response is interesting. Makeig-Jones said he believed that 'other work should give way if necessary'. Since no effort was being given to 1.5m or 60cm, he must have meant that the work to give way would be communications or DF. If so, this is the first indication of thoughts in the Admiralty of reshaping priorities. DSD also said that he now thought he had been mistaken in insisting upon the October 1936 decision that it was necessary to produce good results before asking for unlimited facilities,[20] a chicken-and-egg situation from which Bawdsey did not have to suffer!

The Signal School meeting of 8 September 1937 was to start a chain of events which revolutionized the work on the development of naval radar, events which were to lead to the successful fitting of radar to two large fleet units in a little over a year. Presided over by Captain Murray, it was attended by both old and new DSDs, by DSR (Wright) and Brundrett, who had recently left Signal School to join Wright at the Admiralty, by both old and new experimental commanders, by Cecil, and by all the senior scientific staff concerned with development on 4m and 23cm.[21]

The meeting resulted in action on two fronts – in the Admiralty to provide greater resources, and in the Signal School to give a far higher proportion of effort to radar. For the former, DSR and DSD presented a joint minute to the Controller on 28 October asking for a cruiser for sea trials (*Saltburn* was too small), for more staff, and for more wooden huts and machine tools at both RN Barracks and Eastney. They also said this: 'It is proposed, by internal arrangement, to relieve one of the Principal Scientific Officers in Signal School of some of his less important duties in order that he may devote close attention to the supervision of the type 79X experiments.'[22] These proposals were swiftly approved by the Admiralty.

The action taken at the Signal School is graphically described by Coales:

> Cdr Willett relieved Wylie as Experimental Commander on 15 November and it would appear that the Controller quickly agreed to the proposals put forward by DSR and DSD because early in November a major reorganization of the work was put in hand in Signal School.
>
> To set this in motion, Captain Murray called together the newly appointed Experimental Commander (Basil Willett), Shearing, Ure, and C. E. Horton (then in charge of the DF section) and presumably also Henry Cecil and Cecil Evershed [who had succeeded Brundrett as Secretary]. The importance of this meeting is emphasized by the presence of the Captain of the Signal School himself, who did not usually concern himself with the affairs of the Experimental Department but left them entirely in the hands of the Experimental Commander, the senior civilian officers and the application officers. No record of this meeting has been located but it is known that Captain

Murray told the meeting: 'Signal School has been caught with its trousers down and this must not happen again.'[23]

REORGANIZATION AT SIGNAL SCHOOL: CHANGE TO 7.5–METRE WAVELENGTH

The most important organizational change was that a new 'R' – for RDF – Division was to be set up within the Experimental Department, headed by Horton, who was to take over direction of all the Signal School radar development from Ure. A. A. Symonds was to replace Yeo in charge of the staff working on metric radar at Eastney; J. F. Coales (who had worked with Horton in the DF section since 1930) was to take responsibility for all work on wavelengths less than 1 metre, whether for radar, communications or DF, dividing his time between the RN Barracks and Southsea Castle; and a third section of the new division was set up under A.W.Ross (who had joined the DF section in 1936), heading a small research team for propagation and aerial research at a field station at Nutbourne, some 12 miles east of Portsmouth.

Clearly Horton's first task was to review the progress to date and, with his section leaders, decide on the future development programme. He was no lover of formal meetings or of writing memoranda or papers. His method of working was to discuss the problem with the people involved, decide on a course of action, and leave them to get on with it.[24] Soon, a very fundamental decision was taken by Symonds and Ross to abandon 4m and, accepting that the aerials would have to be larger, develop a new type 79 on a wavelength of 7.5m, a wavelength already investigated by Yeo and Anderson, for which suitable electronic equipment was readily available because the BBC's new television service had been operating on the same waveband since 1936, as well as communications equipment already fitted in ships.

On the centimetric side, Coales (himself only recently let into the radar secret) was able to pick the brains of the leading worker on magnetrons in the UK, E. C. S. Megaw of the GEC Research Laboratories, despite the fact that the latter was not in the secret. GEC had an Admiralty contract to develop naval communications equipment on 60cm, for which Megaw was responsible on GEC's behalf, Coales on Signal School's, so there was every reason for them to discuss centimetric radio design without having to mention radar as such. The outcome of this first of many contacts with Megaw was that, after discussion with Horton and DSR, Signal School in February 1938 discontinued work on 23cm and concentrated on developing radar on 50cm – the wavelength on which the vast majority of gunnery sets in HM Ships in World War 2 were to work.[25]

1938: TYPE 79 SEA TRIALS

If 1937 had been disappointing, 1938 started on a note of optimism. Important technical decisions had been taken and the stage was now set for development at the highest priority of radar on 7.5m for warning of aircraft, and at 50cm for ship detection and gunnery. The team had been greatly strengthened and more was promised. That this optimism was justified is shown by the fact that the Controller felt able to anounce to the Swinton Committee on 13 May that 'two ships of the Home Fleet have been equipped with RDF'.[26] Actually, he was anticipating events, but only by a month or so.

Ross and his team quickly designed an aerial array suitable for workng on 7.5m, the transmitting array to go on one mast, an exactly similar array for receiving on the other, the two to be rotated in synchonism. In passing, it is worth noting that this aerial array was so successful as a compromise between the various conflicting needs of the naval environment that it was used throughout the life of type 79/279 (to beyond 1945) and was the basis of design of aerials for later radar sets on shorter wavelengths, type 281 on 3.5m and types 286 and 291 on 1.5m. Considerable credit is due to Ross and his team who hit on a near-perfect design so early in the history of naval radar.

This new 7.5m type 79X was tried out successfully at Eastney and in *Saltburn* early in 1938 and the performance was promising enough to start manufacturing in Signal School's workshops at the highest priority two 'Chinese copies' of the laboratory set 79X, to be known as type 79Y. In contrast to the slow progress in the previous three years, both these radar sets were at sea and fully operational in the cruiser *Sheffield* and battleship *Rodney* only a little more than six months after the decision to make them had been taken.

At the same time as the decision to produce 79Y was made, work started on a redesign of the set to incorporate all the lessons learned to date, later called Type 79Z, for fitting in the Signal School's promised trials cruiser, *Dunedin*, whose refit at Portsmouth was due to complete in mid-1939. (In the event, it was fitted in *Curlew*.)

Getting the radar to sea in such a short time was achieved by the production process that was in general use at Signal School to procure W/T equipment between the wars. The researcher worked up a laboratory lash-up until it had an acceptable performance (experimental model 79X). This was examined by an engineer from Signal School workshops, accompanied by a draughtsman who took a few sketches to record vital dimensions. From this, the workshops provided two or more 'Chinese copies' of the lash-up (development models 79Y and 79Z), but improved mechanically to be suitable for shipborne use – in which the Signal School production engineers had more experience than anyone else in the UK. During the time the two 79Zs were being made in the workshops, draughtsmen drew a complete set of manufacturing drawings and these were issued to the chosen manufacturer to be worked to rigidly, absolutely no flexibility or interpretation being allowed (type 79 production models).

The same process was used with equal success in 1940 for type 281 with two Signal School production models, and type 271 in 1941 with no less than twenty-five in view of the tremendous urgency in getting 10cm radar into escorts.

Though equipment produced in this way was often accused of not being, what today would be called, 'user friendly', the overwhelming advantage was that something which worked was got to sea in an incredibly short time. As was learnt later in the war, and on all equipment since the war, any modification or 'improvement' introduces problems which take time to solve and, in the war, time was at a premium.

VALVES AND THE CVD COMMITTEE

Whereas with 7.5m the technology was fairly well established by 1938, with 50cm everything had to be new. An 'instant' radar set was not possible and Coales and his group had first to work on many fundamental problems. Back in 1937, Brundrett had remarked that the directional accuracy of the 23cm gear was almost good enough for the direct control of guns and it was already established that radar was significantly better at measuring range than the optical rangefinder.[27] Coales knew that gunnery applications would be of the greatest importance for 50cm radar, for which a compact aerial system would be an enormous advantage, so during 1938 much effort was put into investigations to achieve this. And all the other components – transmitter, receiver, CRT display – had to be designed more or less from scratch.

Some time in September 1938, Coales was visiting the GEC Research Laboratories at Wembley in connection with a centimetric communications set they were developing, when one of their engineers, R. le Rossignol, showed him with great pride a small triode valve which he had just designed. Coales realized that, with a fairly small modification, this could be the ideal valve to increase the power of the 50cm transmitter then being designed. But if GEC were asked to modify the valve, it would be obvious that the Navy needed it for something other than a CW communications transmitter. And secrecy did not allow the real reason to be mentioned to anyone in GEC.

The matter was of great urgency because, if successful, the modified valve could well increase the peak power at least ten times. Wright was approached and eventually gave his permission for Megaw and selected individuals at GEC to be let into the radar secret. But he knew that other valve manufacturers must be brought in some time, so in due course, the CVD organization was set up under the control of the Admiralty who, as we have seen, looked after the valve development needs of all three fighting services.[28] CVD stood for Communications Valve Development, a title adopted for security reasons to disguise the real purpose, which was radar.[29]

This rather technical matter has been treated at some length because of the importance of the CVD and of the 'micropup' valve – used extensively during the war by all three services – both of which stemmed from it. It was an example, too, of the difficulties which arose from the very necessary secrecy.

THE RADAR DIVISION MAKES PROGRESS

Whereas all in the new R Division were by now imbued with a sense of urgency not so evident before the reorganization, the same could not be said of all Admiralty departments. Remarking upon the long delay in the Admiralty's reply to C-in-C Portsmouth's proposals about the *Dunedin* trials, Horton dryly comments on the inconsistency of their Lordships' pronouncing that the trials were of 'great importance' while at the same time ruling that they were only to be conducted 'as opportunity offers'.[30]

On 22 February, Signal School had sent to DSD a five-page memorandum by Willett, Cecil and Horton. Following a succinct review of progress at Bawdsey and Signal School on long-range detection of aircraft, it discussed other possible uses to which radar could be put in the Fleet – in gunnery, in the detection of ships, RDF in aircraft – and made a brief mention of problems of enemy interception, identification, jamming and conflict with W/T communications. It pressed for an increase in scale of effort by the Admiralty and also by the Air Ministry because of the importance of ASV for aerial reconnaissance and shadowing. The concluding paragraph brilliantly sums up current thinking and future hopes:

> RDF is a subject in its infancy. Its potential importance can only be estimated, but enough has been done to show that a certain performance is possible. At the present time, *the work proceeding on behalf of the Navy is tentative and restricted.* If the full value to the Fleet of this important development of science is to be made available, *the scale of attack must be increased.* There is good reason to believe that the subject is now receiving attention abroad, and since it is impossible to confine any discovery or invention to one nation for long, it is of vital

> importance that, at any given moment, this country should be ahead of any other in the practical application and use of this new scientific aid.[31] [Author's italics]

Preserved with the Signal School copy of this paper is a copy of a cutting from *The Daily Telegraph* of 1935, telling of the test off the New Jersey coast of a highly secret 'ray' unofficially claimed 'to be capable of detecting ships 50 miles at sea and aeroplanes at great altitudes . . . it indicates the position of any ship up to distances of 50 miles so accurately that it can be used to direct gun fire.'

By March, the Signal School reorganization was beginning to take effect and on 25 March, what seems to have been the first full meeting of the Naval Staff on the subject of radar (including the Naval Air Division for airborne radar) took place, presided over by the Assistant Chief of Naval Staff (ACNS) and attended by Willett, Cecil and Horton. Development priorities were agreed upon – first, long-range air warning; second, rangefinding for surface gunnery, to include what was later called surface warning radar; and third, continuous short ranging on aircraft for anti-aircraft fire.[32]

As a first and immediate action – or as immediate as bureaucracy and the Treasury permitted – the meeting proposed that 'any available RDF equipment should be fitted in the principal Fleets forthwith'.[33] After Admiral Henderson (by now Sir Reginald) had visited Eastney to see the situation for himself,[34] the Board of Admiralty on 17 May sent a Most Secret letter to the Commander-in-Chief, Home Fleet, informing him of developments in radar, the following paragraph from which is worth recording:

> The detection of aircraft is evidently not the only tactical application of wireless echo location of interest to the Fleet. Priority is also being given to the development of the locating of surface vessels and to the achievement, in the more distant future, of an accuracy of continuous close range location of aircraft sufficient for AA control and searchlight purposes. Further there is a possibility that the apparatus fitted in ships for aircraft detection may also provide short ranges of aircraft with an accuracy appreciably higher than that of existing rangefinders.[35]

In the same letter, the Commander-in-Chief was directed to nominate two ships to be fitted with Type 79Y during the summer leave period. He chose the battleship *Rodney* and cruiser *Sheffield*.

In a minute of 31 May summarizing the proposals made at the March Naval Staff meeting, DSD had pointed to the very large expected demand on the Valve Section from the RAF's Home Chain stations, perhaps as high as 10,000 valves a month in wartime – 'there is a very real and grave risk that the whole RDF defence system, on which such high hopes are placed, may be brought to a virtual standstill in the moment of greatest emergency.'[36] As we have seen, some measures were taken in August – just a month before the Munich crisis – but they were considered again in November when the Swinton Committee called upon the Admiralty substantially to enlarge the Valve Section at Portsmouth to meet the needs of all three Services, who should bear the cost in proportion.[37] Shortly afterwards, the CVD Committee under DSR (Admiralty) was set up as already mentioned.

Then, on 16 August 1938, Their Lordships approved the immediate addition of more staff, including one scientific officer, one assistant and four silica glass workers for the Valve Section. They approved £5,000 as a contingency fund for the purchase of special apparatus (see Tailpiece on p. 29), and £10,000 for the employment of approved firms to develop specialized equipment. Any new building work was to take into account the projected transfer of the whole Signal School to a new site (plans being very tentative at this stage), and was to be restricted to immediate requirements.[38]

The remarks by the Admiralty's Civil Establishments Branch on DSD's hitherto cautious approach to staffing radar development are worthy of record:

> During the last two years it has been necessary to go to the Treasury three times for sanction to increase the number of people working on RDF and the last was only a month or so ago. It seems a pity it is not possible to lay down a definite programme or to foresee requirements such as those set out in this paper and thus avoid constant piecemeal applications for increases . . . CE is not really in a position to offer any useful criticism as it has no knowledge with which to test the word of the scientific authorities that certain research is necessary and that so much staff is required for it . . .[39]

TYPE 79 OPERATIONAL AT SEA

While the war clouds were gathering, only to recede somewhat after Munich, *Sheffield* and *Rodney* were being fitted with type 79Y – the first operational ships of the Royal Navy to have this new-fangled RDF. In October, *Sheffield* reported the following detection ranges on aircraft:

30 miles at 3,000ft
48 miles at 7,000ft
53 miles at 10,000ft[40]

In the autumn of 1938, the Valve Section produced transmitting valves capable of much increased power and as a result of this and of many lessons learnt from the trials in *Sheffield* and *Rodney*, type 79 was redesigned so that an experimental installation at Eastney Fort East developed 70kW transmitter power (compared with 20kW for type 79Y) and in May 1939 aircraft ranges of 40 miles at 3,000ft and 70 miles at 10,000ft were obtained. At the same time, the production of a quantity of this redesigned equipment was set in train by the method described on p. 21 above. The Admiralty had decided that the set was to be fitted in certain selected capital ships and cruisers (not flagships or aircraft carriers) and 36 ships were named. A total order for 40 sets (to cover reserves) was placed under conditions of great secrecy with Aeronautical and General Instruments Ltd. (AGI), a comparatively small firm in south London.

1939: TOWARDS WAR

Redesign of the 7.5m set to prevent interference between the various sets in the Fleet, and the manufacture in Signal School workshops of two further development models took place between the autumn of 1938 and early 1939, when J. D. S. Rawlinson, who had some fifteen years' experience with the development of radio equipment both ashore and afloat, took Symonds's place in charge of the long-range warning team. These sets, designated type 79Z, were completed by May 1939 and fitted in the anti-aircraft cruiser *Curlew* and heavy cruiser *Suffolk*, both operational in September. The first deliveries of the production models from AGI were also received in September.[41]

Meanwhile, *Rodney* and *Sheffield* had taken part in the annual Combined Fleet Exercises of the Home and Mediterranean Fleets off Gibraltar in March. In one exercise, *Rodney*, who was not herself transmitting, detected *Sheffield's* radar 100 miles away with a bearing accuracy within 2°, and this was confirmed by subsequent analysis. This danger of radar giving away the Fleet's position was to have a profound effect on the thinking of senior officers in the early stages of the war. In another exercise, both ships had their first experience of air strikes coming from several directions simultaneously: *Sheffield* seems to have coped better than *Rodney*.[42]

The sea trials of Type 79Y in *Sheffield* had shown that, rather unexpectedly, radar on 7.5m gave useful ranges on surface vessels, but the results were unreliable and detection at any given distance was not a certainty. So detection of surface craft was put next in priority to the Naval Staff Requirement for the detection of aircraft. To meet this, it was necessary both to reduce the wavelength and to increase the power.

While there was a good chance of meeting the need by the work on 50cm, success was by no means a certainty, so work was put in hand on a metric set, similar to Type 79 but with higher power and shorter wavelength. Once more, success in the first instance depended on valve technique, but work aimed at this target – higher power at shorter waves – had been going on continually and early in 1939 it was seen that several hundred kilowatts was to be had at about three metres. Work was therefore started on a set at 3.5m, eventually to be called type 281.

Work on 50cm had been proceeding apace. By June, the transmitter was giving an output power of 1.2kW, almost certainly a 'world first' in generating more than a kilowatt on 50cm. In the same month, sea trials took place in the destroyer *Sardonyx*, ships being followed out to 5 miles and low-flying aircraft to 2.5 miles.

Now that radar was at sea, there had to be people to operate and maintain it. Being the responsibility of the Signal Branch, it was logical that the maintenance and most of the operation should be done by Telegraphist ratings. However, young non-specialist seamen ratings were brought in to help with the watch-keeping, chosen almost at random – literally 'detailed off at Both Watches'. (Get a naval friend to explain this!)

* * *

All this had been achieved in spite of the limited priority still accorded to radar by the Naval Staff. Even as late as July 1939, despite the war clouds that were massing on the horizon, the Signal School's 'Memorandum on expanded effort on RDF' had to remind the Admiralty that radar was still supposed to take second place in the priority of work in HM Signal School, communication equipment for new-construction ships taking first place.[43]

This last fact was one of those laid before the 2nd Naval Radar Panel Meeting which met on 27 July under the chairmanship of ACNS. The panel was told that by September, *Rodney, Sheffield, Suffolk* and *Curlew* would be fitted with type 79, and that the battleship *Valiant* and three more anti-aircraft cruisers would be fitted by the end of the year. They decided on the names of ships for the next 30 sets due off the production line early in 1940, giving priority within that group to AA cruisers, but no aircraft carriers were mentioned. (In the event, it was the dates the ships came into dockyard hands that decided the matter.) All new construction and reconstruction large ships (which did include the new *Illustrious* class aircraft carriers) should be prepared for fitting. The priorities of Staff Requirements laid down in December 1938 (which governed the priorities for work at Signal School) were amended so that short-range air and surface warning (which was to include accurate range and bearing for fire control), and identification, came immediately after long-range air warning, while accurate high angle fire control dropped to third place. The very large staff and accommodation increases asked for in the Signal School memorandum were agreed.[44]

Then, on 10 August, came the following in a letter from the Secretary of the Admiralty: 'I am to acquaint you that Their Lordships have decided that the development and supply of RDF for naval purposes is to be extended and accelerated and given the highest priority'.[45] At last!

Less than a month later, on 3 September 1939, Britain declared war upon Germany.

A PREWAR SUMMING-UP

During the prewar period, naval weaponry against air targets was concentrated entirely on gunnery, unlike the RAF who were concentrating on fighter interception of attacking aircraft. There seem to have been no plans for fitting air warning radar in aircraft carriers until well into 1939, though airborne radar in naval aircraft was always considered necessary.

It takes longer to scramble an aircraft and to reach operational height than it does to close up a gun's crew, and thus the RAF needed longer aircraft warning ranges than the Navy. The first preliminary Naval Staff Requirement of a warning range of 60 miles (p. 11 above) was based on allowing guns' crews to close up at warning, thus avoiding the need to remain at action stations for long periods unnecessarily. Providentially, this range was achieved by the first prototype operational radar, type 79Z, despite the naval limitations of comparatively small aerials and power available at the comparatively short wavelength of 7.5m. The RAF Chain Stations were able to achieve the greater ranges needed because of the higher power available at the longer wavelength of 25m and the higher gain from the much larger aerials possible ashore.

One of the most striking things about the prewar period is the contrast between the limited rate of progress in the first two years and the remarkable achievements thereafter, achievements which allowed the Royal Navy to enter World War 2 with operational radar. The situation in late 1937 was not good: metric radar was still very much in the experimental stage and as things stood seemed unlikely to achieve even the minimum requirements within a reasonable time, while centimetric radar was even more experimental. Yet only months later, not only had most of the significant shortcomings of the metric radar been removed, but two ships of the Fleet had been fitted and were reporting the detection ranges of aircraft at more than 50 miles. And in parallel the 50cm work had been put on a realistic basis for development, resulting in the extensive fitting of fire control radar from 1940 onwards. The two people who probably did most to stimulate the small naval radar team of scientists and engineers to achieve this were Horton at the Signal School and Wright at the Admiralty, aided on the naval side by Willett and Cecil.

TAILPIECES

• Some time in 1938, a young naval stores officer in Portsmouth Dockyard was going through demand notes from HM Signal School when he came across one for electric kettles. The good book said these were not allowed by the warrant, so he returned the form marked 'not entitled'. Very soon after, an irate inspecting officer telephoned. Had our friend not noticed that the demand came under Vote 6, the Admiralty Experimental vote, which should be met without question?

Visiting the Signal School on duty a month or so later, our friend was not entirely surprised to discover that, unlike staff in other establishments at the time, scientific, technical and administrative staff at that part of Signal School were able to supply freshly-made tea and coffee more or less on demand.[46]

But there has been a riposte to this implied slight on the integrity of Signal School at that time. There were already plenty of other ways of making tea – with Bunsen burners, etc., we are told. Obviously, the kettles were required as an artificial load for transmitters in the laboratory.

• There was a policy which prevailed until the end of 1939 that flagships should not have air warning radar fitted because it would interfere with communications. Some unnamed cynics had another explanation: they said that the Fleet Flagship *Nelson* was to have been one of those chosen for the trial in 1938, but that this was vetoed by the Commander-in-Chief because radar demanded the top position on both masts – so where could the admiral fly his flag? *Rodney* and *Sheffield* were neither of them normally flagships.

3

1939–40: The Phoney War – But Not at Sea

Sep 3 War with Germany
Dec 13 Battle of River Plate

On 3 September, the day war was declared, the only two radar-fitted ships that were operational, *Rodney* and *Sheffield*, were with the Home Fleet, based on Scapa Flow in Orkney. Of the next two being fitted, the AA cruiser *Curlew* was completing her refit at Chatham, reaching Scapa for 'work-up' in October (then to be employed principally in escorting East Coast convoys), while the heavy cruiser *Suffolk* was completing her refit at Portsmouth; she sailed to the Mediterranean early in October but returned to join the Home Fleet at Scapa in November. The battleship *Valiant* was to complete her refit at Plymouth in November, the AA cruiser *Curacoa* at Chatham in December.

Air warning radar cover of Scapa Flow was provided by the RAF CH station at Netherbutton near Kirkwall. Initially, this was a temporary mobile station, connected by telephone lines with the Fleet flagship's buoy. On 5 September, Netherbutton reported an aircraft at 70 miles, very high. When sighted and identified as a Heinkel 111, *Sheffield* opened fire and could claim to have fired the first naval shots of the war.

NAVAL RADAR AT WAR

The first radar-assisted naval skirmish took place on 26 September. The submarine *Spearfish* was in trouble off the Horn Reefs, unable to dive, and units of the Home Fleet were sent to extricate her, with the battleships *Nelson* and *Rodney* and the aircraft carrier *Ark Royal* in support. *Spearfish* having been rescued, all ships were withdrawing back to Scapa when *Rodney's* radar reported two or three large groups of aircraft at about 80 miles and closing. She found it relatively simple to track those groups and the Commander-in-Chief in *Nelson* was kept fully informed by flag signals, but *Ark Royal*'s Skua fighters were not flown off. In fact, this main

3. The battleship *Rodney* in 1940. Fitted with the 7.5m WA type 79Y in 1938, she was one of the first two British ships to have radar, with the transmitting aerial at the top of the mainmast, the receiving aerial on the foremast abaft the bridge.

(*IWM*)

bomber force attacked British cruisers and battlecruisers some distance away, and *Hood* was slightly damaged. *Rodney* felt that her radar reports were not being taken seriously enough by the Command.[1]

However, she failed to give any warning of the single Dornier aircraft which dropped a bomb alongside *Ark Royal*, fortunately without damaging her. (This incident led to the first of many German claims to have sunk that famous ship.)

> Single bombing aircraft were detected at about 20–25 miles range. No difficulty was experienced in following them in, but within a range of 10 miles, reports of them by flag signals served no useful purpose owing to their high rates of change of range and bearing.'[2]

So reported Captain Syfret of *Rodney* after the action, drawing attention to the communication problems which were to become so important as the war progressed.

On 14 October, the old battleship *Royal Oak* was torpedoed by a U-boat while at anchor in Scapa Flow, with the loss of many lives. On the 16th, two squadrons of Ju88s attacked ships lying in the Firth of Forth, and the next day, while the main body of the fleet was at sea, a raid of similar strength took place at Scapa, damaging the base ship *Iron Duke* (an even older battleship) so that she had to be beached.

These events caused the Commander-in-Chief temporarily to abandon Scapa as a base, so the Home Fleet had to wander between Loch Ewe, the Clyde, and Rosyth until March 1940 when Scapa's defences were deemed to have been sufficiently improved.[3]

VICE-ADMIRAL SIR JAMES SOMERVILLE

The day after war was declared, the new Controller of the Navy, Rear-Admiral Bruce Fraser (Sir Reginald Henderson had died suddenly in April) had appointed Vice-Admiral Sir James Somerville for special service inside the Admiralty to be responsible for the coordination and development of radar in the Navy.[4] Only a few months earlier, when serving as Commander-in-Chief, East Indies, Somerville had been invalided home with suspected pulmonary tuberculosis. By the inflexible rule of the Service, he was placed on the Retired List, despite certificates from eminent civilian medical men of a complete recovery with no signs of the disease remaining.

Fraser could not have chosen a better person to progress radar in the Navy. As a signal specialist, Somerville had himself been the Signal School's first Experimental Commander from 1917 to 1920 and he had had a brilliant career thereafter. Up to the outbreak of war, RAF needs had been paramount and, in that atmosphere, funds for naval radar and weapons had been difficult to obtain. Somerville, however, had the knowledge, the drive and the necessary seniority both to press the Navy's case externally, and to galvanize the Navy itself, the dockyards and those concerned in industry. And he had the fullest confidence of the new First Lord of the Admiralty, Winston Churchill. Much of the credit for the remarkable speed with which ships were equipped with radar during 1940 and 1941 must go to Somerville – in Watson-Watt's words, 'the foster-father of naval radar'.[5]

Since 1936, there had been a strong feeling in the Air Ministry – largely generated by Watson-Watt who now had the title of Director of Communication Development – that all radar research should be centralized at the Bawdsey Research Station, and that radar work by other Services should be restricted to development specific to the Service concerned. This view was accepted by the War Office but, as we saw on p. 12, the Admiralty had insisted that both research and development for any seagoing application should be done by HM Signal School at Portsmouth.

By August 1939, Watson-Watt once again felt his position threatened – by Admiralty proposals to extend the laboratories at Eastney, and by the Army's plans to move their research from Bawdsey to the new Air Defence Experimental Establishment at Christchurch. (In 1941, ADEE was to become the Air Defence Research and Development [ADRDE]; in May 1942, when it moved from Christchurch to Malvern, it became the Radar Research and Development Establishment [RRDE].) Contending that these proposals cut across agreed policy, Watson-Watt urged that a major policy decision be sought in favour of completely unified control and execution of research, development and production for the three services – to make the most effective use of staff and of the limited number of firms, and the earliest possible application in all fields – proposals which were agreed in Air Ministry up to the level of the Secretary of State. Then, on 1 September under the threat of air attack, all staff except those manning the CH station left Bawdsey, most of the Air Ministry scientists going to Dundee to form the Air Ministry Research Establishment, while the Army scientists went to Christchurch.[6] (AMRE moved from Dundee to Swanage in Dorset in May 1940, where it became the Telecommunications Research Establishment [TRE], moving in turn to Malvern in Worcestershire in May 1942.)

This, then, was the situation that faced Somerville on his first day at the Admiralty on 5 September. On the 8th, he represented the Navy at a Treasury inter-service committee, where Air Marshal A. W. (later Lord) Tedder (Director General, Research and Development) gave the Air Ministry view that the radar research and development establishments for all three services should be grouped in the counties of Angus and Fife in Scotland. On the Admiralty's behalf, Somerville objected strongly, but a further meeting on the 13th gave rise to two proposals of interest: (a) that F. W. Brundrett of DSR Admiralty should act as a 'London recruiting bureau' for all new staff, whether civilian or military, to work on radar for all three fighting services;[7] and (b) that radar research for all three services (except research on valves for which the Admiralty was already responsible) should be conducted by the Air Ministry under Watson-Watt's superintendence, but that development should be done by the three services independently.[8]

Proposal (a) was implemented immediately and Brundrett continued to oversee the recruitment of all scientific staff for government service, later being responsible to Sir Maurice Hankey for this part of his duties. Proposal (b) was finally agreed on paper in August 1940, but by that time it had been overtaken by events – the separation of the three radar establishments was complete and all were doing research as well as development. It was too late to change this state of affairs and there was in fact no centralized radar research during the war, though the establishments freely exchanged research results amongst themselves. It is debatable whether the policy of a central research and separate development establishments would have worked even if it had been implemented.[9]

RECRUITING SCIENTISTS AND ENGINEERS

Before the war, Brundrett, despairing of any inter-service agreement, had acted on his own and by the war's outbreak had recruited no less than 100 men to the Navy's research establishments, inventing new grades to accommodate them. The intake included some of the best and brightest young men from the universities as well as several well known in radio circles. Of the men then recruited and later additions no less than eight eventually became Fellows of the Royal Society and three Fellows of Engineering – *FRS*: H. Bondi, J. F. Coales, C. Domb, T. Gold, F. Hoyle, R. Keynes, M. H. L. Pryce, R. J. Pumphrey; *F.Eng:* R. Benjamin, J. F. Coales, P. Trier. These newcomers and the old hands were welded together into a most successful team by C. E. Horton, whose great achievement was to give technical leadership but, more important, to create what in naval circles is called a happy ship. From that band of people not only did many achieve high academic distinction but, of those who stayed in the scientific civil service, a significant number reached the highest grades.

A year later, in August 1940, the Secretary to the War Cabinet, Sir Maurice Hankey, was made responsible for meeting Britain's ever-increasing demand for radio engineers. Realizing that large numbers were required not only for research and development but also for the armed services generally, Hankey drew up special training schemes for what came to be called 'the Hankey boys'.[10]

If it had not been for the far-sighted vision of Brundrett and Hankey, it is doubtful if radar could have played the decisive role that it did. This was a new weapon to the services; not only were substantial resources needed to develop it but highly skilled manpower was to be needed to use and maintain it. Time did not allow years of 'debugging', so sailors – as well as soldiers and airmen – had to keep in operation equipment which a few years before would not even have been found in scientific laboratories.

SOMERVILLE'S IMPACT

Somerville set about his task with great vigour. During September he did a great deal of travelling to acquaint himself with the current situation – to Bawdsey on the 9th and 16th (of which more anon) and Christchurch on the 11th, the two earlier visits in company with Watson-Watt (with whom he remained on good terms despite their disagreements in committee); to Portsmouth on the 6th, 18th and 30th, visiting not only the Signal School and Eastney but also the Gunnery and Torpedo Schools; to GEC's Research laboratories on the 20th; to Chatham on the 22nd, when he saw *Curlew*'s new type 79 air warning set; to the Home Fleet at Scapa Flow on 26–28 September, calling on the evacuated Bawdsey scientists at Dundee on his return journey; to Bath on 4 October to call on the recently evacuated Admiralty technical departments. And all this had to be fitted in with paper work in London – he had a desk in the Signal Department (see *Admiralty organization* in Glossary) – and many meetings at Air Ministry and Admiralty, where the newly-appointed First Lord was acting with equal vigour – and stirring up the whole Admiralty organization. On most of these visits, Somerville was accompanied by Commander H. F. Lawson of the Naval Ordnance Department, who had been let into the radar secret a few months earlier.[11] (Lawson had been Commander (G) at the Gunnery School before coming to DNO. He became executive officer of *Prince of Wales* in March 1941, and was lost when she was sunk.)

Churchill visited Portsmouth on 21 September.

> I was struck [he wrote to the First Sea Lord] by the concentration of research and development in wireless, including RDF, at the Signal School at Portsmouth. I feel that we are running a great risk in leaving all this valuable plant and work in such a vulnerable place. Would it not be possible to separate the research work and some of the

4. Vice-Admiral Sir James Somerville on board his flagship when in command of Force H, with his Chief of Staff, Captain N. J. W. William-Powlett.

(IWM)

development work from the instructional functions carried out by the School, and from that portion of the development work that must be carried out in the vicinity of ships?.[12]

As we shall see in the next chapter, the dispersal was not fully implemented until 1941, though steps were already being taken to move the part of the Valve Section concerned with centimetric wavelengths to the H.H.Wills Physics Laboratory at Bristol.

Churchill also intervened directly in Watson-Watt's attempts to 'poach' Horton from the Signal School. In about May 1939, a new post of Assistant Director (Communications Development) was advertised by the Air Ministry, to assist Watson-Watt in the planning and direction of radar research for all three services. As this meant a considerable promotion, Horton was persuaded to apply, with the Admiralty's reluctant concurrence:

> Horton stood out head and shoulders above the other applicants, [Sir Kingsley Wood, the Air Minister, told Churchill on 21 September] and on 12th July we wrote to the Admiralty enclosing the letter to Horton offering him the appointment. This letter to Horton had not reached him on September 12th and he had been informed that the letter was mislaid in the Admiralty . . . We recognise the seriousness to Signals School of the loss of Horton, but with a full sense of responsibility for RDF progress as a whole, I ask you to be so good as to release Horton to us for the common good.[13]

Happily for the future of naval radar, Churchill (advised by Somerville and Wright) resisted the blandishments of his Cabinet colleague:

> The Admiralty is, as you know, a rather newcomer [*sic*] into this field. We have the greatest need to develop RDF both in respect of recognition signals and of range-finding. Several ships have already been fitted, and a large-scale application is already in progress at the utmost speed. The process is no longer one of research but of application. Dr Horton is, I am assured, literally the key-man in this . . . In these circumstances I hope you will be so kind as not to press me upon the point.[14]

Very soon after, Horton was promoted two grades, from Principal Scientific Officer to Superintending Scientist, about the same time as Basil Willett, the Experimental Commander, received the acting rank of Captain.

Somerville's visit to Bawdsey on 16 September to assess the performance of the Army's new 1.5m Coast Defence set against HM

Submarine H.34 was described by Professor J. D. Cockcroft, who was present with W. A. S. Butement, one of the designers of the set:

> We, from Rye, went back to Bawdsey to study coast defence and there met the redoubtable Admiral Sir James Somerville. He turned up one day with a tame submarine with a wireless set to control the same, and proceeded to spot it on the CD set. He nearly lost it to one of our bombers but fortunately it survived and he went away full of enthusiasm.[15]

Somerville's diary entry was laconic:

> Then to Bawdsey & carried out RDF tests. Not very conclusive but Butement & Cockcroft quite optimistic. H.34 got bombarded at 1045 by one of our aircraft. Luckily no damage. Should have been obvious.[16]

Later, Somerville became very involved with inter-service air and surface radar coverage, first of the anchorage at Scapa Flow, then of the Fair Isle passage between Orkney and Shetland. He was appointed Inspector of Anti-aircraft Weapons & Devices on 1 January 1940, where he continued to dominate naval radar policy matters until June when he was appointed Vice-Admiral Commanding Force H, based on Gibraltar.

HM SIGNAL SCHOOL AT WAR

In Portsmouth on 1 September, everyone was at fever pitch. The Reserve Fleet was being mobilized and reservists recalled. The Naval Barracks was seething with men being issued with uniforms and there were hammocks and kitbags everywhere. A honeycomb of air-raid shelters was being made under the parade ground to house 4,000 men, while families who lived in the city were moving to stay with friends outside, or in such temporary accommodation as they could find. Old friends who had retired from the Navy reappeared in uniform and everyone was seized with the urgency of the situation. Massive air raids were expected and – though only the select few could know this – the need for radar became paramount.[17]

Up to the actual declaration of war, recruitment for radar work in the Signal School's Experimental Department had, as we have seen, been in penny numbers, grudgingly, one or two at a time, and many promising lines of research had to be forgone.

> Suddenly the flood gates opened and people literally poured in from all quarters; from the universities (laboratory technicians, graduates

> and professors), from Industry (Baird television, Philips, Marconi, and so on), from other Government departments, from Public Utilities. This intake and its distribution was orchestrated by F. S. Brundrett (later Sir Frederick) and was a remarkable achievement. It also showed enormous adaptability (and restraint) on the part of the few permanent staff whose job it was to let the newcomers into the great secret and absorb them effectively into the task – whilst still keeping the vital momentum going without faltering.[18]

The above graphic description is by S. T. Wright, one of those newcomers who joined in December. He continues:

> No training was possible. I, for example, who had been despatched from the Post Office Engineering Department Radio Branch, was sat down on my first day in J. D. S. Rawlinson's office at Eastney and given the only copy of the somewhat dog-eared handbook for type 79Z to read before it was carefully locked away again. After that, you were on your own and expected to get on and defeat Hitler with a soldering iron (provided). But all went very smoothly and efficiently, so that people who had in fact joined only a week or so before I arrived, seemed to me to have had *always* worked on RDF.[19]

It was realized that, with the expansion of radar requirements, Signal School workshops would not be able to make all the development models which would be needed, hitherto always made 'in house' without calling upon the manufacturers. In September, therefore, a standing contract was awarded to Messrs. Allen West, a Brighton electrical firm already well known to Signal School, to do such work on a continuing basis. In effect, Allen West's Moulscombe factory became part of the Signal School as far as development work was concerned.[20]

THE NEED FOR SURFACE DETECTION

At the outbreak of war, the 7.5m air warning type 79 was already at sea in the Fleet and forty more sets were on order for fitting in large ships of the Fleet in 1940 – in existing ships whenever they came into dockyard hands for a long enough period, as well as in the ships due to join the Fleet from builders' yards, laid down from 1936 onwards as part of the rearmament programme – the *King George V* class battleships, *Illustrious* class aircraft carriers, *Fiji* class heavy cruisers, *Dido* class light cruisers and *Hecla* class depot ships.

Type 79 needed special offices to be built in any ship to be fitted and also required alterations to masts – or sometimes new masts – to carry the aerial system. It was thus an eight-week job to fit.[21] The converted C and D class AA cruisers received priority for radar fitting, the only one never to be so equipped being *Calcutta*, who could never be spared for a long enough period before she sailed for the Mediterranean in August 1940; she was sunk off Crete in June 1941.

The need for a set to detect surface craft and low-flying aircraft, for both warning and gunnery purposes, was stated in the very earliest days of naval radar, but wavelengths shorter than 7.5m were necessary and the technical problems were prodigious and only limited research and development effort could be spared before 1939. By the outbreak of war, the design of what was at the time termed type SS – Ship to Ship, as opposed to type 79 which by inter-service agreement was classed as type SA, Ship to Aircraft[22] – was still awaiting a decision on the wavelength to be used.

The preferred wavelength for what Somerville called the 'Combined Wireless Rangefinder and Lookout Set'[23] was 50cm, on which Coales and his team had been working since 1938. However, sea trials of a 50cm set with a single rotating frame aerial in *Sardonyx* on 14 October 1939 proved disappointing so far as surface detection was concerned as they did not achieve the hoped-for ranges to make it a viable surface warning set, so the idea of a combined surface warning and rangefinding set was abandoned. Instead, Somerville, Willett and Horton decided that the immediate solution to the SS problem (only a temporary expedient until individual radio-rangefinders for the various armaments became possible) was that the metric sets must have rangefinding ability. It was resolved:

- To add to existing 7.5m sets (SA – type 79) a ranging panel adapted from the Army's GL Mark I, and design a new receiver to deal with the shorter pulse lengths needed. This would give ranges to an accuracy of 50 yards out to 14,000 yards, but bearing accuracy would be no better than ±10°.
- To make the new 3.5m set (SS – type 281) serve a triple function: (1) as a long-range warning set, (2) as an AA rangefinder, and (3) as a rangefinder for the main armament, accepting the fact that it could not perform these functions simultaneously. Beam-switching (see Glossary) was to give a bearing accuracy of about ±½°, and the shorter wavelength would make it marginally better than type 79 in detecting low-flying aircraft and surface craft.[24]

THE GENESIS OF TYPE 281, THE HOPED-FOR ALL-PURPOSE SET

These proposals were translated into action with commendable speed by Rawlinson and his team at Eastney. The modified type 79 did sea trials in the AA cruiser *Curacoa* in March 1940 and by the end of the year all existing type 79s except those in *Rodney* and *Sheffield* had been modified and renumbered type 279. As for type 281, Somerville chaired a meeting at the Admiralty on 15 February 1940 to decide on where the prototype should be fitted. Signal School had said that, if the requirement for accurate ranging to 10 miles were to be met, type 281 could not be fitted in cruisers of the size of the *Dido* class then building, owing to the increase in aerial weight aloft as compared with type 279. No, said Somerville, it was the small cruiser not the battleship that urgently needed such a set, 'not only for gunnery purposes but also for shadowing'.[25] The prototype must go in *Dido* herself, with mast height reduced if necessary, accepting reduction in performance. (The topweight restriction did not apply to later sets, for which a much lighter aerial pedestal was designed.) And so it was, though it was agreed that the next three ships of the class, *Naiad, Phoebe* and *Bonaventure*, should have type 279 with full-height masts.[26]

So work on type 281 pressed forward with the highest priority at Eastney, breaking new ground by reason of its high power because fresh developments in silica valve techniques (maturing early in 1940) had brought 1000kW within reach. Another notable advance was the incorporation of beam-switching, giving accurate bearing determination. The same accurate ranging equipment was included as in type 79, except that it was integrated into the receiver panels instead of being an add-on unit.

The first experimental type 281 was operating in the laboratories in May 1940 and the first development model was erected, running and on test at Eastney in time to be able to plot the first air raid on Portsmouth which took place about 6pm on 11 July and was plotted in from the coast of France and from west of the Isle of Wight from 5.30 onwards. Nearby Royal Marine guns' crews watching the aerials train on the raiders were heard to marvel at these 'death rays' which dispersed the raiders. By that time, Signal School had had several unofficial reports of these 'rays' stopping motor car engines.[27]

The first prototype made in Signal School workshops was released for sea trial in August and fitted in *Dido*. Even by June, however, experimental results had given sufficient promise to justify the risk of ordering in quantity and in that month the Admiralty placed an order for 36 sets.[28]

CARLISLE AND TYPE 280

Another possible solution to the Navy's SS problem was to make use of the considerable work that had already been done at Bawdsey in producing the Army's GL (Gun Laying) anti-aircraft radar. During the

5. The AA cruiser *Carlisle* at Malta in 1942, fitted with the 3.6m WA/GA type 280 radar, developed from the Army's GL Mark II. Transmitting and receiving aerials were at the top of the masts.
(IWM)

last quarter of 1939, *Carlisle*, completing her conversion to AA cruiser in Devonport Dockyard, was fitted with the 3.6m type 280, based on the Army set GL Mark II 'wireless rangefinder', with aerials specially designed by Ross's party in Signal School.

At the time it was unique as it provided warning cover to a maximum of about 60 miles, with a bearing accuracy of about ±2°. Furthermore, once the target had been sighted and the gun director trained upon it, the radar could 'see' the target and radar ranges (significantly more accurate than optical ranges) could be passed continuously to the gunnery system, being plotted automatically on the range plot of the High Angle Control System (see p. 4 above). Warning capability was of course lost while gunnery ranging was being carried out.

On 1 February 1940, *Carlisle* sailed for Malta to work up and to try out various different aerials designed at Nutbourne. She had no trouble in the detection of aircraft, getting good range and bearing information provided there was no interference from land echoes or the target was not too low. Good polar diagrams of the vertical lobes and side lobes were calculated and confirmed by actual results, so estimation of target height became quite good, giving the HA Director an elevation bracket to search in. Coordination between the radar operators and the DCT (director control tower) became very efficient, especially when the system was modified so that the director trainer could follow the radar trainer as well as the other way round.[29]

Carlisle served with distinction in the Norwegian campaign, in the Mediterranean and in the Red Sea, type 280 thoroughly proving its worth, certainly being better for gunnery purposes than type 79/279. However, the Navy did not continue with type 280 (except for four Auxiliary AA ships in 1940) and it was replaced in *Carlisle* by types 281 and 285 in 1943.

RADAR FOR GUNNERY AND FIRE CONTROL

Soon after the outbreak of war, the Director of Naval Ordnance (DNO) had asked Signal School whether a radar could be developed to provide accurate ranges on dive-bombers from a maximum range of 2.5 miles down to half a mile, to give the moment for the firing of a 'barrage' by ships' main armament. The *Sardonyx* trials having suggested that this performance might be possible, development on 50cm was continued as fast as was possible with the very small team available. It was soon demonstrated that the required performance could be achieved by fitting an aerial array of two Yagi 'fishbones' – one for transmission, one for reception – to the current pom-pom director.

This development is described by Captain Stephen Roskill, the gunnery officer who was the main author of the *War at Sea* volumes of the official history of the Second World War, writing here in 1979:

> After returning from *Warspite* as a Commander, I was appointed to the Naval Staff (first to DTSD, later to Gunnery Division) and was the representative of the Staff on Admiral Somerville's committee on radar development. Although I cannot give a precise date to it I think it was after one of that Committee's meetings early in 1940 [actually, it was a month or so earlier than this] that I asked Mr Horton, a Signal School scientist, if it would be possible to mount the radar antennae on gun directors instead of at the top of the masts. Horton asked, 'If it were possible, would there be a Staff Requirement?', and on my replying emphatically in the affirmative he asked me to come to the Signal School at Portsmouth for a meeting at an early date to discuss the possibilities. There were precisely five of us at the meeting – Captain Basil Willett of the Signal School in the chair, Commander H. F. Lawson of DNO's Department, Messrs Horton and John Coales (Signal School scientists) and myself.
>
> That meeting marked the genesis of the family of 50cm sets known as type 282 (short range AA), 284 (main armament) and 285 (long range AA) . . .[30]

Mounting the radar aerials on the gun directors so that they rotated with them seems to have been a new concept for the Royal Navy, though the German Navy's Seetakt were already so mounted.

Development of the 50cm systems now proceeded at the highest priority. Trials at the Gunnery School's AA Range at Eastney in February 1940 were based on the two-fishbone aerial array on the pom-pom director mentioned above and were successful enough for the two officers attending from the Admiralty, Roskill from DTSD and Commander W. K. Edden, Lawson's colleague in DNO, to give a verbal go-ahead for orders to be placed for 200 sets, a decision officially confirmed in April, five months before the design took final shape. So type 282 was born.[31]

In early 1940, there was not a single 50cm component of which the design had been finalized, yet plans were laid for the production of two hundred complete equipments within six months. Coales's group was increased to about twenty but there was not sufficient space for them in the RN Barracks, so a disused school in Onslow Road, Southsea, was requisitioned, to be the group's headquarters until 1942. The Signal School had always had the closest relations with what today would be called the electronics industry. In January 1940, therefore, arrangements were made for the design and manufacture of the transmitters to be undertaken by GEC, the modulators and generators by BTH, the receivers by Marconi, and the CRT displays and consoles by A.C. Cossor, all under the close supervision of the Signal School, where the aerial system and feeders were to be designed.

THE BATTLE OF THE RIVER PLATE

Two operational matters deserve mention before concluding this account of the first six months of the war – often called the 'Phoney War', though this description certainly did not apply at sea.

In the Battle of the River Plate on 13 December 1939, none of the three British cruisers engaged had radar, but the German pocket battleship *Admiral Graf Spee* had the earliest German Seetakt radar, the 60cm FuMG 38G, fitted in 1938 as a radio-rangefinder, with a mattress aerial fixed to the main armament director tower on the foremast just above the main optical rangefinder, making it easy to compare ranges by the two methods.[32] Its nominal maximum range was about 8 miles. Later models (with wavelength increased to 80cm) had beam-switching for accurate bearing measurement, but *Graf Spee*'s did not. During the battle, her gunnery was extremely accurate and, though her first salvo was 600 yards short, she began to hit *Exeter* at about 9.5 miles. How much this owed to radar is not certain as it was at the FuMG 38G's extreme range. A postwar account by *Graf Spee*'s Artillery Technical Officer, Lieutenant F.W. Rasenack, seems to imply that radar was not an important factor.[33]

However, her subsequent scuttling in the approaches to Montevideo on 17 December gave the British an opportunity of learning something about German naval radar, the very existence of which was not then known, even if suspected. 'A peculiar fact is the rangefinder on top of the superstructure, which was continuously in use at sea and kept revolving all the time by motor.'[34] – so reported the British Naval Attaché on 21 December after examining British merchant seamen prisoners of war from the *Graf Spee*. The Naval Intelligence Division and DSD were alerted. Could this be radar?

A study of prewar photographs of *Graf Spee* revealed that some time in 1938 a new cylindrical structure (referred to later as the radio house) had been built on top of the existing rangefinder house which surmounted the rotatable gunnery control tower on the foremast, and that, in some photographs only (so probably dismountable) a canvas-covered 'mattress' was affixed to the side of the radio house facing in the same direction as the rangefinder. Further interrogation of the British ex-prisoners and a study of photographs taken after the scuttling showed this was some form of radio apparatus, very likely radar.

6. The forward superstructure of the German pocket battleship *Admiral Graf Spee* in 1939, showing the aerial of her 60cm FuMG 38G radar mounted on the main armament director tower above the optical rangefinder. This photograph was taken by a *Graf Spee* officer before the Battle of the River Plate and the film was handed to a local chemist in Buenos Aires to be developed. The chemist happened to be British and passed a copy on to the British naval attaché.

(HMS Collingwood)

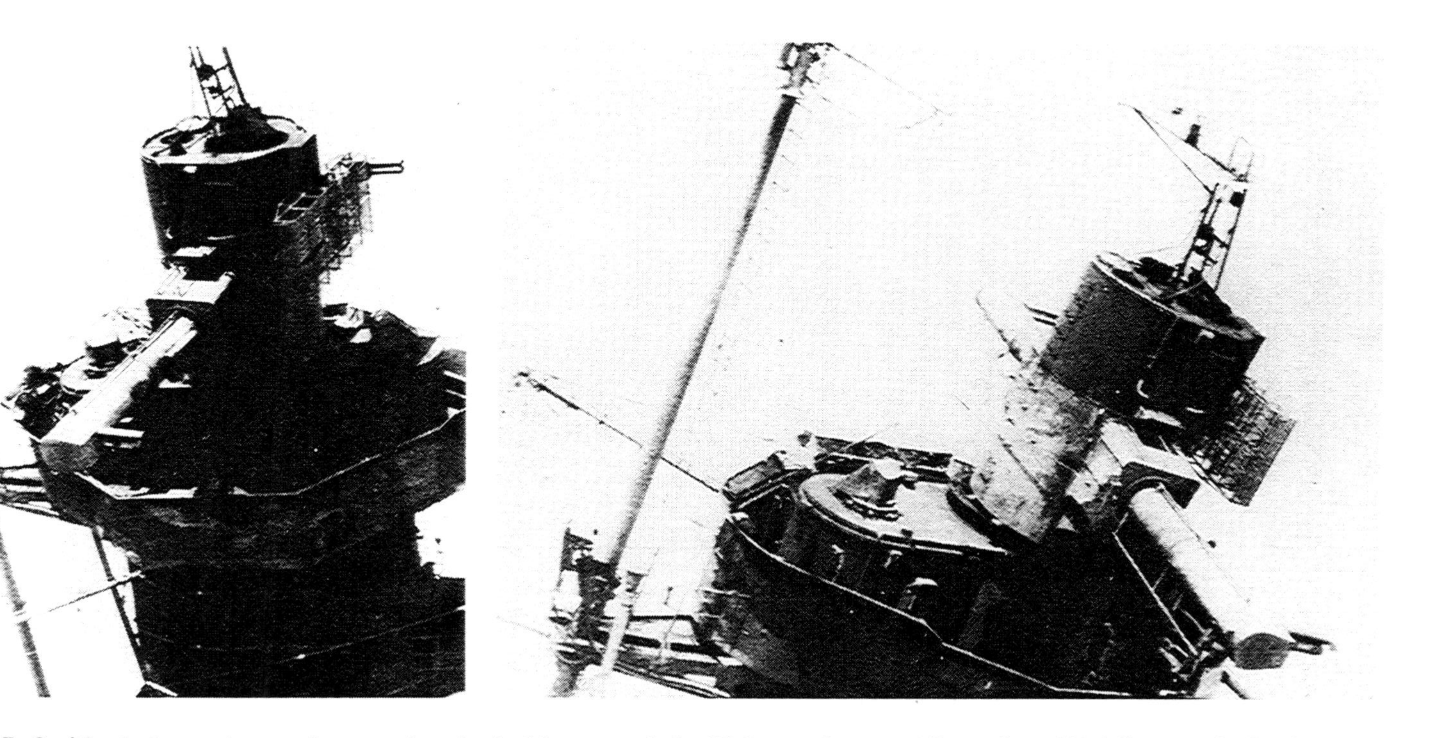

7. *Graf Spee*'s forward control tower after she had been scuttled off Montevideo on 16 December 1939 following the Battle of the River Plate, showing the FuMG 38G radar aerial.

(NMM)

On 14 January 1940, the British Minister in Montevideo, E. Millington-Drake, told the Foreign Office that the British Naval Attaché 'had received instructions to endeavour to make careful examination of it [the wreck] with a view to finding one particular naval secret.'[35] However, he said, the wreck was being closely guarded by the Uraguayan Navy and clandestine examination was impossible; the only way that Admiralty representatives could get on board would be for the wreck to be purchased by someone sympathetic to the Allied cause.

On 7 February, a British radar expert reached Montevideo, posing as the representative of Messrs. Thos. Ward Ltd, shipbreakers of Sheffield. This was L. H. Bainbridge-Bell, an Air Ministry scientific officer who had been one of Watson-Watt's original team at Slough, Orfordness and Bawdsey, seconded to the Signal School since the first week of the war. He had left London on 31 January, studied more photographs of the ship at the Deuxième Bureau in Paris, and travelled by air from Marseilles.

On 20 February, the Foreign Office told Millington-Drake that HM Government would provide £20,000 for the purchase of the wreck. Julio Vega-Helguara, overtly representing a Montevideo engineering firm but actually working on Britain's behalf – Millington-Drake called him a 'reformed gangster' – was already in negotiation with the German Minister, who happened to be a friend, and a price of £14,000 was agreed on the 23rd with the proviso that the Germans be allowed to remove or destroy military secrets.[36]

Vega took possession of the wreck on 1 March but it was not until the 6th that Bainbridge-Bell managed to get on board at low water, climb the foremast, and inspect the radio room and mattress aerial array. By the end of his second visit on the 7th, despite considerable fire damage inside the radio room and earlier souvenir-hunting by the 'natives', he had found enough evidence from the undamaged aerial array itself and from the remains of a cathode ray tube indicator to be able to signal to London implying that this was indeed a radar operating at a wavelength of about 57cm.[37] He also took some 'samples' of the scrap on offer, some of which happened to be elements from her radar aerial and CRT indicator.[38]

After three more visits, Bainbridge-Bell reckoned that he had found out all that could be discovered at that time and that he should make for home. 'He is very satisfied with the results,' Millington-Drake told the Foreign Office, 'considering the damage by explosion and fire and given the inevitable tardiness of his investigations and the systematic pilfering that has undoubtedly taken place by naval personnel, a number of objects having been admittedly taken by the Inspector General of Marine and placed in a depository in the Arsenal.'[39]

Meanwhile, 100 photos of the ship's interior taken by Señor Vega went off to England by air mail, and arrangements were being made in London for more representatives of Thos. Ward Ltd. to travel to Montevideo, two of whom happened to be M.K. Purvis, Admiralty naval constructor, and Lieutenant G.P. Kilroy, a torpedo specialist from HMS *Vernon*, who arrived on 29 March – but that is another story.

An interesting postscript arose nearly two years later when the Treasury wrote to the Admiralty claiming that the £14,000 paid for the wreck, as well as Thos. Ward Ltd.'s expenses, 'had gone down the drain finally'.[40] DSD made the following indignant reply:

> The examination of the *Graf Spee* by [the Signal School] representative was most valuable in establishing the use of RDF by the enemy. It also provided sufficient technical detail in this matter to guide us in the search for enemy RDF in general and in the devising and preparation of equipment for countermeasures.[41]

We now know that three other German ships were fitted with the 60cm FuMG 38G in 1937–8: the light cruiser *Königsberg*, torpedo boat G.10, and trials ship *Strahl*.[42] With hindsight, it seems surprising that, with no less than four German ships fitted before the war began, the Royal Navy should not know for certain that the Germans had radar in ships until Bainbridge-Bell's visit. Was this perhaps yet another result of the obsessive secrecy about radar – even the fact that there was such a technique – so that Naval Intelligence did not know to look for it? The Air Ministry had appointed a scientist, R.V. Jones, to work with Air Intelligence in September 1939: the Admiralty did not follow suit until E.M. Gollin (whose first job at Signal School had been to replace Yeo when the latter started working on radar) began to work with Naval Intelligence about 1942.[43]

SCAPA FLOW

By the time the Home Fleet reckoned Scapa Flow was safe enough to return to in March 1940, the mobile RAF CH station at Netherbutton had been replaced by something more permanent. However, there were considerable teething troubles and the fleet in harbour often had to rely on its own resources. During the forenoon of 16 March 1940, there had been a large number of false alarms from ashore; then, in the evening, the new fleet flagship *Rodney* was straddled by two bombs without warning, and *Norfolk* was hit – before a shot was fired and just at the moment that the Gun Operations Room ashore reported that the plot was clear. All of which led to some pretty searching enquiries.

Rear-Admiral M.W.St.L. Searle, who as a Commander was Fleet Gunnery Officer in *Rodney* at the time, continues the story:

> About a fortnight after the first attack on the Fleet, a further attack was attempted. On a clear and sunny afternoon, the AA cruiser *Curlew* was berthed near *Rodney* and hoisted the flag signal reporting the detection of aircraft approaching from the south, distant 50 miles. The notice for air defence was shortened to 5 minutes (yellow flag) by the Fleet Gunnery Officer [Searle] and a messenger sent for the Captain of the Fleet in the absence at the Admiralty of Chief of Staff. The Captain of the Fleet (Capt. H. R. G. Kinahan, newly joined since 16 March) came up to the signal bridge and approved Red Notice. About nine twin-engined aircraft in a loose line-abreast formation were soon sighted to the south'ard and were met with a very impressive volume of AA gunfire; the self-destroying 2–pounder pom-pom projectiles made a massive wall of black bursts in front of the formation, which turned together to the westward, jettisoning their bombs over Hoxa and Swin Sounds, and made off to the SE. This incident was probably not reported to the Admiralty: no damage had been suffered; no aircraft had been shot down. But some confidence in Air Defence was restored.[44]

This restoration of confidence was largely due to *Curlew* and her type 79 radar, berthed at the flagship's buoy at Scapa and connected by telephone to the Naval authorities ashore as well as to the RAF plot and Army gunners. Lieutenant John R. Hodge, a regular RNVR officer, had been put in charge of her radar when he joined in Chatham in September 1939:

> I informed the captain that I hardly knew how to fit an electric light bulb, let alone advanced electrics. However, he told me that nobody else knew anything much about the machine either and that it was very secret.
>
> We sailed round to Plymouth with Hartley [Lt-Cdr, from Signal School] on board. He and I were closeted in the radar hut night and day trying to get the hang of it. On arrival in Plymouth, we asked for, and got, the RAF to fly all over the place at different heights and directions. From these trials, we made what was probably the first crude graph of the height of approaching aircraft. This was later corrected with a detailed study of attacking aircraft from which we also judged the number of planes in the attack. We had a very good idea of the numbers, height and speed of possible attacks together with probable direction of attack which we passed to the gunnery officer. Working on these crude beginnings, it was amazing how accurate we were able to get.
>
> Armed with this very meagre knowledge, we went to war. . . . From now on we found ourselves a very popular ship with the fleet and

> convoy-commodores, as word got round that we were able to 'foretell' when to expect an air raid!
>
> In Scapa we soon devised a special plotting board to enable us to plot both at sea and in harbour. It was from this board that we worked out the speed of approaching planes.[45]

For the episode at Scapa in March 1940, Hodge was awarded the MBE, the first British radar officer (for so he had unwittingly become) to be so honoured.

4

1940: Norway and the Fall of France

Apr 9	Germany invades Denmark and Norway
May 10	Germany invades Belgium and Holland
Jun 10	War with Italy
Jun 19	*Verity* trials (type 286)
Jun 25	Fall of France
Sep	Radar reinforcements for the Mediterranean – 'Hats'
Sep 28	*Dido* trials begin (type 281)
Oct 28	Italy invades Greece
Nov 11	Attack on Taranto – 'Judgment'
Nov 11	*Usk* trials at Swanage (10cm)
Nov	*Southdown* trials (type 285)
Dec 2	Adml Tovey succeeds Adml Forbes as C-in-C Home Fleet
Dec	*King George V* trials (type 284)

The 'Phoney War' came to an end on 9 April 1940, when Germany invaded Denmark and Norway. For the Navy, the Norwegian campaign from April to June 1940 provided the first real trial for the Navy's newly radar-fitted ships, all eight of whom took part. Ships already fitted at the beginning of the campaign were *Rodney, Valiant, Sheffield, Suffolk, Curlew, Carlisle* and *Curacoa. Coventry* completed her radar trials just in time to take part in the evacuation of Narvik in June. The four AA cruisers in particular learned the limitations of operating close inshore, radar results being very severely restricted by the high cliffs and surrounding high land in the fjords. *Curacoa* was severely damaged and *Curlew* sunk. This description of the latter's last moments, by Mr F. E. Butcher, who was a telegraphist and radar operator on board at the time, reminds us how secret radar then was:

> A high-flying aircraft dropped a bomb which struck us on the starboard side aft, below the waterline . . . After about an hour, she capsized. My departure from the radar scene [he became a normal telegraphist after this] is fixed as a picture in my memory of Jimmy

8. The AA cruiser *Curacoa* in 1942. Several WW1-design 'C' and 'D' class 6in cruisers were rearmed with eight 4in AA guns in 1938 and 1939, having high priority for the fitting of the earliest WA type 79 radar, with aerials at the top of each mast. *Curacoa* had the first type 279 (type 79 with accurate ranging) in January 1940. By the time this photograph was taken, she had additional radar: two 50cm GA type 285 mounted on the HA directors on the foretop and aft, and 10cm WS type 271 forward of the mainmast. She was sunk by a collision with RMS *Queen Mary* on 2 October 1942.

(IWM)

Green, the killick sparker [Leading Telegraphist] in charge of the office, going into it and smashing everything to pieces as much as he could with a heavy hammer, in an effort to make it as difficult as possible for the enemy to gain information about it.[1]

THE SECRECY OF RADAR

A further security precaution which was taken after she sank was to drop depth charges to destroy the aerials which, if recovered, would disclose the wavelength being used.[2]

While the very high level of secrecy about radar in the early days had the virtue of denying information to the enemy, its continuance once radar became operational – and particularly after the outbreak of war – was a great hindrance to the proper use of radar in the fleet. During the first year of the war, the majority of those at sea were ignorant even of radar's very existence, let alone its capabilities and limitations. And this applied not only to the junior ranks, but to senior officers as well, unless it so happened that they had recently served in a staff appointment where there was a need to know about it. For example, when Rear-Admiral J. G. P. Vivian hoisted his flag in *Carlisle* as Rear-Admiral AA Ships before the landings at Aandalsnes in April 1940, he knew nothing about radar, although it was fitted in three of the AA cruisers he commanded. Of course, only a few hours' briefing put him completely in the picture – but he was a privileged person.[3]

Another aspect was the matter of 'radar silence'. Instilled into many generations of naval officers since the Battle of Jutland was the knowledge that, if you used your wireless, the enemy would know where you were. Keep wireless silence until you know he has seen you! The same applied to radar: during the earliest fleet exercises where radar was used in March 1939, *Rodney* had detected *Sheffield*'s type 79 a hundred miles away.[4] During the early part of the war, many restrictions were placed upon the use of radar at sea – say one sweep every five minutes. But from the earliest days, these regulations were often broken by individual commanding officers who realized that in many circumstances the advantages of early warning greatly outweighed the risk of detection by the enemy.

As the war progressed, the restrictions on the use of radar at sea were considerably relaxed, even with metric radar, while the use of centimetric radar – much more difficult for the enemy to detect and DF – was permitted almost without restriction. By 1944, radar silence was imposed only under special circumstances such as the approach of an invasion fleet. It is interesting to reflect, however, that by the Falklands war in 1982 (see the Epilogue) the wheel had turned full circle because of the developments in electronic warfare and radio-controlled missiles. Once more, radar and radio silence was the order of the day for much of the time at sea.

THE NORWEGIAN CAMPAIGN

In the main fleet on the British side, the few radar sets fitted probably did little to affect operations except for air raid warnings. Nevertheless, though none of the carriers had radar, the concept of fighter direction could be said to have originated in this campaign, as discussed in more detail below. On the German side, *Scharnhorst* and *Gneisenau* had Seetakt radar and it was lucky for *Renown*, who had no radar, that, during the action off the Lofoten Islands on 9 April, she managed to score an early hit which put *Gneisenau*'s Seetakt and gunnery control system out of action. The carrier *Glorious* was not so lucky; she was sunk on 9 June by *Scharnhorst*, who opened fire at 14 miles using radar ranging.[5]

Though no British ship yet had dedicated gunnery radar, the Norwegian campaign did provide the first realistic test of the HACS and short-range anti-aircraft gunnery systems, underlining their particular weaknesses – the poor performance of the optical coincidence rangefinder against aircraft, and the human difficulties of operating these systems in fast-moving situations afloat. The welcome given to 'radio-rangefinders' as soon as they became available late in 1940 is therefore not surprising.

Two extracts from the Midshipman's Journal of R.D. (now Vice-Admiral Sir Roderick) Macdonald are of interest. He was serving in the battleship *Valiant*, just fitted with type 279 air warning set. No excessive secrecy here!

> Tuesday 9th April [at sea off Norway] . . . We continued South to cut off a second enemy force, which was reported steaming west from the Skaggerak. At midday we turned north, the enemy seaplane now shadowing us ahead.
>
> Our RDF now showed its worth. At 1415 it detected nine aircraft flying low forty miles to the east. AA action stations were closed up. Five minutes later a further group was reported forty miles to the south. At 1436 aircraft were sighted to the east and the cruisers, disposed around the horizon, opened fire on the starboard quarter . . .
> Sunday 9th June [at sea: troop convoys from Norway] . . . In the afternoon we were expecting to meet *Glorious*, and when RDF reports of aircraft approaching came through, we thought that they were probably friendly. However as they were coming in fast astern, the sun

being on our port quarter, and we could not see them, we went to 'Repel Aircraft' stations. Hardly had this been sounded off and before they had closed up, a bomber flying out of the sun flew in a medium altitude dive down the ship, casting two bombs which fell twenty-five yards off the port beam.
[Pencil note in margin by Captain H.B. Rawlings] We put the helm hard-a-port on RDF.
[Macdonald continuing] This plane disappeared ahead before anyone could open fire on it. At the same time two aircraft were sighted on the starboard bow flying round towards our stern . . .
[at 2100] . . . the convoy consisted of seven transports and in company *Ark Royal, Southampton* and *Coventry* [newly fitted with radar] . . . There were so many of our aircraft in the air that the RDF was rendered of little use . . .[6]

THE BIRTH OF FIGHTER DIRECTION

Fighter direction was one of the possible functions of radar which seems to have been given scant attention by the Naval Staff before the war, although the RAF had already given it much thought. The main defence of the fleet from air attack was to be by guns, so radar's primary function was to give suitable warning and provide accurate range information for gunnery purposes. In the fitting of type 79, existing aircraft carriers had the lowest priority, though the *Illustrious* class carriers then building were to be so equipped. It was the Norwegian campaign that proved that fighter direction was possible at sea.

It is agreed by all the founder members of what came to be known as the Fighter Direction Branch that the chief credit for the invention and development of fighter direction at sea belongs to the then Lieutenant-Commander Charles Coke, a specialist observer in the Fleet Air Arm, who was Air Signal Officer in the carrier *Ark Royal*, which had no radar. Coke relied upon sighting reports from the fleet and on reports from the radar-fitted *Sheffield* or *Curlew* when in company, sent by flag signal until it was certain the enemy knew our position, then by W/T (Morse). He sat in the corner of the bridge wireless office with a telegraphist beside him writing down incoming reports, which Coke plotted on a Bigsworth board – a portable plotting board used by observers in the air – before himself passing messages by W/T to the fighters. (Observers were all experts in Morse.)[7]

In the earliest days, he passed the enemy's position, course and speed to the aircraft, leaving the fighter's own observer to work out a course to intercept; this came to be known as the 'Informative' method of fighter direction. Soon, however, he found that with radar he could track the fighters as well as the enemy, so was able to take the momentous step of telling the observer the course ('vector'), speed and sometimes the height at which to fly, which the latter then shouted to the pilot by voice-pipe. This was the 'Directive' method, evolved by Coke who hardly left the bridge wireless office during the 24 hours of daylight off Norway in May and June 1940. Though this achieved some success, the trouble was that the whole process took some four minutes, so it is not surprising that the success rate varied.[8]

So began fighter direction in the Royal Navy, whose specialist Fighter Direction Branch was to grow from zero in 1940 to 800-odd fully qualified officers in 1945, using highly sophisticated equipment, the most important of which was radar.

NAVAL AIR WARNING RADAR ASHORE

Germany invaded Belgium and Holland on 10 May 1940, a campaign which culminated in the evacuation from Dunkirk from 26 May to 4 June and the Fall of France on 25 June. In the RDF Division of the Signal School in Portsmouth at this time, there was a tremendous sense of urgency. For Rawlinson's group at Eastney, the 7.5m long-range air warning type 279 (by which name the modified type 79 was now known) was in production and much of the group's efforts were taken up in assisting in the fitting and trials of sets being installed, fifteen becoming operational between May and December 1940.

Largely for training purposes, two type 79 sets were in operation ashore near Portsmouth by July 1940, one at Fort Wallington on Portsdown Hill (manned by survivors from *Curlew* and *Curacoa*), one on the sea shore at Eastney Fort East. Both were in operation during the raids on Portsmouth that month, as was the first development model of type 281 (see p. 41 above). Fort Wallington formed part of the RAF's chain and came to be of particular value when two nearby RAF stations were put out of action by air raids.

WARNING RADAR FOR SMALL SHIPS: TYPE 286

For Rawlinson's group, already heavily committed in the final development of type 281 at Eastney, a most urgent new commitment arose. Immediately after the evacuation at Dunkirk, Britain found herself under grave threat of invasion and the Admiralty produced an urgent, almost panic, requirement to Signal School for radar, for both surface and air warning, to be fitted to destroyers and light craft whose duty it would be to intercept any seaborne invasion force.

Some months earlier, a naval Walrus aircraft had been drawn up on a slipway at Lee-on-Solent, facing seawards about 20 feet above sea level, in order to demonstrate to senior naval officers the capabilities of the RAF's 1.5m ASV (air to surface vessel) Mark I radar. The demonstration was a success and the visitors were able to see the echoes of many ships in Spithead and the Solent out to a range of 5 miles or more.[9] No suitable naval set was available for the Admiralty's new requirement, but fortunately the ASV Mark I was in fairly free supply and, as demonstrated at Lee, could satisfy the need if an aerial light enough for the top of a destroyer's mast could be designed. The decision to proceed was taken on a Friday. By the following Sunday week, A. W. Ross's group had produced the first aerial array, a wooden framework carrying one transmitting aerial facing ahead and two receiving arrays angled to port and starboard – known variously as the 'canary cage' or the 'bedstead'. This non-rotating aerial covered an arc from ahead to just abaft the beam on both sides of the ship, but could only give rough bearings without swinging the ship. On 19 June, the first sea trials of the radar that came to be called type 286 took place in the destroyer *Verity* at Spithead.[10]

Happily, the threat of invasion receded but the requirement for a small-ship warning radar set remained – for the detection of U-boats and E-boats, particularly the latter, which at that time were a serious menace to traffic on the East Coast and English Channel. Sets were obtained from the Ministry of Aircraft Production and aerial systems were constructed in Royal Dockyards. As might be expected from those stirring days of 1940, written records of what ships were fitted, and when, are scanty, but we do know that, despite difficulties in finding staff to fit and test the sets, thirty-two ships had been fitted by the end of the year,[11] the first few with ASV Mark I (type 286), later ships with the better engineered ASV Mark II (type 286M). One of the very earliest to be fitted was the destroyer leader *Malcolm*, at Harwich.[12]

50-CENTIMETRE GUNNERY RADAR GOES TO SEA

In Coales's gunnery radar team at Signal School, everyone was working late and at weekends to get the development of the 50cm set completed so that design details could be finalized and large-scale production started. In May, impressed with the results of the close-range trials of what developed into type 282 with fishbone aerials on a pom-pom director at the Eastney AA Range in February, DNO asked whether 50cm radar-rangefinders could be provided for high-angle and main-armament directors as well. Since these director control towers were physically larger, the obvious approach to achieving higher performance in the very short term was to use larger aerial arrays.

For the low-angle main armament, it was decided to use two parabolic cylinders – a configuration that came to be known as the 'pig trough' – up to 21ft wide and 2.5ft high, fed by a line of half-wave dipoles, to produce a beam only 5° wide by 38° vertically for the accurate tracking of surface targets. (Type 282's twin-fishbone aerial gave a beam 38° wide and 35° vertically.) The two pig-trough aerials were so long that they had to be mounted one above the other for transmission and reception respectively, but, with an aerial power gain (see glossary) each way of seven times that of type 282, a tracking range increase by a factor of about 2.6 on a similar target could be expected.

Trials at Southsea Castle with a fishbone aerial system larger than could be fitted on a pom-pom director had already obtained echoes from the Nab Tower at 8 miles and a pig-trough system of parabolic cylinders should do even better. Thus encouraged, an experimental pig-trough array – but only 10 feet wide – was mounted on the main armament director of the battleship *Nelson*, using a type 282 transmitter and receiver to complete the system.

Sea trials of this experimental aerial took place in *Nelson* in June, during which ranges of 18 miles were obtained on ships in convoy, 11.5 miles on the destroyer *Ambuscade*.[13] So good did these results appear to be, that an immediate decision was taken to initiate a programme of fitting all cruisers and above, and orders for the basic set were increased from 200 to 900. In July, it was decided that the first prototype of the new set, type 284, should be fitted in the battleship *King George V* then building, completing in December. It was to displace the planned HF/DF set in the Admiral's Chart House.[14]

It was subsequently discovered that these results were over-optimistic, the ranges obtained during the trial being significantly greater that the normal because of anomalous propagation – familiarly known in later years as 'anoprop' – the channelling of centimetre waves near the sea surface which often occurs during anticyclone conditions, a phenomenon similar to optical mirages. That this could occur was not known at the time.

The experimental radar was removed after the trial and *Nelson* was fitted with a production model of type 284 in 1941.

To apply this radar set to high-angle (HA, or anti-aircraft) gunnery meant that aerials had to be fitted to a variety of directors of different shapes and sizes, and this required further development work. Trials with other aerial systems had been carried out and ranges up to 6 miles on an aircraft were obtained with an aerial system that could be mounted on an HA director. The first prototype of what is now called type 285, with a 'fishbone' aerial attached to her rangefinder-director (dual-purpose HA/LA) went to sea in the new 'Hunt' class destroyer *Southdown* in September 1940, and a decision was taken the same month

9. The Australian destroyer *Nestor* in 1941. The 1.5m WC type 286M fixed 'bedstead' aerial can be seen at the top of the mast, and the 50cm GA type 285 Yagi ('fishbone') aerials on the rangefinder–director abaft the bridge.

(*IWM*)

that all destroyers should have either the 1.5m type 286 or 50cm 285, the former in the older V, W, A, B, C, D and Lend-Lease American destroyers, the latter in E class and later destroyers which had rangefinder-directors.[15] As we shall see, this policy was considerably modified when the 10cm type 271 appeared.

The various applications proposed for the high-angle set, type 285, required close cooperation with DNO, the organization responsible for fire-control directors, and this proceeded at the highest priority. Several combinations of fishbone aerials were necessary – a pair for pom-pom directors, and three transmitting and two receiving – or sometimes two transmitting and three receiving – for the rangefinder-directors fitted for the secondary HA/LA armament in large ships and the main armament in destroyers. Though in April 1940 no single component of the 50cm radars had been designed for manufacture, such was the cooperation between Signal School staff (numbering about twenty), DNO and Industry that manufacturing drawings were completed and production lines set up so that the first sets of equipment from about ten manufacturers came together in time for type 285 to be installed in *Southdown*, and type 284 in *King George V*, both in November 1940 – of which more anon.

FORCE H AND THE MEDITERRANEAN FLEET

Italy's entry into the war on 10 June and France's capitulation on 25 June fundamentally altered the strategic situation, particularly in the Mediterranean. In May, as the Italian attitude became more threatening, reinforcements had been sent to Admiral Sir Andrew Cunningham's Mediterranean Fleet at Alexandria, and these included the radar-fitted AA cruiser *Carlisle* already mentioned. A month later the Admiralty was faced with the urgent need to replace the lost French maritime power in the western basin of the Mediterranean, for which purpose the famous Force H was hurriedly formed at the end of June under Vice-Admiral Sir James Somerville (tuberculosis notwithstanding), to be based on Gibraltar. Force H included the radar-fitted battleship *Valiant* and later the cruiser *Sheffield*, as well as the carrier *Ark Royal* without radar.

The following reminiscences by Admiral Sir Rae McKaig give a good idea of how radar was used in Force H in the early days. He joined *Sheffield* as a midshipman in May 1940:

> This installation [type 79] came into its own when we joined *Renown*, *Ark Royal* and the 8th Destroyer Flotilla at Gibraltar later in the summer as Force H, and contributed signally to the effective use of *Ark Royal*'s fighters against the Regia Aeronautica in the western

Mediterranean, and to [air] warning for *Sheffield* herself against attack from Vichy French bombers after a contretemps at Nemours on the Algerian coast.

In the spring of 1941, I was transferred temporarily to the *Renown*, the Force Flagship, and was able to see the system working as a minor cog in the organization. *Sheffield* passed radar information to *Ark Royal*, who had none, by flag signal for the direction of fighters. My station was on the bridge of *Renown* where, with the aid of chalk and circular blackboard, I displayed the 'air picture' to the Admiral, either by reading *Sheffield*'s flag hoists myself, or listening to the Yeoman of Signals on watch. Admiral Somerville took a keen interest in the result and occasionally interjected command guidance or requests for information by Aldis Light from *Ark*. At this stage the Italian mode of attack was by high-level 'pattern' bombing which gave detection ranges of 70 to 80 miles and sometimes 100-plus miles, if my memory serves me right.

At that stage, RDF was the responsibility of the Communications Branch, and the installation in *Sheffield* was in the hands of a saturnine and taciturn Chief PO Telegraphist assisted by the RDF operators who, being 'Hostilities Only' ratings in a largely 'regular' Chatham ship's company, were treated as people not worth knowing, the more so as they always appeared tired and wan at sea, and because entry into their 'office' at the foot of the mainmast was not only strictly forbidden but said to be bad for the health. Later an RDF officer joined, presumably for maintenance, rather surprisingly a Canadian of the RCNVR.[16]

CARLISLE AT ADEN

Meanwhile, *Carlisle* had passed through the Mediterranean in May 1940, and was sent to the Red Sea for convoy protection there if Italy entered the war, a precaution in case it should become necessary to route shipping for the Middle East via the Cape of Good Hope rather than through the Mediterranean. She reached Aden some two weeks before Italy declared war, weeks spent in training and, more particularly, in devising and practising a plan for the air defence of Aden with Gladiator fighters from the RAF base (where there was no radar) and the Army AA battery of 3.7in guns. The effort was not wasted. In the evening of 10 June, many of *Carlisle*'s officers were dining at the Cable & Wireless mess in Aden when, during the soup course, a messenger handed a note to the mess president to tell him that Italy had declared war and that all officers were ordered to return to their units immediately. Next morning at 4.45am a group of aircraft was detected by *Carlisle*'s type 280 at 40 to 45 miles, apparently in formation at 5,000 to 6,000ft, flying straight and level – in fact twelve SM79s. In the early morning light, the Gladiators were directed with success against this perfect target by Lieutenant F.C. Morgan, *Carlisle*'s gunnery officer, and the ships and army also put in perfect gunnery practice. Nine aircraft were destroyed or damaged, and three landed in the desert and surrendered. Certainly none returned to Eritrea, and the Italian forces never approached Aden again.[17]

THE MEDITERRANEAN

Illustrious was the first British aircraft carrier to be fitted with radar. (The second was *Formidable*, commissioning in October 1940.) As Admiral Hezlet has said in his *Electron and Sea Power*, this was probably the most important step taken with this new device since it was introduced into the Navy.[18] Available radar sets had been fitted in ships building and whenever they came into dockyard hands; if any priority was given it was to anti-aircraft cruisers. But aircraft carriers in the Fleet were considered far too busy to be taken away from operations to be given the priceless aid to the air defence of the Fleet. Having commissioned in April 1940, *Illustrious* and her air squadrons 'worked up' off Bermuda, during which she tried out the 'informative' method of fighter direction mentioned above.[19]

At the end of August, additional reinforcements were sent to Admiral Cunningham – from Force H the battleship *Valiant*, and from the United Kingdom *Illustrious* (flying the flag of Rear Admiral A.L.St.G. Lyster), the AA cruisers *Coventry* and *Calcutta*, all but the last fitted with radar. They gathered on 29 August in Gibraltar where *Ark Royal* was in harbour, so Coke was able to brief *Illustrious*'s Commander (Operations), George Beale, on the fighter-direction techniques evolved so far in *Ark Royal*, particularly on the single-letter code which greatly shortened W/T messages to fighters.[20] As we shall see, *Illustrious* was to carry forward the fighter-direction torch with resounding success.

A joint operation by the forces of Admirals Cunningham and Somerville, known as 'Hats' – Hands Across The Sea – was mounted to pass the reinforcements through the Mediterranean from Gibraltar to Alexandria. The eastern and western fleets met south of Sardinia and the whole operation was completed without interference by the Italians. 'This operation,' said Admiral Cunningham, 'which was the first attempt at passing forces on a large scale between Cape Bon and Sicily passed off smoothly and with remarkably little incident . . . The use of RDF and the Fulmar fighters of HMS *Illustrious* placed an entirely different complexion on the air-bombing menace compared with what had previously been experienced . . .'[21]

Other radar-fitted ships to join the Mediterranean Fleet at the end of 1940 were *Ajax* with type 279, which came round the Cape, and *Berwick* and *Glasgow* with the fixed-aerial ASV type 286M, which sailed through the Mediterranean early in November in Operation 'Coat', similar to 'Hats'.

While Operation 'Coat' was taking place, aircraft from *Illustrious* and *Eagle* made the very successful raid on the Italian battlefleet in Taranto harbour. The whole operation might have been seriously affected had not the fighter direction of *Illustrious* and the Fulmar fleet fighters proved so efficient. Only three erratic attacks were made by the Italians and eight aircraft were shot down.[22]

By the last quarter of 1940, the Home Fleet had also received substantial reinforcements of radar-fitted ships – the battleship *Nelson* and cruisers *Edinburgh* and *Galatea* fresh from damage repairs; the new battleship *King George V* and new cruisers *Bonaventure, Fiji, Naiad, Phoebe, Kenya, Nigeria,* all with type 279; and, as we shall be seeing, *Dido* with type 281.

ILLUSTRIOUS AND FIGHTER DIRECTION

In *Illustrious,* Beale acted as the ship's fighter direction officer (FDO), sitting in front of a simple spider's-web plot showing the air situation given by radar, 'told' by telephone from the A-scan in the radar office. Though he sometimes carried out interceptions himself, he more often delegated this task to others, including, for example, the Torpedo Officer, or Warrant Observer Teddy Wicks, or – most important for the future of the FD Branch – Sub-Lieutenant David Pollock, assistant to the Admiral's Staff Officer (Operations). R/T (voice radio) was tried instead of W/T (Morse) but the equipment proved unsatisfactory.[23] A high state of efficiency and professionalism was quickly attained, enhanced by the intelligent and cooperative attitude of all on board, from Admiral Lyster and Captain D. W. Boyd downwards. Partly because of her new Fulmar fighters, but chiefly because *Illustrious* had her own radar, she established something near to air supremacy over the eastern Mediterranean, a superiority which she was able to maintain until the Luftwaffe arrived in Sicily at the end of the year.

And she identified serious problems, some of which were to dog fighter direction until the final stages of the war:

- *radar security* – when she first joined the Mediterranean Fleet, Fleet Orders said she was only allowed to transmit on radar for one minute in each hour, and if no unidentified aircraft were detected within that minute the set had to be switched off;
- *radio security* – no communication was allowed with fighters until an unidentified aircraft came within 20 miles;

- *land echoes;*
- *heightfinding* – or the lack of it;
- *the detection of low-flying aircraft.*[24]

TYPE 281 SEA TRIALS

The first sea trials of the new 3.5m air warning set in the new cruiser *Dido* in September/October 1940, conducted by S.E.A. Landale and D.S. Watson from Signal School, were an outstanding success, qualified only by difficulties with the mechanical system for controlling and rotating the aerials. (It was so noisy, many operators called it the 'mangle'.) A Hudson aircraft at 25,000ft was detected at over 100 miles, one at 1,000ft at 25 miles; *Dido*'s sister-ship *Phoebe* was detected at 9.5 miles and a trimmed-down submarine at 2.75 miles. (During *Dido*'s passage from Liverpool to Scapa, Watson, operating the type 281, detected a ship at around 7 miles and it remained on a constant bearing until *Dido*, on radar information alone, altered course to avoid. Was this the first example of radar collision avoidance?)

But there were two other aspects which particularly interested the Naval Staff when the trial results were circulated round the Admiralty. First, the flight of 5.25in projectiles had been followed and splashes seen out to 5,000 yards, making blind spotting for range quite possible which, with a bearing accuracy of ±½°, could have a significant effect on night fighting even though it could not give an elevation datum for the director layer. Secondly, the characteristics of the type 281 aerial gave a well-defined first minimum, making the method of height estimation described in Appendix C much easier than with type 279.[25]

It has to be said that in the event the gunnery capabilities of types 279 and 281 were seldom used at sea – with the honourable exception of the Battle of Cape Matapan in March 1941 – first because of commanding officers' reluctance to abandon all-round warning, secondly because, even by mid-1941, a high proportion of big ships were already fitted with additional radar sets dedicated to gunnery. Indeed, by the end of 1942 no new type 279s were fitted with the ranging panel and many were removed from existing sets. It could not be removed immediately from type 281, being an integral part of the set.[26]

GUNNERY RADAR SEA TRIALS

Let us now continue the gunnery radar story we left on p. 61. Sea trials of the first prototype 50cm sets took place in November and December respectively, achieving the following maximum ranges:

First type 285 on rangefinder-director, HMS Southdown [27]

(3 transmitting and 2 receiving Yagi arrays, 40ft above sea level)

Wooden aircraft, height	200ft	4 miles
	500ft	5.5
	1,000ft	7
Metal aircraft, height	2,000ft	9
	5,000ft	8.5
Heavy cruiser, *Fiji* class		6.6
Destroyer, 'Tribal' class		4.5
Submarine on surface		2

First type 284 on director control tower, HMS King George V [28]

(2 parabolic cylinders 21 × 2.5ft, 90ft above sea)

Light cruiser, *Dido* class	10 miles
Destroyer, 'Town' class	7
Submarine on surface	3.5

The advantages of the higher position and greater aerial gain against surface targets are clear.

Thus, by the end of 1940, the Royal Navy was in a position to fit 50cm gunnery radars widely – type 284 in big ships for the main LA armament; type 285 in many classes of ship for the long-range HA armament, which could also be used for surface targets in destroyers; and type 282 for short-range HA armament, eventually in all classes. This was accomplished using one basic radar set but many variants of the aerials.

In addition, type 283 was to follow in 1942 to enable the main armament of larger ships to lay down barrages against air attack as an additional deterrent.

THE BIRTH OF NAVAL 10-CENTIMETRE RADAR: TYPES 271/2/3

For the detection of small surface objects – in 1940, most urgently needed for the detection of U-boats by convoy escorts – the Navy's Director of Scientific Research, Charles Wright, had said from the very start that what the Navy needed was radar at centimetric wavelengths. The shorter the wavelength, the smaller the aerial needed to produce the narrow concentrated beam required for high-definition surface echoes and, incidentally, the more difficult for the enemy to detect and DF at any distance. The first of these considerations was important not only for the Navy but also for military airborne radar for detection of other aircraft

(AI) and of surface craft (ASV). But when the war began, centimetric radar was not technically feasible.

From 1936, E.G. Bowen had been in charge of the development of airborne radar at Bawdsey. We have already heard of his early success with ASV on 1.5m in 1938 (p. 18 above). In his *Radar Days,* he gives an interesting account of his early discussions with Wright on centimetric radar:

> In the next few years [from 1938], there is no doubt that it was the Admiralty which expressed greatest enthusiasm for changing to centimetre waves. There was practically no interest within Bawdsey itself and anyone talking about centimetre waves was thought of as some kind of crank. In the spring and summer of 1939, I discussed the problem a number of times with Sir Charles Wright, Director of Scientific Research for the Admiralty, during his many visits to Bawdsey. In addition to his Naval responsibilities, he was Chairman of the Committee for the Coordination of Valve Development (CVD), which acted on behalf of the Army, the Navy, and the Air Force. We agreed that the greatest need for a centimetre wave system was in airborne radar and he asked what wavelength was most desirable for this application. I did a back-of-an-envelope calculation . . . The only way of improving [on the 1.5m system] would be to project a narrow beam forward to get rid of the ground returns. I estimated that a beam width of 10° was required to achieve this. Given an aperture of 30 inches – the maximum available in the nose of a fighter – this called for an operating wavelength of 10cm. This agreed with Sir Charles's own assessment of what was required for naval purposes.[29]

Soon after this, the CVD Committee arranged for an Admiralty contract to be placed with Birmingham University's Physics Department under Professor M. L. E. Oliphant for the development of transmitting valves for radar at 10cm wavelength.

Early in 1940, a triple breakthrough occurred that was to revolutionize the detection of surface craft, indeed the whole development of radar. These were the developments of:

- *the first multi-cavity magnetron valve,* by J. T. Randall and H. A. H. Boot of Birmingham University, redesigned by E. C. S. Megaw of GEC Wembley into the first magnetron suitable for use in a radar system;
- *the reflex klystron valve,* by R. W. Sutton of the Signal School's Valve Group at Bristol; and
- *the crystal mixer,* by H. W. B. Skinner of TRE at Swanage.

The first of these made possible a radar transmitter – and, incidentally, today's domestic microwave oven – the second two a receiver, capable of working on 10cm or less.

By an inter-service agreement, the initial development of radar on the 10cm band for all three services was the responsibility of the airborne radar group of the Air Ministry's TRE near Swanage in Dorset, in the first instance primarily to produce an airborne interception set for night fighter aircraft to replace the 1.5m equipment then in use. The leaders of the TRE's highly enthusiastic centimetric team were P. I. Dee and H. W. B. Skinner, and included A. C. B. (now Sir Bernard) Lovell and A. L. (now Sir Alan) Hodgkin. Lovell and W. E. Burcham having got the first 10cm echoes from an aircraft on 12 August, they had a 10cm set working in a field with a clear view over Swanage Bay and had obtained echoes from land and surface craft by early autumn. Throughout 1940, Coales in the Signal School had responsibility under Horton for research and development on all wavelengths less than 1m, and was in almost daily contact with Megaw at GEC who kept him informed of the magnetron developments. Some time in September or October, he visited Skinner at Swanage to see how work on 10cm was progressing,[30] thinking that 10cm might – as it eventually did – replace 50cm as the wavelength for the Navy's gunnery radar. On returning to Portsmouth, Coales reported the result of his visit to Horton and this presumably resulted in the chain of events recounted below.

The Battle of the Atlantic was going from bad to worse from the Allied point of view, and radar which could detect submarine periscopes by day and surfaced submarines by night was a most urgent need. Lovell's diary tells how on 28 October 1940 he gave a demonstration to 'two naval commanders who opened their arms at 10cm for submarine detection. Of course we are enormously keen on it.'[31] The names of the two commanders are not known but it seems likely that one was from the Signal Department and the other from the Anti-submarine Warfare Division. Certainly, Captain G. E. Creasey, Director of Anti-Submarine Warfare, was with Willett at Swanage at some stage, as is evident from his letter of a year later, quoted on pp. 113–14 below. Dee's diary for the same day shows that Willett and Horton had visited TRE before this:

> 28 October 1940 – From Leeson House we recently demonstrated 10cm equipment viewing ships in Swanage Bay. Horton and Willett were at first very defeatist about this saying that from the Naval point of view the war would have to be won on 3.5m. Skinner and party however are strenuously advocating fitting a 10cm equipment to a destroyer for anti-submarine work . . .[32]

THE SWANAGE TRIALS

Defeatist or not, Horton and Willett acted with great speed. Some time in late October or very early November, in order to achieve what would now be called an efficient transfer of technology, S. E. A. Landale was charged with taking a small team from Rawlinson's Eastney group (comprised J. Croney, C. A. Cochrane and C. S. Owen) to Swanage on 'temporary loan' to Dee's group to assist in the construction of a centimetric radar apparatus. When this team left for Swanage, there was still no specific Admiralty commitment to develop a radar on 10cm wavelength, but it was charged to send back information to Eastney on techniques and circuits which had been used successfully at TRE with the new centimetric devices.

By early November, the TRE group had three sets of apparatus available for field testing, one of which was a 'Chinese copy' worked on by the Signal School staff. Though differing in detail, the three sets of equipment were necessarily very similar, all being based on a magnetron with a nominal output power of 5kW peak power, on a 'Sutton' tube local oscillator, and on Skinner's hand-made silicon/tungsten crystal mixer. The aerials – Lovell's speciality – were 3ft-diameter circular paraboloids with the transmitting and receiving dipoles at their foci, giving a pencil-like beam. Initially the test site was at Leeson House, an empty girls' school at Langton Matravers, 250ft above sea level, a situation eminently suitable to testing with a view to ASV application. Available for testing this equipment against a ship target was the 92-ton launch *Titlark*, which had presumably been a round-the-bay cruise boat in happier days.

The first test against a ship target on 7 November was with a TRE-built set with the *Titlark* as target. However, it was the second test on 11 November which was of particular naval interest, the Signal School copy apparatus being used and the submarine *Usk* laid on as target. Among those present were Willett and Horton. Though no naval records of this important trial have so far been found, representatives of the Naval Staff from London almost certainly attended also.

Usk, surfaced stern on, was seen out to about 7.5 nautical miles, strong signals up to 6 miles, conning tower only between 4 and 4.5 miles.[33] Lovell's diary says, '. . . gave a superb submarine demonstration which seemed to shake Willett and Horton pretty much'. Skinner's report on this and other trials at Swanage is reproduced in full in Appendix A.

There was then almost a month's gap in the test programme of the Signal School copy apparatus and it was probably during that time that it was built into the 'Admiralty trailer'. The earlier trials had not been entirely realistic for shipborne equipment, first because Leeson House was 250ft above sea level, secondly because the thin pencil-shaped beam of the circular paraboloid aerials meant that the target would be lost every time the ship rolled more than a degree or so – which convoy escorts do all the time. This latter problem called for an aerial which had beam narrow in the horizontal – to retain good bearing discrimination and high aerial gain – but wide in the vertical so that the target was not lost when the ship rolled. Lovell proposed the solution which was destined to become one of the standard forms of aerial for naval radar – the 'cheese' aerial with which he had been experimenting since June. This consisted of a section of a parabolic cylinder closed top and bottom by parallel plates, arranged with its focal line vertical and in the aperture. The rectangular aperture so formed was fed at the centre by a dipole.

With the proof on 11 November that radar on 10cm could detect something as small as a submarine at long range, all doubts vanished as to the urgency to get the set to sea. It only remained at Swanage to be satisfied that with a different aerial form and at low aerial height there would still be adequate performance. While the new cheese aerial of aperture 6ft × 9in was being made, TRE in their continuing test programme tested their paraboloid aerial equipment, first at a height of 60ft from Peveril Point, then at a height of 20ft from the beach near Swanage pier. On 18 December, the trailer aerial was changed from paraboloids to cheese and a further test was run with the *Titlark* from the beach. Having proved that a range of some 3 miles was still possible from this unrealistically low aerial height, the trailer was towed to Portsmouth on 19 December – just in time for the lucky members of the team to have a very short Christmas leave.[34]

Let us sum up this account of the beginnings of naval 10cm radar by quoting from a report of July 1941 which, if not drafted by Captain Willett himself, was at least signed by him:

> The principal urge in this direction [towards 10cm] came from the Air Force requirement for an RDF set for night fighters, the very short wavelength being required to enable an aerial system of sufficient gain to be installed in the very limited space to be had in such aircraft. At the same time wavelengths of this order held promise of improved results against surface targets.
>
> In the summer of 1940 it seemed that 10cm RDF was a long-term project with no immediate future, but thanks principally to the perfecting by the Signal School Laboratories at Bristol of a new type of valve for use in the receiver and to greatly improved performance obtained by the GEC from the magnetron as a transmitter, the art of 10cm moved rapidly forward in the hands of the Air Ministry Research Establishment at Swanage. By November 1940 they were able to demonstrate some very striking results from an equipment on shore directed against one of HM submarines as a target.

A new group was at once formed in Signal School to take advantage of this successful research and at the end of December this party returned to Signal School with a replica of the Swanage laboratory apparatus. Events then moved rapidly and to such effect that within three months an adequately engineered equipment was operating for trial in a corvette in the Western Approaches.[35]

THE TIZARD MISSION TO THE USA

Quite early in 1940, Sir Henry Tizard, realizing that Britain might soon come near the limit of her productive capacity, particularly in electronics, had made the bold suggestion that Britain should disclose her scientific secrets to the United States in return for help on technical and production matters. Opposition to this proposal having been stifled by the fall of France, a scientific mission bearing a 'black box' containing Britain's secrets arrive in Washington in September 1940.

In that black box – actually a legal deed box – were details of the jet engine, rockets, predictors and, above all, radar in its many forms. And the most important item in the box was a sample of one of the first production magnetrons from GEC, Wembley, in the charge of E. G. Bowen, one of Watson-Watts' earliest collaborators, whom we have already met.

Led by Tizard himself, the mission included Bowen, Professor John Cockcroft from the Army research team, and representatives of the three armed services.

Cockcroft has said that the magnetron increased the power available to US technicians by a factor of 1,000. President Roosevelt later described the black box as 'the most valuable cargo ever to reach our shores'.[36]

TAILPIECE

- Though the RAF lorry towing the trailer containing the Signal School centimetric radar returned immediately from Portsmouth to Swanage, the trailer itself remained on the beach at Eastney Fort East. Many months later, the Navy heard that the senior storekeeper at TRE was still trying to discover how a large trailer could possibly have been lost from the inventory, seemingly without trace.

5

1941(1): Matapan and the Sinking of the *Bismarck*

Jan	Luftwaffe at Sicily airfields
Mar 11	Lend-Lease Bill signed by Roosevelt
Mar	German U-boat attacks intensified
Mar 17	First U-boat sinking due to type 286 (*Vanoc*)
Mar 25	*Orchis* trials (type 271)
Mar 28	Battle of Cape Matapan
Mar 30	German counter-offensive in N Africa
Apr 6	Germany invades Greece and Yugoslavia
May	Luftwaffe move from Sicily to Balkan airfields
May 24	*Hood* sunk
May 27	*Bismarck* sunk
May 27	British withdraw from Crete
June 22	Germany invades Russia

HOME FLEET OPERATIONS

At the beginning of 1941, Britain was almost alone in the war against Germany and Italy. The danger of invasion had receded and the Battle of Britain had been won, but the Battle of the Atlantic was critical, and was to become more so as the year progressed. Ground forces were in contact only in Libya and Egypt but the Navy was heavily engaged in the defence of shipping against submarines, surface craft and aircraft, particularly in the Atlantic and in the Mediterranean where Malta was under siege.

The principal task of the Home Fleet under Admiral Tovey, working mainly from Scapa, was to prevent German heavy units from breaking out into the Atlantic to attack British shipping. On 22 January 1941, the German battle cruisers *Gneisenau* and *Scharnhorst* sailed from Kiel under Vice-Admiral Lütjens to attack our shipping in the Atlantic. The British Admiralty received intelligence of this move in time for Admiral Tovey in *Nelson*, with *Rodney*, *Repulse*, eight cruisers, and ten destroyers, to reach an area about 120 miles south of Iceland to intercept the German force and to bring them to action if found.

Just before daylight on 28 January, the cruiser *Naiad* sighted and reported two large vessels, almost certainly enemy, which at once turned northwards and increased speed so that contact was lost, never to be regained. In fact, we now know that the 81.5cm FuMo 22 Seetakt radar in the German ships had detected *Naiad*, her sister ship *Phoebe*, and a destroyer screen, six minutes before *Naiad*'s sighting. Lütjens decided not to open fire on the cruisers because of the danger of a dawn destroyer attack; moreover, he thought it likely that a heavy British force might be in the vicinity, so decided to turn away.[1]

About half the big ships in the British force (including *Naiad* and *Phoebe*) were fitted with type 279 radar, designed for long-range air warning, with relatively poor performance against surface targets. (The following month, *Valiant* reported detection range on a carrier off Alexandria at 6 miles, cruisers at 5 miles, destroyers at 2 miles.[2]) In any case radar silence was the normal fleet policy until the enemy was sighted. (An hour and a half after the sighting, *Naiad* detected and held what seemed to be a surface echo, which then faded.) If action had been joined, *Birmingham* might have been able to use her 50cm surface gunnery set, fitted a month earlier.

> It was all very unfortunate [says the Naval Staff History]. Our main force was excellently disposed for intercepting the raiders, but the few minutes grace afforded to the Germans by reason of their better equipment enabled them to withdraw without being brought to action. It will be noticed that there was no aircraft carrier with the Home Fleet. Had one been present, it might have made all the difference.[3]

Happily, 'better' British equipment – for the detection of surface objects – was on the way, as proved, first by *Suffolk*'s use of a 50cm gunnery set in the *Bismarck* episode four months later, then by the advent of 10cm radar during the months immediately following.

THE MEDITERRANEAN

The Mediterranean Fleet under Admiral Cunningham was based at Alexandria, ready to engage the Italian fleet should it present itself, and to support General Wavell's operations in North Africa. At the beginning of 1941, it had four ships fitted with type 279 air warning radar – the battleship *Valiant*, the carrier *Illustrious*, the cruiser *Ajax*, the AA cruiser *Coventry* – and *Carlisle* with the 3.6m type 280. The cruisers *Glasgow*, *Berwick*, *Orion*, *Manchester* and a few destroyers had the fixed-aerial 286M radar, stocks of which had been sent to Alexandria (with ship-fitting staff) by the Signal School. In May, stocks of surface gunnery sets were sent for fitting in capital ships and cruisers in Alexandria.[4]

At the other end of the Mediterranean, Admiral Somerville commanded Force H, strategically based at Gibraltar and able to conduct operations either in the Atlantic or the Mediterranean. The only radar-fitted ship permanently in Force H was the cruiser *Sheffield* with her old type 79Y.

On 10 January 1941, the Mediterranean Fleet's control of the air in the eastern Mediterranean came to an abrupt end when *Illustrious* was severely damaged by aircraft from 'Fliegerkorps X', the Luftwaffe's specialist anti-shipping force, which had arrived in Sicily earlier in the month. Whereas Italian aircraft had posed a threat which could be contained, the arrival of Luftwaffe units completely changed the whole strategic situation.

This happened just as Operation 'Excess' was being mounted, primarily to pass through the Mediterranean an important military convoy with stores for Malta and the Piraeus, coordinated with convoys to and from Malta and Alexandria. Covered by Force H from Gibraltar, the first convoy was escorted by the new AA cruiser *Bonaventure* (fitted with type 279) as far as the Sicilian narrows where she turned over her charges to units of the Mediterranean Fleet, returning to Gibraltar with Force H. The main body of the Fleet under Admiral Cunningham joined the convoy soon after. They had been shadowed all the way from Alexandria and, on 10 January when just west of Malta, sustained heavy air attacks, concentrated on *Illustrious* which, though her radar gave plenty of warning of the approaching German dive-bombers, had no time to recall fighters then pursuing Italian torpedo-bombers, or to fly off more. She was hit by six heavy bombs and very badly damaged. *Valiant* detected the second attack on *Illustrious* and was able to contact three ex-*Illustrious* Fulmars from Malta and to direct them by wireless to attack and damage three enemy bombers.[5] The following day, *Gloucester* and *Southampton* were subjected to similar attacks, the latter catching fire and having to be sunk.

Though not a single ship from any of the four convoys had been lost, the sinking of *Southampton* and destroyer *Gallant* (by mine), and damage to other ships, particularly *Illustrious*, was a heavy price to pay. For Admiral Cunningham, the arrival of German dive-bombers presented a potent new factor and his only aircraft carrier, *Eagle* – with a small hanger and no radar – was too old and slow to operate with the Fleet. His most urgent needs were fighters and more radar-fitted ships.

The Admiralty took immediate action by sailing the new carrier *Formidable* to take *Illustrious*'s place in the Mediterranean, but the former had to go round the Cape and did not reach Alexandria until 10 March, having been delayed in Port Said while the Suez Canal was cleared of magnetic mines. While there, she met *Illustrious* (on her way to damage repairs in the USA), so that Commander Beale was able to brief *Formidable*'s Commander (Operations), Philip Yorke, on the progress *Illustrious* had made in the techniques of fighter direction.

THE BATTLE OF CAPE MATAPAN

Formidable arrived just in time to take part in the Battle of Cape Matapan on 28 March 1941. The British ships taking part were organized in two groups: Force A under the Commander-in-Chief in *Warspite*, comprised three battleships, *Formidable*, and nine destroyers, while Force B under Vice-Admiral H. D. Pridham-Whippell in *Orion*, had four cruisers and four destroyers. In Force A, the only radar-fitted ships were *Valiant* and *Formidable*, both with type 279. In Force B, *Orion* had type 286M and *Ajax* type 279. The Italians are said not to have known that the British ships were fitted with radar.[6]

Except to say that *Ajax* made enemy reports based on type 279 surface echoes during the forenoon of 28 March, and that *Formidable*'s aircraft made several torpedo attacks on the Italian ships, we need not concern ourselves with the early part of the battle. At 8.14pm, when the cruisers were some way ahead of the battlefleet, both steaming towards the supposed position of the Italian forces, *Orion* with her fixed-aerial type 286M detected the echo of a ship some 6 miles ahead. Ordinary Seaman A. B. Craig subsequently received the DSM and *Orion*'s captain's citation explains the situation well:

> Craig . . . while acting as RDF operator, shewed the courage of his convictions by continuing to report the presence and whereabouts of an enemy ship that was not visible. The position of the enemy was reported and the squadron was manoeuvred entirely on the information obtained from his reports.[7]

Admiral Pridham-Whippell concurred, adding that Craig's work had been vital to the operation, and that his interpretation of the radar screen had been very exact, as subsequent events showed.[8]

The following account from a slightly different aspect was written some years after the event by Commander R. L. Fisher, who was Admiral Pridham-Whippell's Staff Officer (Operations) during the action:

> *Orion* had an early type of ASV radar which, if I remember rightly, you could use by pointing the ship at the target. The only person who could work it was the flag lieutenant. As we came up on the enemy he

> was in the ASV office calling ranges and bearings to me in the plot, and I pretty soon bowled out that the nearest ship was lying stopped, and told the Admiral . . .[9]

In view of the operator's subsequent medal, it seems probable that it was a joint effort, with the operator at the set and the flag lieutenant at the voice pipe. The stopped ship turned out to be the cruiser *Pola,* torpedoed earlier by one of *Formidable*'s aircraft.

The role played by *Valiant*'s type 279 radar in the action which followed is described in her own report:

> Shortly before this action, telephonic communication between the 15-inch transmitting station and RDF office, and [also] range transmission from the gunnery attachment RBL.10 to a range receiver in the 15-inch transmitting station had been fitted by ship's staff.
>
> Before sighting, the RDF was kept sweeping with a view to locating enemy ships, the warning set being used switched to Short Range. The first ship to be picked up was at a range of nine miles, and this ship was later sighted at a range of 4½ miles on the bearing indicated by the RDF. This is now known to have been the *Pola.*
>
> After sighting the *Zara* and *Fiume,* the guns and RDF were put on the *Fiume,* the right hand ship, and fire was opened, the AFC [Admiralty Fire Control] Table being tuned so that True Range was set to RDF Range. This broadside was seen to hit.
>
> Fire was then shifted to the left-hand cruiser, the *Zara,* and the RDF also shifted target. Five broadsides were fired at this cruiser, the True Range being tuned throughout to RDF Range. All these five broadsides were seen to hit.[10]

Subsequently, an officer from the Naval Ordnance Department commented somewhat cynically that, in view of the fact that the range was very short and *Warspite* and *Barham,* neither of whom had radar, also obtained hits with their initial salvos, 'no conclusions as to the value of RDF can be drawn from this report'.[11]

GREECE AND CRETE

In early May, Operation 'Tiger' was mounted, the main purpose of which was to pass through the Mediterranean a special convoy with tanks for the Army of the Nile, but opportunity was taken to send reinforcements for Admiral Cunningham – the battleship *Queen Elizabeth* with newly-fitted type 279 air warning and types 284 and 285 surface and air gunnery radar; the cruisers *Naiad* and *Fiji* both with type 279, the latter with type 284 also; and six destroyers, some of which had type 286M.

Of the fourteen large ships involved in the operations connected with the evacuation of Greece and Crete, nine had air warning radar, two had only the fixed-aerial type 286M, and three – *Warspite, Barham* and *Gloucester* – had no radar at all. But, as Admiral Hezlet points out,[12] while radar could ensure that ships were not surprised by dive-bombers, it could not help against sustained and heavy attacks when ammunition was running short. What was needed was fighter cover, which the RAF could not provide because of the grave situation in the Western Desert, while *Formidable* had too few serviceable fighters even to defend herself. In the Greek, Crete, Syrian and Western Desert campaigns up to June 1941, British naval forces suffered grievous losses due to air attack, mostly by dive-bombers from Fliegerkorps X transferred from Sicilian to Balkan airfields. *Gloucester, Fiji, Calcutta, Terror* and nineteen smaller ships were sunk, *Warspite, Barham, Formidable, Naiad, Orion, Dido, Perth* and seven destroyers very severely damaged.[13]

THE BATTLE OF THE ATLANTIC

The Atlantic was a theatre of war even more important to the future of Britain. As Churchill has said, never for one moment could we forget that everything happening anywhere in those years – on land, at sea, or in the air – depended ultimately on the outcome of the Battle of the Atlantic,[14] so named by him after the First Sea Lord had reported, early in March 1941, exceptionally heavy sinkings. 'We have got to lift this business to the highest plane over everything else,' Churchill told Pound, 'I am going to proclaim "The Battle of the Atlantic".'[15] And he forthwith set up the Battle of the Atlantic Committee of the War Cabinet. 'We must take the offensive against the U-boat and Focke-Wulf wherever we can and whenever we can. The U-boat at sea must be hunted . . . the Focke-Wulf and other bombers . . . must be attacked in the air and in their nests'.[16] For the next year, the Battle of the Atlantic had the highest priority for Britain's resources.

The principal reason for this crisis was a change in U-boat tactics from individual submerged submarines attacking convoys by day – when the British asdic-fitted escorts had the upper hand – to surface attacks by night by so-called 'wolf-packs' coordinated by shore control. The change caught the Navy unawares and unprepared. Asdic – 'sonar' to the Americans – could not detect a submarine on the surface: what escorts must have was the kind of radar which could. And this was centimetric radar.

10. The destroyer *Sardonyx* (of which the author was First Lieutenant) before and after a gale in the Atlantic in June 1941. The weight of the non-rotating 'bedstead' aerial of WC type 286M proved too much for the wooden mast during heavy rolling. A new mast and aerial were fitted but the same thing happened again in October. The much lighter rotating type 286P aerial (see Fig. 20) was then fitted and lasted until the end of the war.

(Author)

By early March 1941, some ninety or so convoy escorts had been fitted with the fixed-aerial 1.5m type 286M. But, though much better than no radar at all, this set was not good enough. The first version, type 286M, had a fixed aerial at the top of the mast, similar in principle to that fitted in contemporary aircraft. To obtain an accurate bearing of an echo, it was necessary to point the ship at the target, though a reasonable estimate of the relative bearing could usually be made without altering course. A more serious limitation was the confusion caused by back and side echoes, (see Glossary; see also anecdote on p. 126) particularly for convoy escorts trying to find a 'rogue' echo among the many friendlies. But the very size and weight of a 'bedstead' at the masthead was also a grave disadvantage. The author was First Lieutenant of the WW1 destroyer *Sardonyx* working with Atlantic convoys when type 286M was fitted in Londonderry in June 1941. During a gale about a month later, the wooden mast and aerial went by the board (Fig. 10). Not two months later in another Atlantic gale, exactly the same thing occurred to the replacement mast and aerial.

These limitations were evident from the very first ship-fitting of type 286M in mid-1940 and the Signal School team at Nutbourne under A. W. Ross began work on the design of a light-weight rotating aerial, a scaled-down version of the type 281 aerial. This was successfully tried in the new destroyer *Legion* in February 1941 when a merchant ship was detected at 9 miles, a destroyer at five, and a surfaced submarine at one and a half.[17] Production of the new aerial began in June and by September, thirty-seven ships had been fitted (including thirty convoy escorts), most replacing existing fixed aerials – of which there were then 210 ships so fitted – thereby converting from 286M to 286P.[18]

On 17 March 1941, the first sinking of a U-boat directly attributable to radar took place – when the destroyer *Vanoc* (Commander J. G. W. Deneys), fitted with the fixed-aerial type 286M, rammed and sank U.100. On 17 February, the Admiralty had appointed Admiral Sir Percy Noble Commander-in-Chief of a new Western Approaches Command, with headquarters at Liverpool, and one of his first actions was to set up escort groups containing a mix of destroyers, sloops, cutters and corvettes, who would 'work up' and train together and remain as self-contained units. One of the first of these, the 5th Escort Group under Commander Donald Macintyre in the destroyer *Walker*, was escorting the homeward-bound convoy HX.112 south of Iceland when they were attacked by a pack of five U-boats. Initially, the only radar-fitted ship in the Group was *Vanoc*.

At 10.30pm on 15 March, *Volunteer* detected the sound of a torpedo and ten seconds later a tanker in the convoy blew up. In the flames, *Volunteer* sighted the torpedo track and, calling *Vanoc* to her aid, searched down the track, but nothing was seen or heard. *Vanoc* rejoined the convoy with the aid of her radar but *Volunteer*, with no radar, lost touch in the fog and never rejoined.

The following night, the wolf-pack attacked again, four more ships being torpedoed. At 12.50am on 17 March, *Walker* sighted and attacked a U-boat some 5 miles astern of the convoy. Forty minutes later she detected another U-boat by asdic and this was hunted by *Walker* and *Vanoc* until it surfaced out of sight and contact was lost by both ships. At 2.50am, while *Vanoc* was proceeding at 15 knots to take station on *Walker* for an organized search, her radar operator reported a contact on the starboard side. The echo was quite unmistakable and could only have been a submarine surfacing, as it arrived suddenly on the screen at half a mile range. Turning towards, *Vanoc* sighted, rammed and sank U.100 before the latter's gun was even manned.[19] U.100's captain, Joachim Schepke, one of the ace U-boat commanders, did not survive. While *Walker* was circling and protecting *Vanoc* picking up survivors from U.100, she gained asdic contact very close astern of *Vanoc* and attacked with six depth charges. U.99 surfaced and surrendered. *Walker* picked up 41 survivors, including Captain Otto Kretschmer, an even more distinguished U-boat commander.

As we have said, type 286 was better than no radar at all – but not good enough. It was to be the 10cm type 271 – together with many other innovations, tactical and technical (such as HF/DF, discussed later) which eventually allowed us to beat the wolf packs. But it was a close-run thing.

However, before discussing the crash programme to design, produce and fit type 271, we must make brief mention of another grave threat to our shipping in the Atlantic brought about by the occupation of Norway and France by the Germans in 1940 – the basing of long-range Focke-Wulf 'Kondor' bombers in Norway and on France's western seaboard. On 21 October 1940, the Canadian Pacific liner *Empress of Britain*, on passage from the Middle East, was bombed and set on fire by a Focke-Wulf some 70 miles north-west of Donegal Bay. She was taken in tow, escorted by two elderly destroyers, *Broke* and *Sardonyx*, but was sunk two days later by U.32. The author was then serving in the latter ship, one of the Signal School's pre-war radio trials ships, which was also escort to convoy HX.112 in the March 1941 episode with *Vanoc*.

These aircraft posed a threat not only from their bombs but also from their ability to tell the U-boats the position of Allied shipping. The convoy routes to and from Gibraltar were particularly vulnerable in the gap between cover from shore-based fighter aircraft from Gibraltar and those of the RAF's 19 Group from home bases.

There was only one solution – the convoys must carry their fighter aircraft with them. The Admiralty made their first plans on these lines – albeit only stopgap ones – before the end of 1940. The old seaplane carrier *Pegasus*, the Bank Line *Springbank*, and the banana boats *Ariguani* and *Maplin* were converted to fighter catapult ships, RN-manned, equipped with two or three fighters, initially Fulmars, later Hurricanes. *Springbank* and *Ariguani* had started their naval life as Auxiliary AA Ships, so were already fitted with the modified Army GL radar set, type 280; *Maplin* was given type 79, *Pegasus* type 286P. Ready for service in April 1941, a Hurricane from *Maplin* secured the first success on 3 August by shooting down a Focke-Wulf 400 miles from land.

In addition, the Admiralty arranged for catapult equipment to be fitted in thirty-five merchantmen, to be known as Catapult Aircraft Merchantmen (CAMs), with RAF-manned Hurricanes and either RAF GCI radar or type 286P. The first launch took place on 31 May, and the first action with a Focke-Wulf on 1 November.[20] But the fact remained that, unless there was a friendly airfield or carrier deck within range, aircraft catapulted from either type of ship were lost after each launching and the pilot might or might not be rescued after a parachute descent. The advent of the Navy's first proper escort aircraft carrier will be discussed in the next chapter.

As the reader will have seen, the naval radar type numbers have begun to proliferate. To tell our story properly, we cannot avoid using them. To help the reader, there is a ready-reference list on p. 340 immediately before the alphabetical index.

THE BEGINNINGS OF RADAR COUNTERMEASURES IN THE ROYAL NAVY

When the German coastal batteries opposite Dover became operational in the late summer of 1940, British convoys attempting the night passage of the Dover Straits were subjected to savage attacks, and losses mounted. Senior officers who were familiar with British radar developments began to suspect German use of radar for maritime surveillance and fire control.

On 11 November 1940, Captain B. R. Willett, the Signal School's Experimental Captain, sent for N. E. Davis of the civilian staff (a senior television engineer seconded from the Marconi company) and informed him of the situation in the Dover Straits, and that German use of radar for fire-control purposes was suspected. Could he please investigate?

Davis travelled to St Margaret's Bay near Dover, taking with him a monitor receiver tunable over much of the decimetre waveband. When German guns next opened fire on a convoy during the dark hours, he set watch and almost immediately detected transmissions on 83cm – from a Seetakt radar. The problem of the control of the German long-range guns was solved! Having made notes of the radar signal's characteristics, Davis identified two radar sites near Cap Griz Nez and Boulogne respectively, and returned to Signal School.

This led immediately to the formation of a new section at Signal School, with himself as leader – the Radar Countermeasures (RCM) Section, whose primary task was to develop means to prevent enemy coastal radars from ranging on Channel shipping, as well as devising a system of monitoring enemy radar developments.

Davis's initial task was to develop an effective jammer transmitter covering the 80cm waveband used by the Seetakt radar. His deputy, O. E. Keall (also seconded from Marconi), set to work on developing monitor receivers. First, Davis visited the GEC Research Laboratories at Wembley to outline his requirements for a transmitting valve and GEC soon responded. He then designed a tunable transmitter using a pair of these valves, giving an output power of about 25 watts. A high-frequency sine-wave form of modulation was chosen for the jamming signal initially, because if similar jamming was used against British radars in retaliation, this could have been filtered out from our radar displays. Later in the war, the form of jamming modulation was changed to 'noise' – similar to that in an ordinary radio receiver off-tune when the volume is turned up high – because such a signal could not be filtered out from the radar display.

The section rapidly produced six experimental jammers, plus the associated aerial systems. Meanwhile, Keall's group constructed monitor receivers for the alignment of jammer operating wavelength with that of the intercepted enemy radar and a motor-tuned monitor receiver was designed to simplify the process.

By mid-February 1941, experimental jammers and associated monitor receivers were installed in the Dover and Folkestone areas. The stage was now set for active operations.

THE FIRST OPERATIONAL TEST

The account of the first operational test of the Navy's experimental radar jamming system is best given in Davis's own words, taken from a covering note to his official report:

> I was at the Sandgate site and an urgent call came through from the Flag Lieutenant [a specialist signal officer], Dover Castle. He informed me that a convoy passing down-Channel was under very heavy gunfire accurately directed. It was at night and demonstrated the effectiveness of the enemy radar. 'Can you do anything to help the ships, Davis?' 'Yes, we are ready to jam.' Then, contrary to Admiralty procedure he said, 'Well, for Goodness sake, jam!' And we did.
>
> The gunfire immediately ceased and our ships proceeded southwards towards Dungeness, unmolested. The Flag Lieutenant came through and said I could lift the jamming, which we did, and this brought a few more shells from the German guns, but our convoy was now out of their range. This incident was a full justification of our policy to jam.[21]

For fear of enemy reprisals against British radars, Admiralty policy before this incident had been not to jam, merely to listen. The policy on jamming was reversed the following day: on such occasions, it was decided, one should jam.

THE DEVELOPMENT OF NAVAL 10–CENTIMETRE RADAR

Let us now continue the story of the development of the Navy's 10cm radar, which we left on p. 70 with the Swanage group under Landale arriving at Eastney just in time for Christmas leave in December 1940. Even before they arrived back, development and design for an operational set had been progressing urgently at Eastney. Despite the many problems posed by setting up experimentally to work in a new frequency band and at pulse lengths appreciably shorter than any so far used in naval radar, and despite the need to design circuits around components already in production or at least in pre-production stage, the first set of units from the laboratory were ready in January 1941 to replace the equipment brought from TRE in the trailer on 19 December. Furthermore, the laboratory workshops, aided by laboratory staff, were committed to manufacture the first twelve copies of these designs.

Speed in development and the ability to produce fast in quantity dominated the design of the new set which therefore had to be as simple as possible. The essential limitation imposed by the new frequency band was that the length of cable between the aerials and transmitter or receiver had to be as short as possible. (In 1940, waveguides were known but not sufficiently developed for use operationally.) The solution adopted was to mount the transmitter and crystal mixer stage of the receiver on the back of the aerials. The low power available from the local oscillator, the Sutton tube, could not feed the crystal mixer through more than 20 feet of coaxial cable, and so the aerials had to be mounted directly onto the roof of the 6–foot-square radar office immediately below, which contained the receiver, display unit, mechanical aerial rotation gear, and a chair for the operator. To make the aerial assembly weathertight, the office was surmounted by a lantern-like structure of teak pillars framing flat perspex windows, the sand-blasting of which served to stop casual observation of the dipoles which could reveal the wavelength being used.

The very size and weight of this structure precluded it being fitted immediately in existing destroyers but this did not apply to 'Flower' class corvettes then coming from builders' yards in increasing numbers. On 11 February 1941, as the first Eastney-built set neared completion, Captain Willett convened a meeting of development, design, production and application staff at Eastney. It was at this meeting that Willett, in telephone contact with the Admiralty's Anti-submarine Warfare Division and DSD in London, took the bold decision to place immediate orders for components for 150 sets, and to increase the number of sets to be built in Eastney workshops from twelve to twenty-four, on the chance of operational success. He ruled that no modifications were to be accepted unless absolutely essential or which could be introduced without delaying production. At this meeting also, the decision was taken that the set should be designated type 271.[22]

Meanwhile, the future of centimetric radar generally showed such promise that, at an inter-service conference at the Air Ministry on 3 January, Air Marshal Joubert had been directed to place an order with GEC for 200 magnetrons, the cost to be shared equally between the Royal Navy and the RAF.[23] Presumably this order was increased after the Signal School's February meeting.

This might be a good moment to record tributes by two people closely concerned. The first is by Dr Landale, writing in 1966:

> The principal components of this centimetric equipment were, first, the magnetron developed by Randall and Boot, and I think the great assistance given by the Research Department of the General Electric Company should not be forgotten, where Megaw's enthusiasm proved invaluable.[27] On the receiving side, the detector was a very strange and now seemingly primitive crystal device developed by Dr Skinner under Air Ministry auspices, and the aerial system proposed and made workable by Lovell, now Sir Bernard of Jodrell Bank.[24]

In fact, Megaw contributed a great deal more than mere enthusiasm. He made the magnetron reproducible and developed it into a useful form; in Randall and Boot's original continuously-pumped form, it would have been useless for a seagoing or airborne radar.

The second tribute was written to DSD by Captain Willett on 29 March 1941 in his covering letter forwarding the results of the first *Orchis* trials described below:

> Whatever else happens, I think that Landale & his party have done a wonderful job in taking over from the Research Department [TRE] in December, engineering and improving the technical performance of the equipment & getting it to sea and working by mid-March. I hope one day you will have a chance to tell him so.[25]

THE *ORCHIS* TRIALS OF TYPE 271

The first Signal School centimetric set was completed on 7 March and plans were made for a sea trial in the corvette *Gardenia*. However, she was

11. The corvette *Periwinkle*, sister-ship to *Orchis* in which the Navy's first centimetric radar, WS type 271, was fitted in March 1941. The radar hut surmounted by the wood-and-perspex lantern containing the 'cheese' aerial array can be seen here between the bridge and the funnel.

(IWM)

delayed, so, in the middle of March, three members of the Signal School scientific staff took the set and installed it in the 'Flower' class corvette *Orchis*, then undergoing repairs at Scott's yard in Greenock. The RDF hut, surmounted by the 'chicken house' (as it was dubbed in *Orchis*), was mounted just abaft the bridge in place of the 20in searchlight.

On 25 March 1941, the first sea trials of type 271 took place in flat calm conditions in the Clyde with destroyers and the small Norwegian submarine B1 as targets, attended on the scientific side of the Signal School by Landale, H. E. Hogben and P. Morey; and on the naval side by the Application Officers, Commander A. V. S. Yates and Lieutenant A. E. M. Raynsford. Trials continued on the 26th, again in calm weather. On the 28th, *Orchis* sailed in search of rough weather – to Londonderry with the Signal School trials party embarked.

> It remained calm for a week, [wrote the irreverent Sub-Lieutenant Orton, RNVR] & we played endless games of cards waiting for the weather to deteriorate. At last it did – & both 'boffins' were so sick that they could only just make it to the set. After two days they said they'd had enough, & because I had specialised in science at Charterhouse before leaving in 1937, they turned over to me all the drawings of circuits & layout etc., & wished me luck & said it was all mine now! They couldn't get away quickly enough![26]

The rough weather trials took place on passage from Lough Foyle to the Clyde on All Fools' Day 1941, the Signal School party being landed at Ardrossan that same evening while *Orchis* sailed to rejoin her escort group on 3 April.[27] During the rough weather, *Orchis* was rolling up to 30° and yawing 10° either side of the mean course, but neither of these made holding the target too difficult. The maximum working range on the small submarine surfaced was 2.1 miles as against 2.5 miles in smooth water. Destroyers were picked up about 6 miles, buoys a little over a mile, and high land beyond the end of the scale, at that time 14 miles.[28]

Though the ranges obtained were not quite as good as might have been expected from the early tests at Swanage, the set more than fulfilled its promise as an anti-submarine weapon. In contrast to the fixed-aerial type 286M – convenient for station-keeping but almost useless for A/S work because of the confusion of side and back echoes – type 271 with its rotatable aerial and narrow horizontal beam width was just what the A/S vessel wanted, doing for surface echoes what the asdic did for underwater echoes, capable of using precisely the same procedures for searching and holding, which everyone was used to and could understand. From an underwater warfare point of view, an escort's ability to find – *and* fix, *and* hold – U-boats on the surface at night fully justified the emergency fitting programme that was immediately set in train. And the navigational advantages were of no mean order either. The Captain of *Orchis*, after two months' experience, having bemoaned the fact that he had met no U-boats, concluded his report thus: 'After being in a ship fitted with type 271, night navigation in one without will seem a perilous business.'[29]

It had already been decided that twenty-four models should be built at Eastney and now a hundred more were ordered from Messrs. Allen West of Brighton, copies of Signal School models. (Allen West was by now virtually part of the Signal School. See p. 39 above.) Components for 150 sets had been ordered in March, increased to 350 in May. The initial design was reckoned to be good enough to be repeated immediately with minor improvements (mostly to make production easier) and a production order for 1,000 sets was placed with Messrs. Metropolitan Vickers, for fitting in 1942.

J. D. S. Rawlinson, in charge of the laboratories at Eastney at the time, had this to say in a lecture to the Fleet in 1944:

> This was a period of great enthusiasm and hard work at Eastney. Mechanics from the Signal School workshop [in the RN Barracks] came to Eastney to help in the task. The [display] indicator used was the ASV indicator completely gutted and rebuilt and 100 of these were done in the labs. at Eastney. We were not experts on mass production but what we lacked in method, we made up in keenness and enthusiasm and I think I speak for all when I say that that period was perhaps the most exciting and happiest time in the radar work. This Mark II set [the Eastney and Allen West sets] had a power output of 5–10 Kw and the same when re-designed as Mark III [the 1,000 production models]. The results were about [the same as] those of the *Orchis* set [the Mark I], 4–5,000 yards [2–2.5 miles] on a submarine, 12–15,000 yards [6–7.5 miles] on a destroyer.[30]

Four more Eastney sets came forward in May, the first going to Allen West for copying, the second to Eastney for instructional purposes, the third and fourth to Chatham for fitting in trawlers. These last two were in response to the threat to East Coast convoys from E-boats, the thought being that the trawlers would detect and plot the E-boats and then direct our motor gunboats by radio to attack them. The trawlers *Avalon* and *Norland* were duly fitted and trials were carried out with *Avalon* at anchor off Harwich at the end of May. Sadly, this early attempt at surface force direction from a ship – to be used with such success in Operation 'Neptune' in 1944 – was not a success at that early stage. A PPI (see p. 108 below), special staff and plotting facilities, IFF in our boats, better radio – all of these were needed even to make a start, said the trials officer. Nevertheless, the Commander-in-Chief, Nore, was able to report in

September that the trawlers *Norland* and *Adonis* could detect motor craft at 2 miles and that they would be used in patrols in the vicinity of the Sheringham Light Float in company with MGBs.[31]

HM SIGNAL SCHOOL AND SHIP FITTING

> 1941 opened in HM Signal School with everyone at fever pitch with excitement and enthusiasm, [writes one who was there at the time] and this despite the fact that since the outbreak of war there had been no leave whatever and everyone had worked all the hours they could. Their Lordships soon realised that this could not go on indefinitely, so we were ordered that everyone must have one day off a week and take a whole week's holiday in 1941.[32]

All departments were working to the limits of their capacities, and none more so than the ship-fitting section under Commander C. V. Robinson which was responsible for providing expert radio supervision of all installation work carried out by dockyards and ship-builders. The work was done at all the main ports at home and abroad – always under very heavy pressure. His staff was largely naval but the already hard-pressed experimental scientific staff, under P. Morey, were often called

12. Captain Basil R. Willett (*right*), Experimental Commander HM Signal School 1937, Experimental Captain 1939, Captain Superintendent Admiralty Signal Establishment 1941–3; and Commander C. V. Robinson (*left*), in charge of radar ship-fitting organization at Signal School and ASE.

(NRT Willett collection)

upon to overcome teething troubles, particularly during the early fittings of new sets, not only visiting ships in harbour, but going to sea as well. B.G.H. Rowley was one of the earliest of these, stationed at Scapa from May 1940, to work in ships of the Home Fleet.

During the first six months of 1941, probably the greatest volume of ship-fitting work was in fitting type 286 in large numbers of small ships – and not a few larger ones – both at home and abroad. For long-range air warning, some twenty-one large ships were fitted with type 279 and, starting in March, six were fitted with type 281.

Back in the Signal School extension in Onslow Road, Southsea, Coales and his 50cm team experienced the same sort of stimuli – though probably a month or so earlier – as those described above by Rawlinson for their 10cm colleagues at Eastney. The sets already designed had to be brought into service and teething troubles cured; and, based on that experience, new and improved sets had to be designed and production organized.

At the start of the year, the earliest 50cm sets had just gone to sea – type 284 in *King George V* and *Birmingham*, and type 285 in *Southdown*. By May, eleven more type 284s were at sea (some to take part in the *Bismarck* affair with great distinction) and more than a dozen type 285s. All of this, particularly the early fittings, kept the 50cm team working practically night and day, assisting and supervising the normal ship-fitting staff and conducting trials, with a Signal School scientist, W.F. Drury, joining Rowley at Scapa to help newly-fitted ships 'working up' there.[33]

By far the biggest challenge was the new battleship *Prince of Wales*, the first ship to get a multiple suite of radar, nine out of the ten operating on 50cm – one type 281 air warning, one type 284 on the main armament director, four type 285s on AA directors, and four type 282s for close-range weapons. The technical difficulties in ensuring that they did not interfere with each other were considerable, but they had largely been overcome before her encounter with *Bismarck* on 24 May. Nor were they made any easier by the fact that the radar officer, Sub-Lieutenant Stuart Paddon RCNVR, was the only 'technician' (this was before the days of radio mechanics) and only one of his thirty or so operators had been to sea before.[34]

For types 79 and 279, the most important development in 1941 was the introduction of single-mast working. From the very beginnings of naval metric radar, the need to have separate transmitting and receiving aerials, each on the top of its own mast, caused many problems because of the competing demands for such positions with other radio systems, communications and direction finding. And none more so than in aircraft carriers where the island structure was of a limited size and the need for a good position for the homing beacon aerial was paramount. The first two carriers with radar, *Illustrious* and *Formidable*, had the receiving aerial above the island and a telescopic transmitting aerial rising out of the deck abaft the island, which could be lowered to permit efficient operation of the homing beacon. This proved most unsatisfactory and, late in 1940, work was started at Nutbourne on the development of a diode switch and matching transformers which would permit a single aerial to be used for both transmitting and receiving.

The first experimental model of the single-mast type 279 was fitted in the battlecruiser *Hood* and sea trials were successfully carried out in April,[35] only a month before she was sunk by *Bismarck*. The next single-masted version was fitted in the escort carrier *Audacity*, whom we shall meet in the next chapter.

Type 281 had proved most successful on sea trials despite criticism of the engineering design of *Dido*'s prototype by her radar officer (Sub-Lieut M. Rosenthal [later Hartley], RCNVR), on which C. E. Horton made the following very fair comments:

> Under war conditions, when our sets receive no adequate trials, we must expect some points to arise when first they appear at sea. May I plead that we don't rush either 290 or 271X to sea too soon?[36]

The first production model of type 281 proved equally successful in *Prince of Wales* when she commissioned in March. Meanwhile, the Tizard Mission in the autumn of 1940, and the signing of the Lend-Lease Bill on 11 March, had resulted in very full RN/USN cooperation and the USN technical liaison officer in London, Commander Dow, was in the closest touch. On 5 April, the US Naval Attaché in London sent the First Lord of the Admiralty the following message from Colonel Knox, Secretary of the Navy: 'In order to facilitate and expedite production of equipment similar to your RDF type 281, I would much appreciate any steps which can be taken by you to expedite shipment of sample equipment to the United States.'[37]

As a result, a complete set was shipped in May, but, said the Admiralty, the US Navy could not keep it. Supplies were so short that, while they might have it for study for about six weeks, the set must then be fitted in the battleship *Malaya*, the first of many British ships about to be repaired in US yards. In explaining this to the Controller, DSD said it was the best that could be done without giving the Americans priority over HM Ships: C-in-C Home Fleet, he said, considered the need for tactical warning could only be met by type 281.[38]

THE SINKING OF *HOOD* AND *BISMARCK*

Let us conclude this chapter with some account of the first fleet action in which radar played a really significant part – the sinking of the new German battleship *Bismarck* on 27 May 1941.

When, on 22 May 1941, the Commander-in-Chief, Home Fleet, Admiral Tovey, received the not unexpected news that *Bismarck* had sailed with the cruiser *Prinz Eugen* from Bergen, he had already made his dispositions to intercept the German squadron should it try to break out into the Atlantic. Cruisers were already patrolling the two possible routes *Bismarck* might take, the Iceland-Faröes gap and the Denmark Strait. The battle cruiser *Hood* (flagship of Vice-Admiral L. E. Holland commanding the Battle Cruiser Squadron) with the brand new and un-worked-up battleship *Prince of Wales* – she sailed with civilian dockyard and contractors' men still on board – and six destroyers were already on their way from Scapa towards Iceland. The Commander-in-Chief's own fleet flagship, *King George V*, the carrier *Victorious*, the battlecruiser *Repulse*, four cruisers and seven destroyers sailed from Scapa that night, unobserved by the Germans.

In the Denmark Strait, covering the 25–mile gap between the minefield off Iceland's north-west coast and the edge of the Greenland pack ice, were two 8in cruisers under the command of Rear-Admiral W. F. Wake-Walker, *Norfolk* and *Suffolk*. While *Norfolk* merely had the fixed-aerial type 286M radar, *Suffolk* had very recently acquired the 50cm type 284 with the aerial mounted on (and rotating with) the main armament director abaft the bridge; she also had the air warning type 279.

Though fitted primarily as a radar-rangefinder for *Suffolk*'s main gunnery armament, type 284 was able to provide both range and accurate bearing of surface targets, for which type 279 was not so suitable. (Nevertheless, until 10cm sets were fitted, ships often did use type 279 successfully for surface plotting – as, for example, in the Battle of Cape Matapan – though, because of its wide beam-width, the bearing accuracy left much to be desired and it was difficult to discriminate between two targets at roughly the same range.) However, using type 284 in this way did involve continuous training of the director, which was not advisable for more than a few hours at a time, but Captain R. M. Ellis had decided that risk of breakdown could, with due safeguards, be accepted in order to make tactical use of the gunnery set. He and his navigator had devised a 'drill' to accomplish this and the radar operators, gun director's crew, and the surface plotting team had been practising assiduously on ship and land targets during the previous few weeks. But there was a definite limit on how long it could be used continuously in the tactical role.

> Adjusted arc of sweep of type 284 as requisite to cover bearings on which the visibility was below 8 miles. Type 279 kept listening watch but was not used for transmitting.'[39]

So says *Suffolk*'s subsequent report of proceedings in the narrative entry for 6.10pm on 23 May. The next momentous events are best described in Captain Ellis's own words:

A lookout first sighted her [*Bismarck*] at 7.22pm on May 23, 1941, emerging from mist fine on the *Suffolk*'s starboard quarter, on much the same course and overhauling fast. The radar was being rested, and anyway would not bear so far aft, the ship being on a westerly leg of a to-and-fro patrol roughly parallel to the coast of Iceland and some 30 miles from it, close to the ice and just outside the minefield.

The *Bismarck*'s range was only 14,000 yards [7 miles], lethally close if she sighted us. It was imperative not to be destroyed before getting off our enemy report, so we turned into a nearby fogbank, between two arms of the minefield. Next came getting the radar to work, to plot the enemy's movements, while ourselves manoeuvring unseen into a shadowing position astern of the rear enemy ship, the cruiser *Prinz Eugen* . . . Once astern of the enemy and out of the fog, I found we could hold the enemy on the type 284 out to about 13 miles and occasionally further.[40]

Bismarck had detected *Suffolk* by radar and hydrophone at about the same moment that *Suffolk* had sighted *Bismarck* visually. Just as *Suffolk* was disappearing into the fog bank, she came into sight just long enough for the Germans to identify her as a 'County' class cruiser, but not long enough to open fire. *Suffolk*'s enemy report was intercepted and deciphered by *Prinz Eugen*, and reported to the German command ashore.[41]

Captain Ellis continued:

I learned afterwards that our first enemy report had been picked up only by the *Norfolk*, patrolling about 50 miles to the west. It was to transmit this radio report that everything had to be subordinated in those few moments before *Bismarck* might open fire with the usually deadly accurate first salvos of German gunnery . . . Our first enemy report was not received by any distant ship or station [because of ice on the main W/T leads to the aerials]. It was only after the heat from several transmissions had dried out the insulators that full power emission was achieved, enabling the Commander-in-Chief and the Admiralty to receive our signals, direct or via Scapa W/T . . .

The *Norfolk*, lacking surface radar, went looking for the *Bismarck*, after plotting my first sighting report, and met her quite close in mist and to her extreme peril. This time the enemy was alert.[42]

Bismarck detected *Norfolk* with her forward radar at 8.30pm, opening fire as soon as she was sighted – for *Norfolk* dangerously close – at 6 miles. *Bismarck* managed to get off five salvos, straddling with three, before *Norfolk* disengaged under cover of smoke, miraculously undamaged. She then sent the first enemy report actually received by the battle fleet some

600 miles away to the south-east. The blast from *Bismarck*'s guns had put her own forward radar out of action, so *Prinz Eugen* was ordered to lead the line instead of *Bismarck*.[43]

> It was obvious [continues Captain Ellis] that she [*Norfolk*] could not keep within visibility distance of the enemy in the prevailing weather, and so took up a shadowing position out of sight, broad on the port quarter, conforming to the enemy's movements as deduced from the *Suffolk*'s sequence of signals. Even so, she ran considerable risk . . .
>
> Throughout the night the *Norfolk* and *Suffolk* remained out of visual and radar touch with each other, but in continuous radio contact. Speed around 28 knots. From time to time it was necessary to rest type 284. Though not visible, the enemy was boxed in by the ice and the two cruisers. No reports therefore meant no changes to report.
>
> It snowed during the night and the wind rose. About dawn it cleared to the southeast and southward, and we could see ice-blink to the north and west. But we still could not see any other British ship. The *Norfolk* was still miles out of sight and gun range when the action between the Germans and the *Hood* and *Prince of Wales* began. Of our battleships we could see only the funnel tops and the gun flashes. The morning of this brief, disastrous battle was May 24th. When fire was opened, the *Suffolk* was roughly 18,000 yards [9 miles] astern of the enemy, ready to flank-mark our heavy ships' fall-of-shot . . .[44]

This is no place to tell the full story of the operations that led to the sinking of *Hood* and *Bismarck* and we will confine ourselves to the radar aspects. *Hood* had very recently been fitted with the prototype single-mast type 279M air warning set and with the type 284 gunnery set for the main armament. As mentioned above (p. 89), *Prince of Wales* had type 281 and nine various gunnery sets. *Bismarck* had three rangefinder towers – one abaft the bridge, one on the foretop, and one aft – each accommodating an 81.5cm FuMO 23 radar as well as a large optical rangefinder. *Prinz Eugen* had two larger towers – one on the foretop and one aft – able to accommodate the improved 81.5cm FuMO 27 Seetakt radar, with a 2m x 4m mattress aerial.[45] *Bismarck* had in addition a radar search receiver, though it is not clear how effective this was.[46]

During the approach phase, it was vital that the battle cruiser squadron should do its best to achieve surprise, so Admiral Holland imposed strict radio and radar silence until action was imminent, relying on *Suffolk*'s reports to keep him up to date on the whereabouts of the enemy. However, soon after midnight, *Suffolk* lost touch. At 2am, the Admiral told *Prince of Wales* to search an arc of the horizon with her type 284, but Captain Leach reported that his main armament director could not cover the desired arc and requested permission to use his type 281.

13. The German 80cm FuMO 23 radar mounted on the forward gunnery control tower behind the bridge of the German battleship *Bismarck*.

(Fritz Trenkle)

This permission was refused. Then, at 2.47, *Suffolk* regained contact and a stream of reports began to come in. Though we do not know the situation in *Hood*, *Prince of Wales* at least was able to develop an accurate plot of the enemy's movements. The British squadron sighted the enemy at 5.35am and came into action at 5.53. *Hood* blew up at 6am.

The sudden destruction of the British flagship enabled both enemy ships to concentrate their fire on *Prince of Wales* and she suffered a fair amount of damage, including the shooting away of the type 281 receiving aerial on the main mast and the destruction of the after type 285 office. Then 'Y' turret seized up and Captain Leach decided to break off the action and retire under cover of smoke.

In the official history, after noting that all radar sets in *Hood* and *Prince of Wales* were known to have been in working order before sailing and that the latter's 'modern search radar set [type 281] could also transmit

14. The *Bismarck*'s after FuMO 23 radar. A third gunnery control tower with radar was mounted on the foretop.

(Fritz Trenkle)

ranges to the main armament', Roskill says, 'In the *Prince of Wales* no results were obtained from either of her two sets [types 281 and 284] throughout the action.'[47] The following account by her radar officer, Sub-Lieutenant Stuart Paddon, RCNVR, seems to imply that the fault was organizational rather than technical:

> My position was in the 281 receiving office and I personally manned the display tube. In those days we had no gyro repeats; we had to give the bearings by red or green [port or starboard], but we did have two M-type transmission units, little counter-drums on which we could transmit range. The transmission counter-drum that I controlled had as its counterpart a receiver on the bulkhead of the Transmitting Station – the TS. This had to be read and placed by someone into the calculations which were being carried out on the plot. Unfortunately,

> despite the fact that I transmitted these ranges, no one even knew they were coming in on the counter-drum and no radar ranges were used by our gunnery people. We had had only one previous shoot and had not developed any drill. This was unfortunate because I had three distinct echoes of three ships at 26,000 yards [13 miles], clearly portrayed on the radar screen, an A-scan with a linear blip. I was able to follow them in with complete accuracy and complete detail, religiously giving the range as I have described. The third ship turned out later, on investigation, to have been a supply vessel in company with the *Bismarck* and the *Prinz Eugen*. I have never seen any reference to this supply vessel in anything I have subsequently read . . .[48]

After the loss of *Hood*, the two cruisers and *Prince of Wales* continued to shadow, *Suffolk*'s type 284 radar being mainly responsible for keeping contact. During the evening, there was a brief exchange of fire, a diversion which enabled *Prinz Eugen* to slip away to the south. About 10pm *Victorious* flew off a striking force of Swordfish torpedo bombers and Fulmar fighters which, thanks to ASV Mark IIN fitted in the Swordfish (the first naval operational squadron so fitted), succeeded in locating and attacking *Bismarck*, though the Fulmars, without radar, had some difficulty in finding the carrier on return.[49] Then, at 3.07am on the 25th, *Suffolk* lost touch when *Bismarck* turned south-east towards St Nazaire, not to be located again until 10.30am on the 26th, when a Coastal Command Catalina located her with its ASV Mark II.

Admiral Lütjens seems to have been surprised at *Suffolk*'s success in shadowing:

> Presence of radar on enemy vessels, with a range of at least 35,000 metres, [he reported to Group West soon after he had at last succeeded in breaking contact] has a strong adverse effect on operations in Atlantic. Ships were located in Denmark Strait in thick fog and could never again break contact. Attempts to break contact unsuccessful despite most favourable weather conditions . . . Detachment of *Prinz Eugen* made possible by battleship engaging cruiser and battleship in fog. Own radar subject to disturbance, especially from firing.[50]

Lütjens overestimated, Ellis's report of 13 miles range is actually about 24,000 metres – more probably (and reasonable) in view of the known capability of type 284 in cruisers.

One incident of specific radar interest may be mentioned in conclusion – *Sheffield* succeeded in shadowing with her old type 79Y radar and then in directing a strike by *Ark Royal*'s aircraft, one of whose Swordfish (though not the ship herself) had radar. For an account of the remainder of this operation – which ended with *Bismarck* sinking with her flag still flying during the forenoon of 27 May – the reader is referred to the official history.[51]

These operations have been described at some length because, except for the Battle of Cape Matapan, this was the first occasion on which radar played a really important role in a fleet action. Without our radar, the results would have been very different. Two points concerned with the tactical need for surface warning radar are worth noting – first, mild surprise that the Germans should not have expected our ships to have had such a capability; and secondly, praise for Captain Ellis and his team for devising ways of using gunnery radar for this purpose. In fact, radar sets dedicated to surface warning – the type 271 series – were already in production.

A fortnight later, Admiral Wake-Walker, who had flown his flag in the *Norfolk* during the *Bismarck* episode, sent a personal letter to the Captain of HM Signal School containing an extract from his own official report, saying:

> The results obtained by *Suffolk* were very good indeed and in type 284 we have a set of great efficiency under practical conditions. It is impossible to overestimate the tactical value of a set of such range and accuracy. Its success may be some compensation to those who have worked so hard in developing and fitting the gear.
>
> Coming from me, [Wake-Walker said in his covering letter] who so recently wrote direct to the Controller to try and ginger up the process of fitting – I hope this will be regarded as an amend for that stab in the back! – although really it was a compliment as it showed how much store I set upon the gear'.[52]

Among the many honours and awards announced after the action were two Distinguished Conduct Medals – to Ordinary Seamen (RDF) C. F. Tuckwood and A. J. Sinker of HMS *Suffolk* – 'belonging to the Action and Relief Action (or Defence) crews of the type 284, to whose zeal and skilful operation of their equipment the success of the low visibility shadowing by its aid was due.'[53]

TAILPIECE

• By Captain R. M. Ellis, commanding *Suffolk* during the *Bismarck* episode, from unpublished autobiography, *When the Rain's before the Wind*:

> ... when I took command of the *Suffolk* and found her newly fitted with a centimetric [50cm] gunnery radar outfit, not a soul could tell me a thing about its possible tactical applications, I had to figure that one out for myself, from such basic knowledge as I had, or could add to experimentally at sea ...[54]

• During January and February 1941, *Eagle* lent one of her observers, Lieutenant D. G. Goodwin (with his Bigsworth board), to *Valiant* to direct the former's two fighters during Fleet operations. '*Valiant* was one of the first ships fitted with radar,' wrote Captain Goodwin in 1983, 'but the idea of actually using it to direct our own aircraft (as opposed to merely detecting the enemy) was startlingly new. We hadn't a clue how to do it and it was a shambles – but rather exciting and fun!'[55]

6

1941(2): The Battle of the Atlantic and Pearl Harbor

Jun 22	Germany invades Russia
Aug 10	Atlantic Charter signed; USN convoy escorts
Aug 26	ASE becomes a separate command
Nov 14	*Ark Royal* sunk
Nov 16	First U-boat sinking due to type 271 (*Marigold*)
Dec 7	Japanese attack on Pearl Harbor
Dec 10	*Prince of Wales* and *Repulse* sunk

THE STATE OF RADAR IN THE FLEET

During 1940 and 1941, tremendous efforts were directed towards getting as many ships as possible fitted with some kind of radar. In deciding what alterations and additions had to be made to existing ships, the Admiralty had a system of priority categories lettered from A to D, the most important items being Classification A. In January 1941, priorities became such that a new superior Classification A* had to be created.[1] In June, even that proved not to be strong enough for radar fitting. 'Generally speaking,' said the Admiralty, 'RDF items are considered of the first importance among A* items.'[2]

In 1940 and 1941, some 23 large ships joined the Fleet from builders' yards, which all had air warning radar fitted and those completed in 1941 also had several gunnery sets each and latterly some had centimetric surface warning radar. Whenever an existing ship was in dockyard hands for damage repairs or other reasons, she also had radar fitted at the highest priority. By the end of July 1941, half of the Navy's capital ships and a third of the cruisers had been fitted with long-range air warning radar, and 20 per cent of capital ships and cruisers had radar rangefinders for their main armaments; 210 destroyers and small craft, mostly on convoy work in the Atlantic against U-boats, or the North Sea against E-boats and aircraft, had been fitted with the naval version of the RAF's ASV set.[3]

The tables below show the position at the end of September 1941.

Table 6.1 Large warships fitted with air warning radar, new construction and existing, 1938–41 (type 79 (or 279) + type 281)

Type of ship	*1938–9*		*1940*		*1941*	
	New	*Extg*	*New*	*Extg*	*New*	*Extg*
Capital ships	–	2	1	1	0+2	5+3*
A/c carriers	–	–	2	–	2	1**
Heavy cruisers	–	2	3	1	1+1	2+7
Light cruisers	–	–	3+1	2	2+2	0+2
AA cruisers	–	1	–	3	–	–
Other ships	–	–	1	–	1+1	9
TOTAL	–	5	10+1	7	6+6	17+12

* *Rodney* converted from type 79 to 281.
** *Audacity* was fitted with type 79B, as was *Hood*.

In addition to the above, *Carlisle*, *Alynbank* and *Foyle Bank* were fitted with type 280 in 1940, *Springbank* and *Ariguani* in 1941.

Table 6.2 Ships fitted with surface and combined air-surface warning radar, as at 28 September 1941

Type of ship	*286M*	*286P*	*290*	*271*	*272*	*273*
Capital ships	–	–	–	1	–	1
Cruisers	10	2	1	–	1	–
Minelayers	4	–	–	–	–	–
AMCs, OBVs, etc	8	2	–	2	–	–
Destroyers	177	12	1	2	1	–
Sloops	20	1	–	1	–	–
Minesweepers	–	–	–	2	–	–
Corvettes, cutters	20	1	–	25	–	–
Trawlers, etc	6	1	–	4	–	–
CAM ships	–	15	3 CHL	–	–	–
Submarines	–	1	–	–	–	–
Coastal craft	20	1	–	–	–	–
Shore training	–	1	–	2	–	–
TOTAL	265	37	2	39	2	1

Type 286M – 1.5m wavelength, fixed aerial.
Type 286P or 290 – 1.5m, rotating aerial.
Type 271 – 10cm, cheese aerial (corvettes, etc).
Type 272 – 10cm, remote ditto (small cruisers, sloops).
Type 273 – 10cm, paraboloidal aerials (large ships).

In addition, 9 type 287 (modified type 284) were set up ashore covering controlled minefields in port approaches.

Table 6.3 Ships fitted with gunnery radar as at 28 September 1941

Type of ship	*282*	*284*	*285*
Capital ships	2	7	4
Carriers	1	–	1
Monitor	–	–	1
Cruisers	1	16	7
AA cruisers	1	–	1
Aux. AA ships	–	–	2
Minelayers	–	–	4
Destroyers	–	–	31
Sloops	–	–	5
Shore training	2	1	2
TOTAL	7	24	58

Type 282 – close range weapons.
Type 284 – main armament low angle, large ships.
Type 285 – high angle, large ships; main armament, small ships.
All 50 cm wavelength. Most large ships had more than one type 282 and 285.

Source: Admiralty Letter S.D.O.2053/41 of 1 October 1941 (PRO ADM 220/79).

THE NEW ADMIRALTY SIGNAL ESTABLISHMENT

The need to disperse the activities of HM Signal School at Portsmouth had been under discussion since 1938 and in 1940 Experimental Department extensions were opened in Bristol, Cambridge, Waterlooville, Eastney Fort East, and two in Portsmouth. Then, in the autumn, the air raids on Portsmouth brought matters to a head. In April 1941, the instructional staff of the school moved from the RN Barracks in Portsmouth to a 'stone frigate' known as HMS *Mercury* – in fact a country house called Leydene near Petersfield – while the Experimental Captain and his staff chose the district of Haslemere in Surrey, headquarters being set up at Lythe Hill House, a country mansion about a mile from Haslemere, though the radar teams at Eastney, Onslow Road and Nutbourne stayed where they were for the time being. During 1941, the Admiralty acquired the nearby King Edward's School at Witley for radar laboratories and workshops, and transfer was effected from the Portsmouth area without disturbance to output by the autumn of 1942, though Eastney, Nutbourne (aerials), and Waterlooville (valves) were retained. (Because Eastney became unsuitable for aircraft trials owing to the enemy air threat, a trials station was opened in 1944 at Tantallon near North Berwick in Scotland.)

For a few months after the first move, the Experimental Department remained part of the Signal School, but on 26 August 1941 it became an independent command called the Admiralty Signal Establishment, or ASE, Captain Willett being appointed Captain Superintendent, in command of HMS *Mercury II*. Henceforward, it was the Admiralty Signal Establishment, not Signal School, which was responsible for research and development of the Navy's seaborne radar, as well as communications and DF.

IMPROVEMENTS IN 50–CENTIMETRE GUNNERY RADAR

With a large fitting programme for the original series of gunnery sets now under way, and experience of the problems at sea now accumulating, it was time for the 50cm team to give attention to various improvements which further laboratory developments in 1940 and 1941 were making possible. And this had to be done while continuing to provide support for the sets already fitted in ships and assisting the people concerned at sea, many of whom were meeting radar for the first time. These close contacts between scientists and engineers on the one hand, and users on the other, generated a heavy load on the former but were seen to be vital in ensuring the best possible use of finite resources. Indeed, it has been said that there was never a period, before or since, when there has been such close cooperation with such a fruitful outcome in advancing a new field of technology, as was demonstrated by the mass of new developments which were exploited. Apparently, this kind of close cooperation never developed on the German side. This became quite clear from interrogation of German engineers from Peenemunde who were interviewed by ASE staff late in 1945 and again in 1946.

With so many ideas emerging, each with its own protagonists, it was not long before some rationalization was seen to be necessary to produce a programme which would satisfy the aspirations of the users – and the scientists – as quickly as possible within the resources which were likely to become available on a relatively large scale by the end of 1941. But ideas were already developing rapidly and, on 22 May 1941, a meeting with Captain Willett in the chair took place at Onslow Road to consider future developments for gunnery sets. First, the shortcomings of existing sets were listed as follows:

- Inadequate range accuracy on long-range sets.
- Shortness of range of type 284 and 285.

- Insufficient bearing accuracy for blind fire.
- Interference between multiple sets in the same ship.
- Weakness of many components.
- Time needed to start up.
- The apparatus was not in continuous readiness unless constantly kept running.

It was decided that the priorities for Coales's team should be as follows:

- To make such improvements to types 284 and 285 as were possible without scrapping existing equipments or needing major dockyard alterations; in other words, to get the best out of the sets already fitted and fitting.
- To develop a new version of type 285 incorporating all technical advances to date.
- To develop a new type 282 to meet the requirements of the 'ideal system' for close-range AA fire.

In addition, it was decided to develop an entirely new main-armament radar rangefinding equipment incorporating all experience to date, which it was subsequently decided should be on a 10cm wavelength.[4] Four weeks later, on 23 June, a meeting to discuss ways and means of doing this was held at the newly-founded Gunnery Division at the Admiralty.[5] Subsequent developments on this front will be discussed in the next chapter.

In July, the Admiralty issued formal 'staff requirements' to cover the recommendations of the Onslow Road meeting, the most important being the need to increase range and fit beam-switching for more accurate bearings for both types 284 and 285. Later in the year, requirements were specified for equipment to permit main armament barrage fire against aircraft, and for using ranges from the centimetric surface warning sets for gunnery purposes.[6] Much of this work was already well advanced in development.

Increasing range and range accuracy entailed a number of lines of approach, all of which were followed up. The first seagoing trials of the experimental gunnery radar in June 1939 had achieved a range of 2.5 miles against low-flying aircraft and 5 miles against ships. The peak pulse power of this set was only 5kW and the aerial design was one not taken further. The first production type 282 had an output of 25kW and used single Yagi ('fishbone') arrays, one for transmission, one for reception, which was adequate for the control of short-range pom-pom guns with a maximum range of 3 miles displayed. The type 285 and 284 radars were similar, but had larger aerial arrays and displayed ranges of 15,000 yards

15. Pom-pom directors of the battleship *Prince of Wales*, 1941, on which were mounted the Yagi ('fishbone') aerials of the 50cm GC type 282.

(IWM)

(7.5 miles) and 24,000 yards (12 miles) respectively on a cathode ray tube of the same diameter. Squashing these greater ranges onto the same length on the display had the effect of reducing the accuracy with which ranges could be read off – from the 50 yards of type 282 to 150 and 240 yards in types 285 and 284 respectively, but, by using voice reporting (practicable in surface engagements) and a paper scale on the CRT, this could be improved to 120 yards. Nonetheless, all these accuracies were substantially better than could be achieved by the optical coincidence rangefinder which typically had an error of ±1,000 yards at ranges of 10,000 yards (5 miles).

The range performance of types 285 and 284 in *Southdown* and *King George V* trials (see p. 6 above), although impressive so soon after the technical problems of 1938, fell short of the requirements specified by the Admiralty in 1935 – precise location of aircraft at 10 miles, of ships at 5 miles[7] – particularly in regard to anti-aircraft performance and smaller ships. Thus, urgent items in the improvement programme were (a) to increase transmitter power, and (b) to make more efficient use of aerial arrays without making them bigger.

Peak power of 150kW had become feasible and a reduction in pulse length from 2 to 1 microseconds improved range discrimination between adjacent targets from about 300 to 150 yards. These changes in power and pulse length were first introduced into types 284 and 285 and later in type 282 (converting it to 282P).

Using aerials more efficiently without increasing their size raised for the radar designer new and apparently daunting problems which were solved in a particularly elegant way by introducing common-aerial working – using the same aerial for both transmission and reception whereby, without any increase in overall size, the width of the 285 aerial was doubled for both transmission and reception, giving effectively a four-fold increase in power. This, combined with the six-fold increase in transmitter output, theoretically increased the 285's range by a factor of about 2.2 on a given size of target, the corresponding factor for type 284 being 1.6.

On the strength of these improvements, the displayed ranges of types 285 and 284 were extended to 30,000 and 40,000 yards (15 and 20 miles) respectively. However, with the original ranging system – a CRT display with a separate mechanical cursor driven by the range operator's handwheel, and a separate calibration trace with a series of accurate range pulses – increased range coverage meant increased range errors. And there were also scaling and setting errors in such a system which became worse the larger the range covered by the display, so that doubling this roughly doubled ranging errors, whereas Staff Requirements demanded smaller errors. By expanding the trace in the vicinity of the range marker (which the operator had to keep lined up on the target echo) and by other means, a cyclic error not exceeding 25 yards repeating every 1,000 yards, regardless of the overall range displayed, was achieved. With rate-aided (see Glossary) handwheel controls for setting the marker on the target, the accurate range rates required for the prediction of future range were directly obtained at the same time as very accurate ranges. For the short-range type 282, a simpler solution was possible to achieve an accuracy of 25 yards in 6,000 yards. An accurately linear potentiometer was used to position the electronic range marker, being driven by a rate-aided mechanism set by the operator but included in the computer mechanism. Using this range-rate, the auto-barrage computer assessed the moment for opening fire to achieve near the target a pre-set barrage range up to 4,000 yards without further intervention by the operator.

Later, this concept was extended to the larger guns of warships and a modified system, type 283, was introduced to allow the main armament of bigger ships to contribute to the AA barrage.

Improving bearing accuracy – another high priority of the Naval Staff – also required new techniques, because the width of the beam could not be reduced further beyond the halving in types 282 and 285 which had resulted from common-aerial working. An increase in aerial size had to be avoided, yet a substantial improvement in aiming accuracy was needed to reduce the errors of the order of 2° in type 285 and ¾° in type 284 which resulted when simply aiming the beam at the target to obtain maximum echo strength. Even with the narrower beam resulting from common-aerial working, errors would still be about 1° in 285. In the case of 284, the aerials already occupied the full width of the director and, being placed one above the other, the beam width could not be reduced.

These large pointing errors arose because the tip of the radar beam is rounded rather than pointed and the sensitivity of echo size to sideways movement is poor. However, the sides of radar beams are steep, so if the side of the beam (rather than its centre) is aimed at the target, the sensitivity in azimuth is substantially improved, albeit at the price of some signal strength. Further, if the beam is swung rapidly between two positions where the sides of the beams cross over at the required level, the target echo will remain steady if it is in line with the axis of this 'split' beam system, but will fluctuate if it is off-axis either way. With this 'beam-switching' scheme, aiming errors averaged about 15 arc-minutes against aircraft on 285, and about 5 arc-minutes on 284 against small ships but rather more against larger ships presenting an extended target in bearing. This most elegant solution to the problem was developed by W. A. S. Butement of ADEE originally for use in Army anti-aircraft radar.

It was not practicable to make the beam oscillate in this way by mechanically swinging the large aerial carrier itself, but a similar effect was achieved by rotating a much lighter mechanism based on the geneva cross device used in cinema projectors to shift very quickly successive frames of the picture whilst allowing each picture to remain in place for most of the cycle. In the beam-shifting application, the rotation of a capacitor moved the beam first to the left and then to the right twenty-five times a second in a similar manner. This ingenious approach produced a most effective solution to another difficult problem in improving the gunnery radars, but not without much development work to convert from the light load of a cinema projector to the much heavier mechanism needed to switch some 10kW of power.

For type 285, narrower beams and beam-switching had an additional advantage, better discrimination between targets close to each other. Two aircraft targets of equal size could approach each other within about 4° before becoming confused, rather than 7° with the early system, and large surface targets could be distinguished from smaller aircraft targets by beam-switching because of the fluctuations in echo amplitude in the former. Before beam-switching, echoes from friendly destroyers even as far as 14° off the radar sight line could cause serious confusion with enemy aircraft being followed.

These improvements were largely completed during 1941 and were ready for first fitting early in 1942, except for the barrage set type 283 for larger guns which was not fitted until later that year. Improved sets were indicated by suffixes added to the basic type numbers.

RADAR IN SMALL SHIPS

Before turning to operational matters, let us say a few words about further developments in warning radar in 1941 at ASE, dealing first with the 1.5m sets being fitted mostly in small ships. As soon as the programme to fit the Air Ministry ASV sets was under way in 1940, work had started at Eastney on an all-naval replacement on the same wavelength, designed from scratch to meet seagoing conditions. To be called type 290, it was to have type 286P's rotating aerial and a hoped-for pulsed power of 100kW as against the 286's 7kW. The first set was fitted in the light cruiser *Aurora* in May, just in time for the *Bismarck* operation, and the second in the destroyer leader *Broke* later in the year.

Type 290 never functioned as planned, owing to problems with the transmitter and modulator but, while it was being finally engineered, technical improvements became available which could solve these problems and work was started on yet another 1.5m set to be called type 291, the first experimental model of which did trials in the destroyer *Ambuscade* and, with a special lightweight aerial, in a Coastal Forces craft in February 1942.[8] Eventually, type 291 replaced all versions of types 286 and 290 and remained in service until the early 1950s.

From Table 6.2, it will be seen that type 286 was fitted in a submarine and in coastal forces craft (MTBs, MGBs, MLs, etc). Radar in these specialized classes of vessel, as well as naval airborne radar, will be covered in the Appendices.

THE PPI

Let us now introduce the Plan Position Indicator, or PPI, which had been in operational use in RAF GCI stations for night-fighter direction since late-1940. The earliest method of showing radar information was the A-scan, a range-amplitude display, which showed on a CRT the echoes detected on the bearing on which the aerial was trained, measuring range along one axis, and the echo-amplitudes – a measure of the strength of the returning signal and therefore of the size and nature of the target – on the other. But there was no bearing information on the display itself. To obtain a two-dimensional plan of the echoes around the ship, a separate manual plot – with pencil and paper or the equivalent – was needed, and this was essential to maintain an all-round lookout for air or surface warning purposes in any situation involving more than two or three echoes.

The PPI goes a long way towards overcoming this disadvantage, producing in effect a radar-made map, the centre of which represents the radar itself. A special CRT generates a radial line from the centre of the face which can rotate in synchronism with the warning radar aerial, and, by displaying echoes as bright spots on the trace (instead of blips as on the A-scan), a complete picture of the situation round the ship was generated during each rotation. By ensuring that the picture fades slowly using special phosphors on the CRT, a relatively steady 'plot' can be provided without human intervention.

Proposed by P.E. Pollard and E.G. Bowen in 1938, the PPI was developed at Dundee and Swanage by G.W.A. Dummer and others under the direction of W.B. Lewis from December 1939, the first GCI night-fighter-controlled operations taking place on 18 October 1940.[9]

At the suggestion of the Naval Air Division, the Admiralty in January 1941 asked Signal School (where a model was already in the early development stage) whether it could be fitted in carriers for use with air warning sets for fighter direction. No, said Signal School, not with naval sets, which have a much wider beam than the RAF's 1.5m sets; perhaps later with 10cm sets. 'Although the PPI is no doubt something to strive for,' wrote Commander Cecil, the Application Officer for radar, 'I don't think its priority should be such as to divert a major quantity of experiment and research, etc.'[10] Early in 1942, *Illustrious* asked for an RAF PPI for use with her new type 281 (whose beam was much narrower than her previous type 279) but got much the same answer – the wide beam would create saturation on the display.[11] However, she was later able to 'scrounge' one in Ceylon without official Admiralty knowledge.

In March 1942, a TRE 12in PPI was fitted in the Home Fleet flagship *King George V* to work with type 273. The first PPI to go to sea, it was about twice the size of an office desk and was fitted in the 'Admiral's Plot' (later called the Bridge Plotting Room). Treated as something of a curiosity, it was neither reliable nor very effective and was said not to have been used very much.[12] A 9in TRE model went to sea for trials with type 271 in the destroyer *Watchman* in August[13] and a contract was then placed with EMI for development of a naval version, to be known as the 'JE', which led to the almost universal fitting of PPIs in naval systems.

THE SUCCESS OF 10-CENTIMETRE RADAR AT SEA

By mid-July 1941, twenty-five of the new 10cm type 271 sets had been or were in the process of being installed. By the end of September, twenty-three 'Flower' class corvettes were fitted, two ex-American 'four-stacker' destroyers, three leaders of minesweeping flotillas – who found it invaluable for the precise navigation that is the epitome of minesweeping – and the sloop *Rochester*.[14]

The aerial of the standard type 271 had to have a good all-round view, so the whole assembly – aerial, 'lantern', and radar office – had to be high up in the ship. In many ships, topweight considerations made this impossible, so a start had been made on the design of a modified type 271 where the aerial (with transmitter and crystal mixer fixed on the back and rotating with it) could be placed up to 40 feet from the radar office, driven remotely by a Bowden wire drive.

In August, the prototype for this new set, designated type 272, was fitted in the anti-aircraft destroyer *Wallace*, of the Rosyth Escort Force, with the aerial lantern on a lattice structure abaft the funnel and the radar office below. Given similar conditions, performance was the same as with the standard type 271, but tuning and maintenance was not easy, half the set being up the mast and half below, with the lantern aloft too small to enter more than head and shoulders. Eventually it was extensively fitted, serving as the surface warning set in August 1943, for example, in two fleet carriers, seventeen escort carriers, twenty-two cruisers, six minelayers, two fighter direction ships, eighteen landing ships, twenty-one fleet destroyers, four escort destroyers, and twenty-four sloops.[15]

In passing, it is worth mentioning that an early model of type 271 was sent to the Army at ADRDE, installed in two GL Mk2 trailers with two 2-metre paraboloid aerials, and, manned by the Army, set up on 19 July

1941 on the South Foreland just north of Dover, 330 feet above sea level alongside the naval type 287 set (a 50cm type 284 with a large aerial for use ashore).[16] Connected to the combined services plotting rooms at Dover, it was the first of many naval 10cm sets to be used ashore by both Army and Air Force. By October, Admiral Ramsay and his staff at Dover were unanimous that this radar had revolutionized naval operations in the area, saying it was possible to maintain an efficient watch on the Straits with about a quarter of the patrol staff formerly used. The admiral told Commander Yates from ASE that, when the station was out of action, he felt like a destroyer captain who had lost his binoculars.[17]

The immediate success of type 271 in small ships prompted the decision to fit the same set in large ships, where, owing to their very size and steadiness in a seaway, the set could be made even more effective, first because the aerial could be mounted at a greater height above sea level, secondly because it was possible to use circular paraboloid mirrors giving a highly concentrated 'pencil' beam rather than the small ship's 'cheese' aerial with a wide vertical beam. The big-ship set, dubbed type 273, was identical to type 271 except for the two 3ft circular mirrors (not stabilized in the earlier models) and a slightly larger lantern.

The first type 273 was fitted in the battleship *Prince of Wales* while she was at Rosyth for damage repairs after the *Bismarck* episode. Immediately after she had finished taking Mr Churchill to Newfoundland for the signing of the Atlantic Charter, Landale joined her at Scapa late in August for sea trials of the new set. Working ranges of 19 miles were obtained on the battlecruiser *Repulse*, 18 miles on the small cruiser *Euryalus*, and 14 miles on the destroyer *Lively*. The Commander-in-Chief was delighted. A great advance on previous sets for the detection of surface ships, he said, and not only did it give greater ranges, but it was far easier to use for searching than type 284 on top of the cumbersome main armament director; all Home Fleet ships should be fitted – immediately! He added that finding a site in a big-gun ship not vulnerable to blast from one's own guns was not easy, and the set would be much improved with beam-splitting, an accurate ranging unit, and a PPI.[18]

While 'beam-splitting' (to improve bearing accuracy) was never added, a start was made in 1941 by adding an accurate ranging panel to type 271 in certain classes of ship, thereby making the set even more versatile for gunnery use, etc. This applied also to types 272 and 273. An Air Ministry PPI was fitted in the fleet flagship *King George V* in March 1942 and naval designed PPIs began to be fitted in mid-1943.

In Operation 'Substance' a month before the *Prince of Wales* trials of type 273, Force H, reinforced by units of the Home Fleet, had successfully passed a convoy through from Gibraltar to Malta. Several of the warships involved had air warning and surface gunnery radar but it was the destroyer *Cossack*'s fixed-aerial type 286M which gave the first indication

16. The 6in cruiser *Kenya* in 1941, with WA type 279 at the mastheads, 50cm GS type 284 'pigtough' aerial array mounted on the main armament director behind the bridge, and the lantern of the 10cm WS type 271 (later converted to type 273) forward of and below the bridge. The 273 lantern and aerial array were continually being damaged by gun blast when 'B' turret fired on an after bearing, so type 273 was later replaced in 'Colony' class cruisers by the less powerful type 272 with its lantern (remote from the set itself) on the foremast above the director. *Kenya* was later fitted with three GA type 285.

(IWM)

of enemy surface activity, detecting three Italian E-boats who made an abortive attack in the early hours of the 24 July while passing Pantelleria.[19]

In a similar operation – 'Halberd' – six weeks after the trials, type 273 fully lived up to its early promise. *Prince of Wales* was among the Home Fleet reinforcements, wearing the flag of Rear-Admiral A. T. B. Curteis. On the passage to Gibraltar, she detected a submarine, and at 2am on 29 September, on the return passage after the convoy had reached Malta safely, the leading seaman (RDF) on watch in type 273 detected an echo at about 7 miles which was later confirmed as an Italian submarine. The next night, the same operator detected another submarine.[20] The Commander-in-Chief was delighted to have this confirmation of his earlier enthusiasm for the set. 'There seems little doubt,' said Admiral Tovey when forwarding Curteis's report to the Admiralty, 'that type 273 was instrumental in enabling effective avoiding action to be taken against Italian U-boats on several occasions, and that effective counter-attacking procedure is possible.'[21]

Before the end of 1941, ten capital ships had been fitted with 10cm sets of which at least six were type 273 with unstabilized aerials. We know that *King George V* (probably in June) and *Duke of York* (by the date of commissioning in November) were initially fitted with type 271, *Prince of Wales* and *Resolution* with type 273, but records do not show which of the two sets was initially fitted in *Ramillies, Royal Sovereign, Revenge, Renown, Rodney* or *Warspite.* Certainly those fitted with type 271 would have been converted to type 273 at the earliest opportunity. Cruisers had to wait until 1942 for 10cm radar, except for *Kenya,* fitted in August with type 271 below the bridge, abaft 'B' turret, where the aerial lantern kept being smashed by gun blast, as a result of which later *Fiji* class cruisers were fitted with type 272 with the aerial at the back of the bridge.

RADAR OFFICERS AND RATINGS

As we saw in Chapter 2, the Navy's earliest radar operators were either junior ratings of the Telegraphist Branch or bright young seamen chosen in a rather haphazard fashion, while the sets were maintained by senior telegraphist ratings. In September 1940, a new specialist RDF branch was created for operators – to be known as Ordinary Seaman (RDF), Able Seaman (RDF), etc. – all to be 'hostilities only' ratings.[22] For radar maintenance, a new 'Radio Mechanic' branch was created in May 1941.

The earliest 'RDF officers' were RNVR officers, preferably but not necessarily having some radio experience or with a degree – in biology or history perhaps, most of the physics graduates having been creamed off for research and development work elsewhere. In January 1940, on the initiative of the Canadian-born DSR, C. S. Wright, the British Admiralty made a request to the Canadian Government for the loan of engineering physicists for various duties, particularly radar. The first batch of Canadian physics graduates, twenty-strong, started their radar training in England in May 1940. It was only the timely arrival of these Canadians that saved the Navy from facing a truly disastrous position in regard to radar personnel, and the radar in a very high proportion of the larger British warships was kept working at sea by Canadian radar officers, particularly before 1943. Writing about the Canadian radar officers then at sea to the Acting President of the National Research Council of Canada, C. S. Wright, said this in June 1941:

> I am pleased to be able to tell you that they have turned out well, some of them extraordinarily well. They have now, of course, had a great deal of practical experience, and we are frequently hearing from the ships' officers at sea how valuable these young men are. To be quite honest, I really do not know what we should have done without them, and I am most extremely grateful to you for helping us in this way. I should also like to add that I am quite certain any more you can send us of the same type would be most welcome.[23]

By May 1941, graduates from British universities were beginning to be recruited under the Hankey scheme mentioned on p. 35.

Owing to the widespread lack of knowledge in the Fleet, radar officers afloat were initially the only officers in a ship in a position to advise on the operational use of the equipment, and they were originally trained and appointed to supervise both the operation and the maintenance of the equipment, though by 1941, other specialist executive officers were beginning to play their part in this, radar forming part of their training. A series of weekly three-day courses for General Service officers was started at Portsmouth and Glasgow in 1941.

THE BATTLE OF THE ATLANTIC

The following extract from a personal letter to Captain Willett from Captain G. E. Creasey, Director of Anti-submarine Warfare at the Admiralty, dated 11 September 1941, gives a good indication of the urgency of the situation at that time, and of the value placed upon centimetric radar by the Naval Staff:

> Thank you for sending the list of ships fitted with type 271. I am entirely in agreement with you that when we get this generally fitted we shall have made an immense stride forward and that it puts type

> 286M completely in the obsolete class (in justice to A/S Warfare Division you must admit that we have always believed that this would be so ever since we saw the first short-wave set functioning at Swanage one year ago). [See p. 68 above.]
>
> I am rather horrified to note on your list the very large number of type 271 sets being fitted in ships which are playing no part in the Battle of the Atlantic. I assure you that I appreciate what is *now* being done to fit corvettes, etc., but I cannot help pointing out the large number of escort vessels engaged in the Battle of the Atlantic and what a small proportion of these are now fitted with type 271.
>
> The point is, of course, that the Battle is already raging. The fact that all ships will be fitted at some future date is something to look forward to, but is no help at the present time when we are faced with a desperate struggle to get our shipping through.
>
> Do not, I beg you, be deluded by our successes of the past two months. These have been due to astounding good fortune in dodging danger. This good luck cannot continue and, in fact, has already been broken. We must face the fact that, sooner or later, we shall have to fight our way past the U-boats instead of dodging them. Our ability to do so with success rests largely on getting efficient RDF into our ships. With it we can face the dark nights with assurance; without it, we shall inevitably be facing disastrous losses.[24]

Prophetic words in the last sentence! There were, however, things which Creasey could not tell Willett. The convoys dodging danger was not all due to 'astounding good fortune' but – unknown to the Germans and therefore highly, 'Ultra', secret, even to Willett – was also due to the fact that, since June 1941, we had been able to read the German submarine command's Enigma wireless signals to U-boats, and convoys could therefore be routed clear of danger. And Creasey was also correct in prophesying that the 'good luck' might not continue. German changes in Enigma procedures in January 1942 meant that we were unable to read U-boat messages from January to December 1942.[25] The letter continues:

> Can you wonder that, with this as a background, I have been pressing night and day for ten months now for more energy and speed in getting our escort vessels fitted out with RDF.[26]

Captain Creasey added the following postscript in his own hand:

> On reading this thro' I feel it sounds a bit ungrateful for all you have done and are doing to get RDF in to our A/S ships. Believe me I do

realise & appreciate it and am *most* grateful. If I seem critical it is because of the desperate urgency of our need.[27]

Following *Vanoc*'s success in March, no further U-boat kills could be attributed directly to the fixed-aerial type 286M, but there are three reports of definite U-boat detections with that set, all in the vicinity of Gibraltar convoys: by the sloop *Folkestone* on 2 July in thick fog, contact at 2.25 miles, sighting about half a mile; by the sloop *Deptford* at 2.20am on 11 August, contact 1.5 miles, three attacks after sighting at less than a mile; and by the destroyer *Vidette* on 22 October, contact 3.5 miles, sighting 1.5 miles.[28]

During the autumn of 1941, a start was made in replacing escorts' fixed ASV aerials with rotating ones. One of the ships so converted – from type 286M to type 286P – was the destroyer *Hesperus* (Lieutenant-Commander A. A. Tait). On the night of 14 December on patrol in the Straits of Gibraltar, the radar operator, Leading Seaman J. W. Sheard, reported a contact right ahead, probably less than a mile. Lieutenant Duncan Knight, her First Lieutenant, who was on watch, switched on the 10in signalling projector and to his amazement saw a surfaced U-boat dead ahead, going fast on the same course, which dived forthwith. Knight ran straight over the spot and released a full pattern of fourteen depth charges 'by eye', which it seemed could not miss; but the U-boat survived despite the night-long efforts of *Hesperus* and her consort.[29] A month later, at 1.10am on 15 January 1942 with *Hesperus* covering Convoy HG.78 under almost exactly similar circumstances – Knight and Sheard were on watch again – an RDF contact was obtained at 1.25 miles. Knight immediately rang down for 18 knots and turned towards. At 1.20, the submarine was illuminated by a bridge signalling projector and Tait was able to ram her. At 1.26, U.93 abandoned ship, sinking shortly afterwards, forty prisoners being taken. Tait received the DSO, Knight the DSC, and Sheard the DSM.[30] A few months later, *Hesperus*'s gunnery director abaft the bridge was replaced by the hut for radar type 271, and we shall meet her again in the next chapter.

As we have seen, the centimetric type 271 first went to sea in the corvette *Orchis* in March 1941. The first U-boat sinking directly attributable to that set was that of U.433, detected by the corvette *Marigold* (Lieutenant J. Renwick RNR) sailing independently east of Gibraltar on 16 November. At 10.50pm, she obtained a contact 30° on the port bow at 2 miles, closing. Turning towards, *Marigold* sighted the U-boat at about a mile but it dived before ramming was possible. Asdic contact was gained but lost after the first depth charge attack. At 11.44, contact was regained and *Marigold* attacked again, whereupon U.433 surfaced and abandoned ship. Thirty-eight survivors were picked up.

The radar and asdic operators, Able Seaman T. O'Carrol and Leading Seaman J. Morgan, were both awarded the DSM.[31]

Two other definite centimetric U-boat contacts with type 271 were reported in 1941: the ex-US coastguard cutter *Sennen* with an outward-bound Freetown convoy OS.12 detected a U-boat at 3.3 miles on 24 November, leading to an unsuccessful hunt; and *Marigold*'s sister ship *Bluebell*, with convoy OG.77, detected a U-boat on the night of 11/12 December at 2.5 miles, subsequently sighting her at 1.5 miles.[32]

THE RADIO WAR AT DOVER

By June 1941, an engineered version of Davis's jamming transmitter, designated Type 91, had been designed. Before the first production models were available, the RCM Section had hand-built a number of transmitters which were installed with monitor-receivers and these were installed at a number of sites in the Dover area. Naval ratings were trained in the operation of the jamming network and the individual sites (Z-stations) were organized by Sub-Lieutenant Jennings RNVR on the Dover Command staff. The efficiency with which he used his resources is borne out by the subsequent small losses in convoy shipping using the Straits of Dover.

However, the RCM war was far from stagnant. The Germans were continually trying new tactics to avoid the effects of jamming whilst the British devised new ways to counter them. They first installed radars on widely-spread operating wavelengths, and numbers of Freya, Seetakt and Small and Giant Würzburgs appeared along the French coast during 1941. Since all these radars could be used for coast-watching, a new wide-bandwidth aerial system was designed for Type 91 which did not need tuning. These aerials were directional so that they could be trained on known German radar sites, and so strengthen the jamming signal by beaming. Also, the Section constructed batteries of the Type 91 aerials pointing at the known German radar sites. Ultimately, 157 jammers were installed in the Dover area, covering both horizontal and vertical polarization.

To counter the effects of British jamming, the Germans introduced a tuning system on individual radars so that, when one of these was jammed by a Type 91, the radar could be tuned to an un-jammed wavelength.[33] Sometimes, in an attempt to deny our operators knowledge as to whether the British jamming was effective, the enemy would leave certain radars apparently operating normally, even though they were in fact completely jammed, which necessitated the utmost vigilance on the part of British operators. A similar technique was used by the

British Air Defence radars in Malta in the face of German jamming. The Germans soon gave up jamming, in the belief that it was ineffective.

ENEMY RADAR MONITORING DEVELOPMENTS

In September 1941, an independent enemy-radar monitor station was established at Dover, equipped with a range of special aerials, monitor receivers and signal analysis equipments to cover the wavelength spectrum from 10 centimetres to 10 metres.

B. M. Adkins was appointed officer-in-charge of the Dover Monitor Station, with a staff of four WRNS operators. Daily reports were sent from the station to the Admiralty, to ASE, and to an inter-service body known as the Noise Investigation Bureau (NIB) in London.

RADAR AND FIGHTER DIRECTION

At the beginning of the period covered by this chapter, the Navy had only two large fleet aircraft carriers operational: *Victorious* with the Home Fleet; and *Ark Royal* with Force H. Admiral Cunningham had no carrier in the eastern Mediterranean, where both *Illustrious* and *Formidable* had been damaged and both were now under repair in the United States. Then more disaster struck. On 3 November, the newly-commissioned *Indomitable* ran aground off Jamaica while still working up her ship's company. But an even greater disaster was the torpedoing of *Ark Royal* on 14 November. Now, the only modern carrier operational was *Victorious*.

One of the few good things to come out of this sorry episode was that, for a few days, *Illustrious, Formidable* and *Indomitable* were together in Norfolk Navy Yard, so the respective fighter direction officers were able to compare notes on methods and equipment, and to obtain, through the generosity of American individuals, radio-telephone equipment.

The Crete operations had shown what could happen if fighters and fighter direction were not available to defend the Fleet against air attack and had in particular demonstrated our vulnerability to low-flying aircraft. As an air warning set, type 281 was superior to type 279 in this respect – albeit at the expense of high cover – and the Naval Air Division of the Naval Staff suggested two other reasons why type 281 was to be preferred to 279 for the fighter defence of the Fleet: (a) better general performance and precision; and (b) 281 would work with a PPI (which they hoped might be available by mid-1942) whereas 279 would not.

In November, therefore, the decision was taken that all fleet aircraft carriers should eventually have type 281, the existing type 279s in *Illustrious, Formidable, Victorious* and *Indomitable* being replaced by type

281 as soon as practicable.[34] At the same time, the Naval Staff began to lose interest in type 281's gunnery capabilities as more and more sets wholly dedicated to gunnery were fitted.

There was at that time no single-mast version of type 281, but when in 1943 it did become available, Fleet carriers were able to have the best of both worlds – two long-range air warning sets each, type 79B and type 281B.

But providing the equipment for fighter direction was not enough. Officers had to be trained to do the job, officers who could gain the

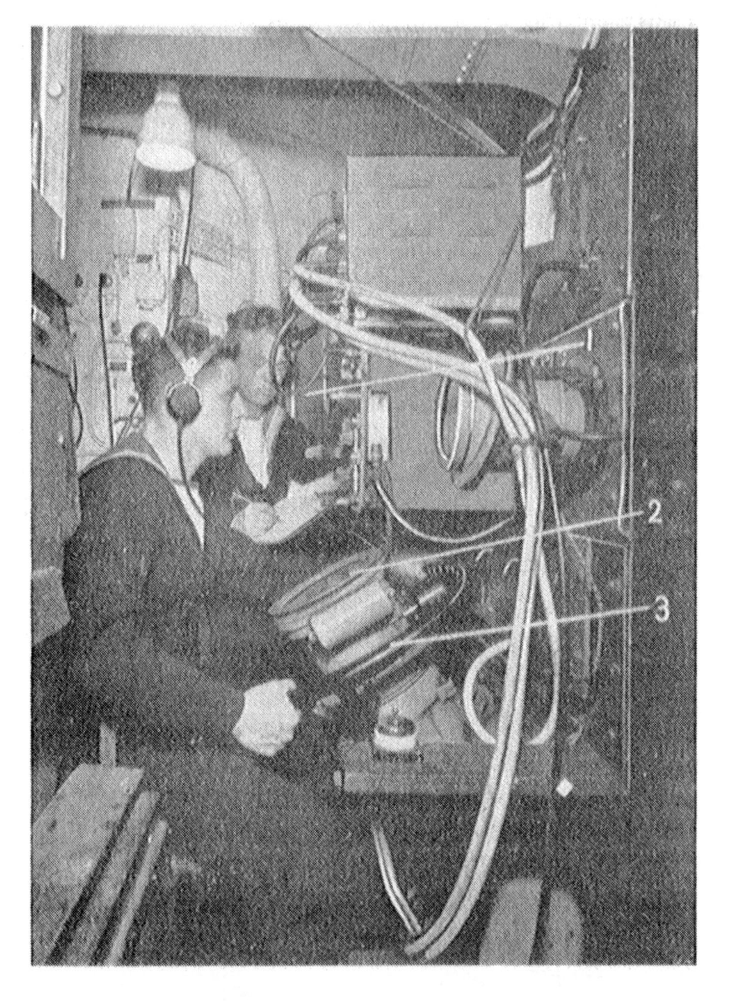

17. WA type 279 receiving office. The operator is training the aerial and watching the A-scan. The second operator is logging the reports made.

(Radar Manual, 1945)

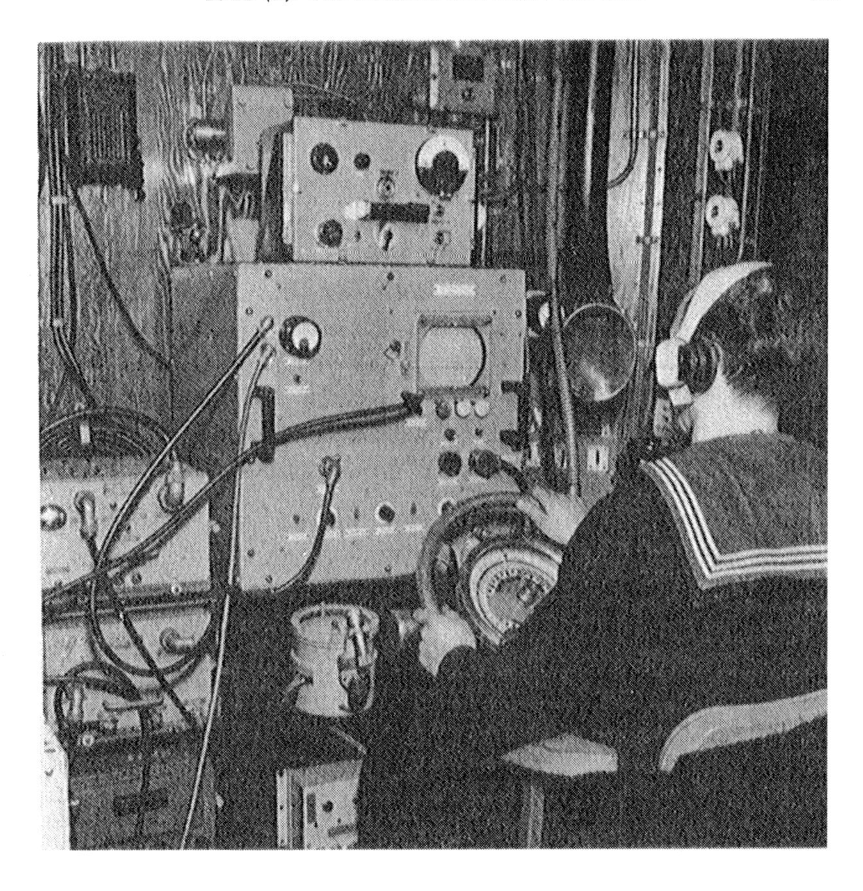

18. WS type 272P radar office in a 'Tribal' class destroyer. Interrogator type 242 is on the left.

(P. S. H. Lawrence)

confidence of the fighter pilots whom they would direct. The setting up of the Fighter Direction Branch comes outside the scope of this book. Suffice it to say here that a naval Fighter Direction School was set up under the direction of Commander Charles Coke (who had done some of the earliest naval fighter direction in *Ark Royal* during the Norwegian campaign) at the RN Air Station at Yeovilton in Somerset. The first course began there in July 1941.[35]

THE ESCORT CARRIER *AUDACITY*

In the last chapter, on p. 80, we told of measures taken to permit convoys to have their own fighter aircraft with them to counter the menace from

Focke-Wulf aircraft when outside the cover from shore-based fighters. But these were only stopgap measures. The long-term solution was that an aircraft carrier should accompany each convoy, not only to provide fighter cover, but also air reconnaissance. The prototype for such an escort carrier was the ex-German prize, *Hannover*, renamed *Audacity*, fitted with a flight deck (but no hangar), carrying up to six Grumman Martlet fighters. For air warning radar, she had the second prototype single-masted type 79. Her fighter direction officer was Sub-Lieutenant John Parry RNVR, a Cambridge don who had just taken the first course at Coke's new fighter direction school at Yeovilton.

She entered service in June 1941 and joined the Gibraltar convoy escorts in September, seeing her first success in the shooting down of a Focke-Wulf while convoy OG.74 was being heavily attacked by U-boats on 20 and 21 September. Two more were shot down early in November with OG.76. On 14 December, *Audacity* began the homeward leg of her second voyage to Gibraltar, escorting HG.76 with the famous 36th Escort Group led by Commander F. J. Walker in the sloop *Stork*. We cannot here tell the whole story of the defence of that convoy, in which five U-boats were destroyed and only two merchantmen sunk: suffice it to say that, with the aid of her type 79, *Audacity*'s Martlets (only four of which were serviceable) shot down two Focke-Wulfs and assisted in the sinking of U.131, before the ship herself was torpedoed by U.751 on 21 December. Sadly, Commander Mackendrick was lost, but Parry survived, being awarded the MBE and receiving accelerated promotion – but only from Temporary Acting Sub-Lieutenant RNVR to Sub-Lieutenant RNVR – for his part in the operations.[36]

Audacity was a sad loss but, as Roskill has said, to her will always belong the distinction of having first closed the air gap on the Gibraltar convoy route.[37]

GUN DIRECTION

The ideal 'fighter interception zone' was later reckoned to lie between 15 and 70 miles from the Fleet, up to 30,000ft. Inside that and below 20,000ft was the 'target indication zone' where theoretically the Fleet's guns should have a primary role in air defence. Like the fighters, the guns have to be organized and directed – and coordinated with the fighter defences if necessary. This in turn implies tactical control of the defences with the object of destroying the enemy, or at least preventing the attainment of his mission.

All gun direction systems by definition involve means by which the sensors which initially detect targets – in our case, radar – can be linked with weapons and their control systems, whilst permitting the

'Command' to influence the progress of the battle in which the ship, or force of ships, is taking part. This process raised many theoretical and practical problems, and in the early days – many years before high-speed digital data-processing and transmitting – even the practical problems were only partially soluble. The early history of gun direction is a struggle to find acceptable compromises between the often incompatible requirements of a fast response by the weapons and adequate control of the battle by the Command. At first, as ships rapidly acquired radar for the first time, individual ships sought individual solutions based on the particular experience of the key personnel concerned, solutions which naturally tended to be extensions of their pre-radar experience. It was not until late 1942, under the growing pressure of the developing anti-aircraft war, that sufficient consensus was reached to allow the specification and development of a specialized radar with the Target Indication Unit (TIU) for gun direction.

At the outbreak of war, the sensors available on most ships were the eyes of the bridge personnel and signal deck, aided by binoculars and sometimes simple mounted visual sights giving bearings of targets relative to the ship's head and, for aircraft, elevation above the horizontal. This system at least ensured that the Command and the 'sensors' were in close proximity and immediate communication. Lookouts would report visual sightings of ships to the bridge with bearings relative to the ship's head; the bridge would then convert these to true (compass) bearings, and the gun director would then be conned onto the target by voicepipe or telephone. The director's range-taker might then provide an optical range of the target.

To enable the Command to see the developing surface situation, visual sightings might then be plotted in 'the plot' which, when augmented by information from other ships sent by radio or visual means, provided ship and force commanders with a tactical picture. Even before the radar warning systems, visual data could be extended over considerable areas when a major force was accompanied by an advanced cruiser screen or there was air reconnaissance. These data too could be added to the plot, though such reports might have large positional errors and delays. Generally however, given the technical limitations of the time, the method of handling surface engagements was not out of keeping with the pace of surface engagements, and permitted the close involvement of the Command which was desired.

So far as the air threat was concerned, a similar situation prevailed. The main sensors were human lookouts with relative-bearing sights, aided sometimes by one master sight with gyro stabilization which gave true bearing directly without reference to a separate compass. As in surface engagements, direction was given in azimuth by voice to directors and their associated weapons. The recently fitted ships' public-address systems for internal messages were also of general value to Gunnery and Air Defence Officers who could if needed inform the whole ship's company of the development of an impending attack. Early war experience soon led to improved visual sights with elevation and azimuth stabilization, allowing the target's position to be passed directly to the weapon directors without the use of voice.

Such was the position before radar. When it arrived, it had urgently to be grafted onto a slow-moving tactical control system which was basically visual and already overloaded. With more data available than ever before, the strain was even greater and, to assist the Command, individual ships developed their own systems of tactical control – typically based on 'home-made' aircraft plots on the bridge or at the Air Defence Position to enable commanding officers to deploy their weapons to the best effect for air defence, a good example of which is provided by the anecdote about Admiral Somerville on p. 62 above. These plots were similar to those employed in the early days for vectoring fighters toward threatening attacks. Such gun control still entailed the use of voice information over broadcast or gunnery telephone networks. Although with practice the team achieved surprising improvements in effectiveness, the fundamental weaknesses of slow voice control and hand plotting would have to be tackled in the end.

As we have seen, by the end of 1941 a limited range of warning radar systems was available, covering aircraft and ship targets, and in various forms suitable for various sizes and classes of ship. In addition, ships had up to eleven radars for fire control, covering surface targets, long- and medium-range anti-aircraft gunnery, and close-range gunnery. Despite the deficiencies in the gun-direction systems already discussed, and, within the limits imposed by the 50cm gunnery radars, it had proved possible to perform sufficiently well against small and poorly coordinated attacks by enemy aircraft to provide a substantially improved deterrent or a spoiling effect in reasonably good visibility, and most early attacks had been of this nature. However, we have seen that, during 1941 in the Mediterranean theatre and later in the Pacific war, developments in air attack techniques were exposing the limitations of the system, demanding urgent action to deal with this much more severe threat.

THE SINKING OF THE *PRINCE OF WALES* AND *REPULSE*

On 7 December 1941, the Japanese attacked Pearl Harbor. Three days later, the battleship *Prince of Wales*, wearing the flag of Admiral Sir Tom Phillips, and the battlecruiser *Repulse* were sunk by Japanese aircraft in the Gulf of Siam. Happily, the radar officers in both ships survived and have given accounts of their experience that day.

19. The battleship *Prince of Wales* arriving at Singapore on 2 December 1941, just eight days before she and *Repulse* were sunk in the South China Sea. The 3.5m WA type 281 aerials can be seen at the mastheads, with 10cm WS type 273, 50cm GS type 284, and two 50cm GA type 285 below the foremast. She also had two more GA type 285 and four GC type 282. Commissioned in March 1941, she was the first ship to have a multiple suite of radar. The ship alongside, *right*, is the cruiser *Mauritius*.

(IWM)

Prince of Wales's radar officer was Sub-Lieutenant Stuart Paddon, RCNVR (later Rear-Admiral, RCN) whom we have already met in the *Bismarck* action. He was very impressed with Admiral Phillips who, within an hour of having embarked with his staff in the Clyde, sent for Paddon and asked to be taken to every radar office in the ship, to sit in the operator's chair, and have explained to him what the operator had to do, all of which Paddon duly did. The Admiral had no previous sea experience of radar, having been on the Board of Admiralty since the beginning of the war, first as Deputy, then Vice Chief of Naval Staff.

We will continue the account in Paddon's own words, starting on the night before the sinkings:

> We remained at what I would call second degree of readiness, not action stations, but sufficiently so that I stayed close to my action station. Some time in the night I was called to my action station in the office of the 284 radar which controlled the main armament. I had no sooner closed up when it was reported that a small vessel had been detected. I assume it had been detected by our 273 radar, but I do not know that for certain . . . All I can say is I do not believe the Navigator would not have had the 273 on. If it was not functioning I would have heard about it very quickly . . . During the night the 281 air warning radar had been shut down by a policy of radar silence . . .
>
> About 10 [a.m. 10th], or thereabouts, we were suddenly called to anti-aircraft stations; I put the air warning radar on and almost immediately we were under attack by Japanese aircraft. I should mention that we had received a report that there had been a landing from Japanese ships, transports presumably, at some place with an unpronounceable name up the Malayan coast, and the Admiral had decided to investigate it. He flew off the two Walrus, one from *Prince of Wales*, the other from *Repulse*, and sent a destroyer in to check out this reported landing. The two Walrus never did return to the ships but flew back to Singapore. No sign of Japanese landings was found at the point of inspection.
>
> In the meantime, we were lying offshore, 20 to 25 miles, waiting for the destroyer to join us, when we suddenly found ourselves under attack. When I put the air warning radar on, we found aircraft approaching from something less than twenty miles. When they broke clear of the land echoes, they showed in groups of nine or ten per group. My air defence station was on the after part of the bridge, aft of the air defence control position, where I watched several attacks come in on us. As I remember, they concentrated on *Prince of Wales* initially, without paying too much attention to *Repulse*. I watched three torpedoes hit *Prince of Wales* on the starboard side, without any apparent effect. Suddenly there was a very heavy shock in the ship, as

if she had been picked up fifty feet and then dropped. When we stopped bouncing around we had a list to port of roughly 15°. I now understand that this was one or possibly two hits on our port shaft and as we were doing a good speed, the high revolution rate of the shaft, on being hit by a torpedo, caused a great deal of destruction in the shaft tunnels. Most of the engine-room and so on on the port side was inundated with water and we lost power throughout most of the ship. This would now be approximately 11.30.[38]

The loss of power having put paid to any further radar duties, Paddon was able to help in many other ways. About 1.20pm, after another high-level attack during which three 5.25in and pom-pom guns were still able to respond, Paddon was on the starboard side of the bridge structure helping to free Carley floats when the ship gave a lurch and started to roll over to port. Paddon was one of the few who managed to escape over the port (low) side, to be picked up by the destroyer *Electra* some ninety minutes later.[39] He was back in Singapore the same night. Eighteen officers and 307 men from *Prince of Wales* were lost, including Admiral Phillips and Captain Leach.

In a report written in Singapore two days later, Paddon said that three of the four type 285 radar sets on the HA/LA directors were in action, following aircraft in from ranges of between 6.5 and 3.5 miles to about 1.5 miles. They remained in action until the torpedo hit on the port side. Because the pom-poms' type 282 sets had low priority for maintenance in Singapore, only one had been operational, and this had given ranges from 3 miles.[40]

Repulse's radar officer was Sub-Lieutenant (Special) G.K. Armstrong, RNVR, who had joined the ship in July 1941 while she was having a short docking and refit at Rosyth, during which a type 284 surface gunnery set was fitted. It was a tight schedule and dockyard workers had to sail with her to Scapa where the installation was completed. Before sailing from the Clyde for the Indian Ocean on 2 August, they embarked six type 286P radar sets, a petty officer radio mechanic and eighteen ratings to man those sets. They installed one of them in *Repulse* while on passage to Freetown. On arrival, an aerial was fitted on a short mast which could be stepped on the top of the Air Defence Position (ADP) and the set was complete and working by the end of the four-day stay in Freetown. With its good aerial height and short feeders from the office immediately below the compass platform, it produced good results. The other five sets, the petty officer and fifteen of the ratings were delivered to the carrier *Hermes* at Mombasa, to be fitted to *Hermes* and other ships in the Indian Ocean.

It was the 286 [says Armstrong] with which we located the Japanese reconnaissance aircraft at 10.40 in the morning of 10 December. After we located this aircraft, it was identified visually from the ADP. If the Admiral in the *Prince of Wales* had broken W/T silence at that time, fighters would have arrived from Singapore in time to save the *Repulse*. She was not seriously damaged until hit by five torpedoes just after 12.20. We operated the 286 continuously, without its normal cooling periods, from 10.15 until the ship sank. We picked up groups of Japanese aircraft at between 25 and 30 miles. It was unaffected by gunfire even though 'S.1' triple 4in [gun mounting] blew our doors open every time it fired on a forward bearing. Even after five torpedo hits, we still had electric power, and I switched off the set in the end, when I was ordered to abandon ship.[41]

So ended 1941, a sad year for the Royal Navy when so many good ships and their companies were lost by enemy action.

TAILPIECES

• By Commander A. E. Fanning, formerly flotilla navigating officer of the 16th Destroyer Flotilla, based at Harwich:

Back echoes – Type 286M, which only faced forward, nevertheless suffered badly from back echoes. One night in the autumn of 1941, while on E-boat patrol off Lowestoft, the Hunt class destroyer *Eglinton* detected an echo at 1.25 miles, classified 'possible E-boat'. Turning towards at maximum speed (about 24 knots), the range closed steadily. At about half a mile, the captain ordered: 'Stand by to illuminate with starshell.' At that very moment, the target (which turned out to be a *Puffin* class corvette also on patrol) opened fire with starshell ahead of herself. She had detected the *Eglinton* astern, as a back echo but, thinking it to be ahead, had given chase. As the range gradually closed, they had assumed they were gaining on their target and, at half a mile, had illuminated with starshell ahead of them. They found nothing ahead but were soon made aware of a 'Hunt' class destroyer rapidly overhauling them from astern.[42]

• By Vice-Admiral Sir Roderick Macdonald:

I joined the destroyer *Fortune* at Chatham late in 1941 after a refit in which type 286P had been fitted. The drafting authorities had cleverly arranged for all three radar operators to be broad Scots. I was the only officer who understood their reports and, during our passage up the East Coast en route for Londonderry was constantly being summoned to the bridge to interpret. This was not only irritating, but hardly contributed to speedy reaction to attack. So I devised a reporting code based on a contemporary comic strip. An aircraft detection became 'Pip', a surface contact 'Squeak', and land 'Wilfred'. This worked, particularly, 'Pip, Pip, Pip, bearing and range so and so.'[43]

- By Mr A.H.G. Butler of Wingfield in Wiltshire:

The pawnbroker's balls – In November 1939, as a Sub-Lieutenant ex *Flying Fox 7* division RNVR, I joined HMS *Westminster*, a WW1 destroyer converted for anti-aircraft work. Some time in 1941, as part of the Rosyth Escort Force, we went into Newcastle for the night. I was 'duty boy' and all the rest of the officers went ashore (without the Captain). When I came into the Wardroom next morning, I found a complete pawnbrokers sign in one of the armchairs and it sculled around in the Wardroom for some time. They had had a lot of fun obtaining it without getting caught.

Fairly soon after arrival at Rosyth, we were fitted with RDF type 286. It would give us range and bearing 45° on either bow; should we need a wider field, we altered course accordingly. A little later a number of sets came on the market, all called by a type number, one of which I remember looked not unlike a glass fluted lighthouse and was christened 'The Perambulating Tomato Hatchery'. So the scene was set for the introduction of 'type 298'.

The First Lieutenant, the late John Hamer, RN, was a 'wag' of some renown and he had the idea of setting up the pawnbroker's balls at the back of the bridge as a mascot. This was done and the Gunner in so doing felt that it should be wired up to the Transmitting Station, and it was not long before we realised that, being spherical and in triplicate, it would not only give us range and bearing, but height as well – and, as we were an AA ship, this was a valuable addition. All this time, the sign remained in its appropriate colours – golden balls suspended from a black bracket – and this was never changed.

So 'type 298' was born. We had the Flotilla RDF Officer to tune it and the Flotilla Gunnery Officer to calibrate it. It was not long, therefore, before the whole Flotilla got the buzz that *Westminster* had a new set. Captains and First Lieutenants were seen walking round the dockyard in the dog watches to sneak a look at the golden balls. New 'types' were at that time so frequent that they really believed the buzz that they had heard – and went away jealous.

The sequel to the story was that, when one day Admiral Fraser came in under the Forth Bridge in the *King George V*, he made a signal to C-in-C Rosyth: 'Understand you have type 298 fitted in one of your escort force. Would much appreciate full particulars.'[44]

This true story was the inspiration for a work of fiction – Edward Hyams's *Silvester* (London, 1951).

7

1942: Malta Convoys and the Invasion of North Africa

Feb 9	*Scharnhorst and Gneisenau* Channel passage
Feb 15	Fall of Singapore
Mar 22	2nd Battle of Sirte
Mar 26	Adml Somerville commands Eastern Fleet
Mar 28–29	*Trinidad*'s action north of North Cape
Apr 5	Eastern Fleet retires to Mombasa
May 4	British landings in Madagascar: 'Ironclad'
May 4–8	Battle of Coral Sea
Jun 3–7	Midway; turn of tide in Pacific
Jun 12–16	Malta convoys 'Harpoon' and 'Vigorous'
Jun 21	Axis capture Tobruk
Jul 5–10	Arctic convoy PQ.17; severe losses
Aug 10–14	Malta convoy 'Pedestal'
Aug 19	Dieppe raid
Sep	Battle of Guadalcanal
Oct 23	El Alamein; Allied offensive in Egypt
Nov 8	Invasion of French N. Africa: 'Torch'
Nov 20	Siege of Malta raised
Dec 22–31	Battle of Barents Sea; *Lützow* and *Hipper*

SCHARNHORST AND *GNEISENAU*'S CHANNEL DASH

On 11 February 1942, the German battle cruisers *Scharnhorst* and *Gneisenau*, accompanied by the heavy cruiser *Prinz Eugen* and a destroyer screen, sailed from Brest after dark, turning up-Channel unobserved by any British forces. Such a move had been expected in Britain and the most detailed preparations had been made to meet this eventuality – or its alternative, breaking out into the Atlantic – to be put into force by the codeword 'Fuller'.

But the British inter-service planners had made two unwarranted assumptions: first that the German force was most likely to pass through the narrows between Dover and Calais during the hours of darkness; and secondly that in any case they would be detected soon after leaving Brest. In the event, the Germans, steaming at 27 knots to pass through the narrows in daylight, were off Le Havre by 8am on the 12th still undetected, largely due to ASV failures on Coastal Command aircraft.

Starting at 9.20, severe jamming of British coast radar indicated that something unusual might be afoot, but this had happened many times before and could mean nothing special. However, the naval 825 squadron of Swordfish recently arrived at Manston in Kent from Lee-on-Solent was brought to immediate readiness. A naval type 271 radar with 7ft paraboloid aerials – known by the Army Coast Defence personnel who manned it as the type NT 271 – had recently been installed on the cliffs at Fairlight, east of Hastings. (An Army NT 271 radar set, also called CD No.1, Mk.IV, just installed at Ventnor in the Isle of Wight, was unfortunately out of action owing to an outbreak of fire the night before, but it probably could not have picked up these vessels hugging the coast, even from its site high on the cliff top.) Being of 10cm wavelength, it was not subject to German jamming. According to an Admiralty source,[1] the first firm indication received in the naval operations room at Dover that this really could be 'Fuller' was a message timed 10.50 from Fairlight reporting a force of large vessels at 23.5 miles range, moving much faster than a normal convoy, off Etaples. Another NT 271 at Beachy Head had picked up a group of ships earlier but their report to the naval HQ at Newhaven and to the RAF seems never to have reached Dover and there was also a considerable delay in the Fairlight report reaching Dover through Newhaven. Incredibly, no direct telephone had been provided between Fairlight and Dover.

We need not concern ourselves here with the controversy which still surrounds events on that morning. Suffice it to say that two Spitfires visually identified the German squadron at 10.42 but that Dover operations room did not receive this most important news until after the Spitfires had landed at 11.09, the pilots having decided to keep radio silence while airborne.

Communications between the Navy, the Army's Coast Defence, and their Coast Artillery colleagues at Dover seem to have been appalling. Coast Artillery with their battery of 9.2in guns on the South Foreland had no warning before 11.30 when the German squadron was detected by an NT 271 radar at Dover itself – nothing from Beachy Head or Fairlight nor news of the Spitfire sighting. Though their 1.5m fire-control radar was being jammed, they had a naval 50cm type 284 (NT 284) which would have been immune – but this was switched off and took 45 minutes to warm up. The guns eventually opened fire at 12.19, firing 33 rounds, with

20. The flotilla leader *Campbell* in May 1942, soon after leading the 16th and 21st Destroyer flotillas' attack on *Scharnhorst* and *Gneisenau* off the coast of Holland. The WS type 271 hut hijacked in Chatham dockyard in January can be seen behind the bridge. The rotatable WC type 286P aerial is at the top of the foremast, below the cross-shaped R/T communications aerial.

(NMM)

three confirmed hits on *Gneisenau*. When eventually operational, the Coast Artillery NT 284 worked only fitfully but the Coast Defence NT 271 passed plots of the enemy force.[2]

Altogether an unfortunate series of events, the only cheerful note being the excellent performance – but not communications – of the Coast Defence NT 271 radars, a tribute to their naval designers.

We will not concern ourselves with the attacks on the German squadron by our Swordfish torpedo bombers, by our MTBs, or by RAF aircraft, but will move on to the destroyer attack in the afternoon. But before doing so, we must go back a few weeks.

In January 1942, while *Campbell* (leader of the 21st Destroyer Flotilla based at Sheerness) was undergoing a week's boiler cleaning in Chatham dockyard, Captain Mark Pizey, Captain (D) of the flotilla and *Campbell*'s own captain, spotted a type 271 hut and aerial lantern lying on the dockside. At this time, priority for allocation of this set went to the Western Approaches and Home Fleet, not to ships like *Campbell* engaged in the E-boat war on the East Coast. The set at Chatham had been allocated to a Home Fleet destroyer which was under repair for action damage and not due to complete for three months or so. Appropriate strings were pulled through the Commander-in-Chief the Nore, and the set was fitted in *Campbell* in two days. She was already fitted with type 286P. No additional operators were available.

Some three weeks later, on the afternoon of 12 February, Captain Pizey found himself leading a force of six WW1 torpedo-fitted destroyers from the 16th (Harwich) and 21st (Sheerness) Flotillas on a dash across the North Sea to intercept the German squadron. It was an afternoon of considerable confusion. No reliable position of the enemy had been obtained since it faded from the Dover radar at 1.12pm, some 10 miles SW of the West Hinder Light Float off Ostende. Visibility was sometimes down to a mile and it was blowing hard. Interception, if it were to be achieved, seemed unlikely to be before 3.15.

In the air, the situation was probably even more confused. During the afternoon, Pizey's force was several times attacked by friendly aircraft, and on at least one occasion found itself escorted by enemy Messerschmidt 110s. *Campbell*'s air warning set type 286P did splendid work during this period, detecting most approaching aircraft in time for the guns to be put on a 'look-out' bearing. Meanwhile, *Campbell*'s new type 271 had 'gone on the blink' and was only restored about 3pm after Herculean efforts by her petty officer telegraphist.

At 3.17, when *Campbell* was about 22 miles west of the Hook of Holland, two large echoes were detected at a range of 9.5 miles (later learned to be *Gneisenau* and *Prinz Eugen*, *Scharnhorst* having dropped astern after striking a mine) and from then on, as the German squadron steamed north-east off the Dutch coast, an excellent radar plot was developed although *Campbell* had only the old-fashioned ARL table with a 5in 'spider's web'. Lieutenant Antony Fanning, the flotilla navigating officer, remained at the plot in the chart house throughout the action, receiving radar reports by telephone. Although the situation became more difficult to interpret on the A-scan (there was no PPI, of course), once the German destroyers came within range, between them, Fanning and the operator (probably the PO telegraphist) always found it possible to identify the two main targets and to provide information for torpedo control.[3]

Torpedoes were fired by all destroyers about 3.47 – by which time they were under very heavy fire from both of the large German ships and from their escort – but no hits were obtained. In the whole operation – from leaving Brest to arriving in Germany – the only significant damage sustained by the German squadron was from mines. Altogether a sad episode for the British forces, the only thing which seems to have come well out of it being the naval type 271 radar, both ashore and afloat.

NAVAL RADAR'S 'PRODUCTION YEAR'

During 1942, radar stopped being something special, something to be marvelled at, something fitted only in special ships. Instead, it became, in all its various roles, an integral part of a fighting ship, as essential as the armament, the propulsion, the communications equipment. Within the Fleet, the secrecy which had surrounded it was swept away, except for future development. In the period covered by the last four chapters, the number of operations in which radar played an important part was limited and we have been able to touch upon most of those. From now on, however, radar played some part in almost all operations, so we shall be able to describe a very small sample only.

By the beginning of 1942, air warning radar had already been fitted in all but one (*Royal Sovereign*) out of thirteen surviving capital ships; in all large carriers; and in thirty out of sixty-five surviving cruisers. (Most of the ships not so far fitted had been abroad continuously since the beginning of the war.) As for gunnery sets, all capital ships but *Valiant*, all large carriers and twenty-seven cruisers had one or more. Since fitting started in August 1941, 10cm surface warning sets had been fitted in nine capital ships and five cruisers. A high proportion of smaller ships – fleet destroyers, escorts, minesweepers – had either the 1.5m type 286 or 10cm type 271.

When the main body of HM Signal School's Experimental Department moved from Portsmouth to Haslemere in April 1941, later to become the independent Admiralty Signal Establishment, members of the RDF Division continued to work in their various offices, laboratories and

21. General view of a cruiser foremast and superstructure, showing WA type 279 at the top of the mast (1), surmounted by interrogator type 243 (2); WS type 272P (3) aerial enclosed in a perspex cylindrical lantern; GA type 285 (6) on the HA Director *bottom right*; and GS type 284 (5) on the main armament DCT *left*, with 'fishbone' instead of the usual 'pig-trough' aerials, as fitted in a few cruisers including *Suffolk* during the *Bismarck* action (later replaced by a pig-trough).

(Radar Manual, 1945)

22. The 8in cruiser *Kent* in 1942, with WA type 281 aerials at the mastheads, that on the foremast being surmounted by the aerial for interrogator type 243 used to provide IFF responses to type 281. The large 'pig-trough' aerial of GS type 284 is mounted on the DCT behind the bridge and the lantern for WS type 273Q hut can be seen on a structure amidships, raising the aerials above the funnels. Two GA type 285s were mounted on the HA directors on either side of the forward superstructure, not easily seen here.

(IWM)

workshops in the Portsmouth area. During the summer and autumn of 1942, they moved piecemeal to King Edward's School at Witley in Surrey, some 5 miles north-east of ASE's headquarters at Haslemere, in time to greet HM the King who visited Haslemere and Witley on 30 October.

Within ASE, 1942 was accounted as 'production year', with the highest priority concentrated on production, supply and fitting. By the end of the year, virtually all capital ships and cruisers had been fitted with long-range warning sets. Gunnery sets (50 cm) and type 271 were each fitted at an average of one a day of each type throughout the year. The Army had also been supplied with 280 sets of type 271.[4]

LOOKING TOWARD 'BLIND FIRE'

Fitting the Fleet with the first generation of 50cm gunnery radar sets was well advanced by the end of 1941 and ASE's Gunnery Section had been working on the improvements to these sets and on the development of certain new techniques which could be added to them without major redesign, consolidated at the 'way ahead' meeting of May 1941 and outlined in Chapter 6 on pp. 102–7. Most of these modifications and additions were ready for fitting in new-construction ships early in 1942, and the converting of existing sets continued at high priority during 1942 and 1943. The general fitting of the accurate ranging panel L.24 (see pp. 105–6) started early in 1943.[5]

The increase in transmitting power, receiver sensitivity, and aerial gain had met the first requirement for increased ranges – battleship to battleship, from 20,000 to 29,000 yards (10 to 14.5 miles) with type 284M; aircraft at 1,000 feet, from 16,000 to 40,000 yards (8 to 20 miles) with type 285M.[6] The 14in guns of the *King George V* class battleships had a maximum range of 36,000 yards (18 miles) so the improvement with type 284M was arguably not enough. In point of fact, the disadvantage was more theoretical than actual because, for various reasons, fire would seldom be opened over about 28,000 yards (14 miles) and, for all cruisers except the 'County' class (with 8in guns), and for destroyers, the radar could match the maximum ranges of the guns. The new ranging panel L.24 ensured that the extension of range performance did not mean less accuracy.

The increase of bearing accuracy with the adoption of beam-switching was also significant – with type 284M, from 45 to 5 arc-minutes on small surface targets, rather more for extended targets (a WW2 destroyer at 5 miles range subtended an angle of about 40 arc-minutes when beam-on, 4 arc-minutes end-on). Type 285M could now measure the bearing of an aircraft to about ±15 arc-minutes, though it could not measure its elevation. With both sets, beam-switching was of great value in assisting the operator to determine which echo, of the several that were usually present on the radar display, corresponded to the target being engaged by the director.

The earliest concept of the use of radar for gunnery purposes was that its role should be to supplement – and maybe replace – the optical rangefinder, to give target range, and equally important, to provide the data to measure 'rate' – the rate of change of range, from which could be forecast the future target position. Indeed, the very term 'radio rangefinder' was extensively used for all fire control radar in the early years. Very soon, however, it was realized that radar could do much more. It could also measure bearing. And it could see the fall of shot relative to the target. Maybe in the fullness of time it could be made to measure elevation. Could it one day replace the human eye as a sensor for the aiming of guns? From these thoughts was born the concept of 'blind fire', the ability to engage an enemy at night or in fog, when the target could not be seen visually.

But to do all this, the radar had to be fully integrated into the ship's fire-control system instead of being merely an 'add-on'. It was intolerable to have the operator who is producing such valuable intelligence shut up in a radar office some distance away from the fire-control centre. One of the modifications was to provide remote director-training units, additional CRTs displaying the side-by-side signals when beam-switching was operating, fitted where they were most needed.

With type 285M, this display was mounted in the director so that blind aiming for line could be done by the director-trainer rather than by the operator in the radar office. This also assisted in complicated tactical situations. In the restricted space of the director, however, it was often difficult to position this display conveniently for the director-trainer, so a later development was the Radar Training Sight (RTS) which allowed an image of the radar display to be presented in one eyepiece of the trainer's binocular sight.

A similar remote CRT was used in type 284M, but this was generally positioned in the below-decks fire control centre for surface fire (the Transmitting Station) and not in the director. With an expanded range display of 4,000 yards centred on the target, this allowed splash spotting in range, and estimates of range 'over' or 'short' to be passed directly to the Fire Control Table, although shell splashes could only be seen out to about half the maximum range of the main armament and misses right or left were difficult to estimate from the range tube.

Although the 50cm sets as modified went a long way towards making blind fire possible as far as range and bearing were concerned, they could not achieve completely blind fire because it was not practicable to convert them to measure the target's angle of elevation owing to the size of the aerial that would have been needed. Nevertheless, for surface actions at least, something approaching it was quickly reached, as is shown in the *Scharnhorst* action described in the next chapter.

With blind fire in mind, the long range air warning set type 281 had been fitted with beam-switching and, though the beam of this set was very wide, ships and aircraft could be tracked in bearing with an accuracy of ±½°. However, to use the set for blind fire on fast-moving targets, the operator had to concentrate on that one target and all-round warning was lost. This was often unacceptable, so type 281 was seldom used for gunnery purposes.

BARRAGE FIRE

The usual procedure when being attacked by aircraft is for the enemy to be detected first by long-range air warning radar, leading to fighter direction operations. This will be followed by long-range high-angle fire from ships. For aircraft which slip through, it is valuable if a curtain of exploding shells can be put up at a distance of, say, 1,500 yards around the ship or convoy being attacked. And if some of those exploding shells are from a battleship's big guns, the barrage should deter all but the bravest pilot from pressing home his attack. Such radar-controlled 'predicted' barrage fire was devised by Captain G.M.B. Langley in *Carlisle* as early as May 1940 during the Norwegian campaign and the Admiralty very soon tackled the problem.[7]

Early in 1941, it was decided to exploit the ability of early radars to give good ranges and range rates to improve barrage fire, without waiting for further advances in blind aiming by radar. A precision ranging panel displaying 6,000 yards with ranging errors not exceeding 25 yards was rapidly developed. This panel, designated L.22, employed an electronic 'step' marker with a high order of setting accuracy, and a highly linear potentiometer controlled by the operator's rate-aided handwheel (see Glossary) to align the marker with the target echo. By this means accurate range and range-rate were immediately available. Panel L.22 was used in conjunction with the Auto Barrage Unit, generally located in the High Angle Control Position below decks. The ranging panel fed the target's range and range-rate into a mechanical calculator in the ABU which then determined the correct moment to open fire to achieve the predetermined future range already set on the shell fuzes. This provided a better control of barrage fire than was possible with the early type 285 sets controlling AA guns and became available for fitting at the end of 1941.

When added to the type 282 sets with improved transmitters and receivers and beam-switching, the L.22 accurate ranging panel converted them to type 282P, corresponding to the 284P and 285P sets which had the new precision ranging panels L.24 for longer range working. During 1942, it was decided to produce a version of 282P which could be mounted on special barrage-fire directors so that large ships could use their surface armament for anti-aircraft barrages in addition to their specialized AA weapons. These sets were designated type 283.

To sum up: by the end of 1941, the likely improvements in 50cm systems had been developed and extensive fitting of this improved equipment proceeded during 1942 and 1943. These sets and the associated weapon system developments had resulted in a series of manually operated weapon systems based largely on (a) existing mechanical computing devices, (b) weapon directors carrying the aerials, and (c) guns with time-fuzed shells. The human operator had little assistance apart from rate-aiding (see Glossary), and only for heavy main armament directors and gun mountings was there power assistance of a rudimentary kind. Nevertheless, reasonable blind fire against surface targets was possible, fair harassing blind fire against aircraft was also possible against low-flying targets, and barrage fire had achieved a useful level of deterrence with the auto-barrage system. Visual fire had been substantially improved by manual radar ranging with rate-aiding which was widely applied.

This outcome was better than might have been expected, since from the beginning 50cm had been regarded as a stopgap – to be used pending the development of components capable of generating adequate power and operating at much shorter wavelengths. In the event, on the gunnery side at sea much of the war was to be fought with the widely fitted 50cm equipments. But in 1942 this certainly could not be foreseen, so, as we shall see in the next chapter, effort was next concentrated on developing a series of centimetric systems to succeed the 50cm equipments, as had always been the intention.

TRINIDAD'S ACTION OFF NORTH CAPE

The account which follows is a good example of what could be achieved even with unmodified 50cm gunnery radar – assisted by type 281. In March 1942, the cruiser *Trinidad*, escorted by the destroyers *Eclipse* and *Fury*, was providing close cover against possible attack by German surface forces on the eastbound Russian convoy PQ.13 (which was generally out of sight of *Trinidad*). The main Home Fleet force of three capital ships, three cruisers, and eleven destroyers was providing distant cover many miles to the west.

Trinidad was a comparatively new ship, having left Devonport dockyard where she was built some six months earlier fitted with type 281, type 284, three type 285s and two type 282s, none of which had been modernized. She had no 10cm surface warning set. The destroyers had type 286.

In the early afternoon of the 28 March, some 90 miles north of the North Cape of Norway, a Junkers 88 dived out of the clouds and dropped three bombs. It was seen in time to put the wheel hard over but had approached undetected because type 281 had been out of action for some time owing to oil freezing in the training gearbox of the receiving aerial on the mainmast. 'In spite of arctic conditions,' wrote Captain L.S. Saunders, 'the two RDF officers had worked magnificently at the main masthead to the limit of their endurance, and the gear box from the masthead had been removed and cleared of oil and had just been replaced when this attack took place. There was no time to refill the gearbox so it was connected without oil and type 281 put into action.'[8] (In fact, the captain got it wrong. The senior RDF officer, Lieutenant (Sp.) R.A. Laws, RNVR, worked at the masthead with a Leading Radio Mechanic, as he considered it unwise that both RDF officers to be up the mast together. It was the cover, not the gearbox itself, that was removed and replaced. The frozen oil was cleared out using a massive blowlamp.[9])

The shadowing aircraft was repeatedly in sight and during the next hour and a half type 281 detected many aircraft. At 2.40, *Trinidad* found a snow squall and managed to escape further attention from aircraft by keeping in the thickest part of it.

In the early hours of the 29th, the outlines of three ships were sighted to the southward which proved to be the British destroyer *Oribi* and two Russian destroyers from Murmansk. At 8.42, type 281 reported a surface contact ahead at 6.5 miles. The 6in DCT was 'put on' by type 281 and type 284 immediately confirmed the contact. Conditions were atrocious – visibility about 2 miles, snow, and spray freezing solid as fast as it came on board.

Five minutes later, type 284 reported there were now three echoes at 3.6 miles. The plot had difficulty in assessing their movements but an approximate course of 300° was reported. At 8.49, three destroyers, obviously foreign, were sighted. Captain Saunders knew they could not be Russian, so gave the order to open fire without further ado. In fact, they were three German destroyers which had sailed from Kirkenes the previous day on the strength of the first air report of the convoy. Continuing the radar story in Saunders's own words:

> The bearing and range from type 284 was now 064°, 4,200 yards [2.1 miles] and 'Alarm Port' was immediately passed to all quarters, followed by 'Open Fire'. Fire was opened at 0850A without waiting to challenge. The main armament opened fire tuned to RDF range. The first salvo fell short and the second salvo scored hits and started a fire with much flame and smoke. The DCT was ordered to shift target to

> the second destroyer and during the first two salvos hits were observed on the target. A further two salvos were fired by the main armament before the target disappeared in the fine snow, but fall of shot could not be observed accurately as the target was hardly discernible. The RDF [Lieutenant Laws was in the 284 office] reported one salvo on the target and one over.
>
> During this engagement, the port 4in battery had also engaged the enemy, controlled by the after 4in Director using ranges from type 285 RDF, but fall of shot was obscured by 6in splashes.
>
> At 0859A type 284 obtained a contact bearing 358°, 8,200 yards [4.1 miles] . . . the plot gave an approximate course of the enemy as 270° . . . *Trinidad* altered course slightly to starboard and fire was opened on a bearing almost right ahead with an RDF range of 2,900 yards [1.45 miles]. Course was [again] altered slightly to starboard in order to open 'A' arcs [see Glossary] and finish off the enemy, the range being 2,600 yards. The 4in armament opened fire with the port Director controlling, using ranges from type 285 RDF as soon as the guns would bear on the target, but 4in fall of shot was in most instances obscured by 6in shell splashes . . . DCT personnel had great difficulty in seeing the target on account of their binoculars freezing over. RDF ranges were used throughout . . . and no rangefinder ranges were obtained . . . Spotting corrections were applied on the strength of type 284's fall of shot reports, which were prompt and frequent . . .[10]

At 9.22, *Trinidad* fired torpedoes at an enemy destroyer and a minute later was herself struck by a torpedo and severely damaged, so that she only made the Kola Inlet under her own steam with the greatest difficulty. Meanwhile, *Eclipse* had finished off the German destroyer engaged by *Trinidad*. It was subsequently discovered that the torpedo which hit *Trinidad* was one of her own, the extreme cold having caused the steering mechanism to behave erratically. She left Kola Inlet six weeks later after repairs in the dry dock at Rosta but was attacked by German aircraft and had to be sunk by a torpedo from one of our own destroyers – but that is another story.

Captain Saunders had the following conclusions on the radar aspects:

> Much depended upon the type 284 in each phase of the brief engagement and it did all that was asked of it:
>
> 1. Tracking from 'first contact' to 'enemy in sight'.
> 2. Tracking and reporting the number of enemy out of sight ahead while chasing. The TS was able to make good estimates of enemy course and speed and changes in both.
> 3. Passing fall-of-shot reports direct to the Control Officer.

4. Ranging throughout. In the first encounter, three targets were reported. During the chase, it was repeatedly confirmed that only one target was seen on the scan.[11]

IFF AND RADAR RECOGNITION

The problem of how to identify echoes seen on the radar display – how to identify friend from foe, or IFF – was one that had been addressed by Watson-Watt's team as early as 1938. Several methods were tried but the one adopted was to fit a small transmitter–receiver, known as a transponder, in friendly aircraft. This received the signals from the locating radar and retransmitted them so that the amplitude of the echoes on the radar display increased perceptibly whenever the transponder made a reply. And it could be made to do more than just identify friend from foe. At the touch of a button in the aircraft, it could send a special signal signifying that the aircraft was in distress, or give some code identifying itself more positively.

And IFF could provide another service – as a navigational beacon. A transponder installed on an airfield or ship – or anywhere where a radio lighthouse was needed – could be picked up by the aircraft's radar, and used as a homing beacon.

The Mark I IFF responded only to the RAF's Chain Home radar, but the Mark II followed almost immediately, responding also to other RAF and Army ground radars of 1940, as well as to the Navy's type 79. Then it was again modified to respond to the RAF's GCI (Mark IIG) and to the Navy's type 286 (and later, 290 and 291), to become Mark IIN.

The Mark IIN system was crude and unreliable but – if the equipment worked and if the aircraft remembered to switch it on – it did provide a limited temporary answer to the aircraft identification problem, and its usefulness in aircaft led to IFF being fitted in ships early in 1942, where it was known as type 252.[12] Thus, a ship fitted with type 291 should in theory be able to identify one operating type 252 – again if the latter had remembered to switch it on. Of course, there were worries at sea that the enemy might be able to set off the transponder, but this risk could generally be accepted.

As a result of the Tizard mission, the United States agreed early in 1941 to adopt the British IFF Mark II for their own armed forces, despite the fact that they already had a rather more sophisticated IFF system under development.[13]

However, the Mark II system suffered from the fundamental limitation that its IFF transponders could only respond to signals on a limited number of wavebands. Modifications had been made to accommodate new radars like GCI and type 286, but there was a limit. For the Navy's

3.5m type 281, for example, ASE had to fit a type 286 with an aerial rotating with the main 281 aerial, specially to trigger the aircraft's IFF and then pass the returned signal to the main type 281 display. Used in this way, the set became an 'Interrogator' and was designated type 241. But only in big ships was there room to do this, and the 10cm band was still not covered. In the next chapter we will meet the hoped-for solution, the Mark III system.

RADAR JAMMERS AT SEA

As early as 1941, Davis had made an experimental installation of a type 91 radar jammer in a warship, the monitor *Erebus*, and successful trials were carried out. It was decided early in 1942 that all RN battleships and cruisers should be similarly fitted, together with the associated monitor receiver, so that German radar, whether ashore or afloat, could be jammed. An omnidirectional wide-bandwidth aerial which did not require tuning was used, whatever the frequency being radiated.

THE CONVOY BATTLES AND HF/DF

In the Battle of the Atlantic, Allied shipping losses rose to unprecedented heights during 1942, largely through U-boat attacks using pack tactics. However, the measures the Allies were taking – increased effort by shore-based aircraft, the advent of escort carriers, the provision of ahead-throwing weapons, to mention a few – were to turn the tide when they became fully effective in 1943.

In July and August 1942, three U-boats were sunk in quick succession by RAF Coastal Command aircraft patrolling the Bay of Biscay to catch U-boats on the surface in transit to and from their operational areas. This was achieved by Wellington aircraft fitted with 1.5m ASV Mark II and the 'Leigh Light', a searchlight slung under the fuselage, which could surprise a U-boat on the surface at night by being switched on to illuminate a target detected by ASV.

But German reaction was swift and this success was short-lived. By autumn, U-boats began to be fitted with a simple search receiver called 'Metox' which could detect the 1.5m ASV transmissions sufficiently far away to permit the U-boat to dive before the aircraft came within Leigh Light range. U-boat sightings in the Bay dropped once more to almost nil and were not to recover until Coastal Command acquired the 10cm ASV Mark III (a development of Bomber Command's H_2S radar) which Metox could not detect. In the two months April to May 1943, of the fifty-six U-boats sunk, over thirty were destroyed by aircraft, roughly half when in transit in the Bay and half when with convoys.[14] In a radio broadcast in the late summer of 1943, Hitler complained that 'the temporary setback to our U-boat campaign is due to one single technical invention of our enemies'.[15] This undoubtedly referred to the use of ASV Mark III over the Bay – a tribute indeed to the magnetron.

But there was one measure which did begin to be fully effective in 1942 – the introduction of shipborne HF/DF. At the beginning of the war, the only effective sensors an anti-submarine vessel had were the naked eye, binoculars and asdic (sonar), while seaborne weapons were limited to the gun and the depth charge. In 1940, another sensor was added – radar, enormously enhanced by centimetric radar in 1941. Then, in 1941 and 1942, U-boat pack tactics gave rise to the development at ASE of yet another sensor – high-frequency direction-finding, HF/DF or, to many of those who used it at sea, 'Huff-duff'.

As we have seen, the Germans drastically changed their U-boat tactics, gradually introducing 'wolf-packs' between October 1940 and March 1941. Patrolling U-boats were ordered by their shore HQs to positions on a line across the expected path of convoys. The first U-boat to sight a convoy reported and shadowed it while the others were ordered to close it and prepare to attack. Then, when ordered by the shore HQs, the pack attacked on the surface at night.

These 'pack tactics' required U-boats to report by radio to their shore HQs when they made or lost contact with a convoy (and to report at other times when not too close to anti-submarine ships). When these tactics were understood by the Admiralty – and when the Admiralty appreciated how much they depended upon HF radio transmissions by U-boats close to convoys – a programme of fitting the necessary direction-finding equipment (HF/DF) in the larger escorts, and in merchant aircraft carriers and rescue ships, was undertaken at the highest priority.

A single HF/DF ship with a convoy could provide reliable detection of a U-boat making a transmission on the frequency the ship was guarding when the latter was within 'ground wave' range (15–30 miles). Though a single line of bearing could not actually fix the position of the U-boat, it was most useful operationally providing one could distinguish between the bearing and its reciprocal, not possible in the earliest shipborne HF/DF sets. It was the invention of a 'sense' aerial to overcome this limitation early in 1941 (by W. Struszynski, a Polish engineer working at ASE) that provided the breakthrough that made HF/DF really useful operationally. However, what was desirable was to obtain a 'fix' of the position of the U-boat relative to the convoy, and this demanded cross-bearings from two – or better, three – HF/DF-fitted ships. Furthermore, because the escorts were never certain which of several frequencies the U-boats might use, additional HF/DF ships made it more likely that any transmission could be detected.

Each HF/DF report enabled the escort commander to send an escorting aircraft if he had one, or anti-submarine ships, to search for and attack the U-boat. The U-boats needed their surface speed to keep up with or overtake a convoy, and even if the searching escorts failed to find and attack the U-boat, they could probably force it to submerge until the convoy was out of sight. If they could be spared for long enough, they might prevent the U-boat ever regaining contact.

It could be assumed that any U-boat transmission within ground-wave range related to the convoy. The Germans prefixed each message with an indication of priority which was self-evident and was known to us. Although escorts could not decipher the text of U-boat messages, an estimate could be made of what they were about by considering their priority and length. For instance, the first patrolling U-boat to sight a convoy would make a very high priority short 'first sighting' report giving only the convoy's position and approximate course; it would possibly follow this within minutes with an 'amplifying' report that was longer and had lower priority; and, while shadowing the convoy, would make periodical short 'in touch' reports. One could similarly guess at the meaning of reports of losing contact with the convoy or of the convoy having altered course. By guessing the contents of the U-boats' messages, the escort commander might estimate what they were doing and might establish anti-submarine patrols or ask for an alteration of convoy course to throw off the shadower or to make it more difficult for the other U-boats to concentrate on the convoy.

To summarize: HF/DF often provided the first warning that U-boats were in touch with a convoy, and also gave information on which to take the initiative against impending pack attack.[16]

The first ships to be fitted with HF/DF outfit FH3 were the fleet destroyers *Gurkha* and *Lance* in July 1941. They had their aerials in the optimum position from a DF point of view – at the top of the foremast – but that meant they could not have radar type 286 as well. The Commander-in-Chief Western Approaches insisted that his ships must have radar, so in September the next four, *Zulu, Lively, Rochester* and *Broke,* had their DF aerials sited aft on a special pole mast, accepting lower performance. By the end of January 1942, twenty-five escorts had been so fitted.[17]

FH3 had been developed by Struszynski's team at ASE with great urgency in 1940 and 1941, making the best use of existing equipment and technology in order to have an effective HF/DF operational at sea in the shortest possible time. Meanwhile, development of an improved equipment, FH4, was under way at ASE and Plessey Co., the main improvement from the operator's point of view being visual presentation on a cathode ray tube instead of the aural presentation (where the operator had to wear headphones) of FH3.

23. The sloop *Erne* off North Africa in 1943, with GA type 285 on the rangefinder-director, WS type 272P lantern on the structure aft, and an HF/DF aerial on the top of the foremast. Some sloops at this date had WC type 291 instead of HF/DF.

(IWM)

The first experimental FH4 went to sea in the ex-US coastguard cutter *Culver* in October 1941, and it was first used in action in *Leamington,* escorting the Madagascar-bound troop convoy, WS.17, in March 1942.[18] However, general fitting of FH4 did not start until 1943, so the most critical convoy battles in 1941 and 1942, and many in 1943, relied upon FH3, earning the admiration of all convoy escort commanders concerned.[19]

Though not actually radar, HF/DF has been given more than a passing mention in this book, not only because the apparatus was developed by ASE, but also because, next to 10cm radar, HF/DF was probably the most important element in the winning of the Battle of the Atlantic. In combination with radar, it eventually made submarine attacks on convoys too hazardous to be attempted. An important part was also played by naval shore-based HF/DF; this again was developed by ASE but is outside the scope of this book.

A classic example of how HF/DF, radar, lookouts, hydrophone effect (using the asdic to listen for underwater noise without transmitting; the term 'passive sonar' used today explains it most graphically) and asdic echo contact were used together (or rather, in succession) occurred with convoy ON.122. At 9.30pm on 23 August 1942, the destroyer *Viscount* and rescue ship *Stockport* obtained an HF/DF fix which the former immediately followed up with the corvette *Potentilla,* who obtained a radar contact at 2.5 miles 45 minutes later. Shortly after, she sighted a U-boat which dived when the range had closed to 0.75 miles, contact then being held by hydrophone effect. Shortly after, both ships obtained asdic echoes. Three attacks were carried out.

Over the next day and a half, many radar contacts resulting in U-boat sightings were obtained by the escorts at ranges between 3.5 and 1.5 miles. The report concludes:

> Further attacks, some the result of RDF contacts, were made by the escort during this day [the 25th] and, as a result of the rough treatment they had received, no further attacks on the convoy developed.
>
> It is clear that the escort was outnumbered by at least one and the heavy scale of counter-attacks developed is most creditable, and shows what can be done with the combined and efficient use of HF/DF, RDF, HE, and the asdics.
>
> Type 271 functioned well in all ships and was responsible for 14 sightings. Type 290 played a small but invaluable part by relieving type 271 of all responsibility for station keeping, thus leaving the 271 free for whole-time search.[20]

For a description of the radar aspects of other A/S operations in 1942, the reader is asked to turn to Appendix B, where a memorandum, issued by the Commander-in-Chief Western Approaches in October 1942, gives a good insight into contemporary thinking.

'All work and no play makes Jack a dull boy.' The point must be made that ships of the Western Approaches Command – destroyers, sloops, corvettes, escort carriers – did not spend all their time escorting trade convoys in the Atlantic. Sometimes they were seconded to other escort duties like Arctic or Malta convoys, or to troop and supply convoys to Madagascar or North Africa. But whether these diversions could really be counted as 'play' is a moot point.

PRODUCTION AND FITTING OF TYPE 271

Let us now return to the radar equipment story. To the great credit of ASE and of many other bodies and individuals, no less than 236 British ships were at sea with 10cm surface warning radar by 4 May 1942 – just over a year after the first type 271 went to sea in *Orchis*. The majority of destroyers and smaller ships not fitted with a centimetric set had the 1.5m RAF-designed type 286 combined air and surface warning set, the majority still with the fixed 'bedstead' aerial though, by the end of the year, many had been converted to the smaller rotating aerial as type 286P. Furthermore, the upgrading of these sets to type 286PQ by the substitution of a naval-designed modulator and transmitter giving greatly improved performance was well under way in those ships not already fitted with the all-naval type 290 (see p. 107) and the sea trials of its successor type 291 – with the same wavelength, aerial and pulse power – took place in the destroyer *Ambuscade* in February 1942.

In the last chapter, we read some of the comments made by the Admiralty's Director of Anti-submarine Warfare in September 1941. Here is part of another letter from Captain Creasey to Captain Willett of ASE, written in July 1942:

> . . . I do indeed believe that the signal world have done wonders in rushing the 271 into service and there is no doubt that people at sea are putting more and more faith in it.
>
> But I know you are not running away with the idea that all is for the best in the best of all possible worlds. I am sure that a lot more can be done to improve the surface warning RDF, both in the way of performance and presentation [the PPI], and I know your people are driving on with this.
>
> I hear there is a much improved HF/DF set in the experimental stage, which sounds a great advance. *Copeland*'s efforts in Convoy HG.84 show what can be done with intelligent use of these DF bearings.

24. The destroyer *Tumult* off Naples during an eruption of Vesuvius, 1944, with the older radar, with aerials for GA type 285 on the rangefinder–director, WC type 291 at the top of the mast, and WS type 271Q aft. Later destroyers, sloops and frigates were given lattice masts (see Figs 34 and 35) so that they could have WS type 272 forward (instead of type 271 aft), later replaced by WS type 276 or WC type 293.

(IWM)

> Your people have done their share in producing the instrument which will do the job. It now remains up to the rest of us to do our side of the bargain by making sure that the gear is, in fact, intelligently used. This sounds easy, but I expect you can appreciate that, in practice, it is proving extremely difficult. The main trouble, of course, is that everybody has an inherent reluctance to study the various orders sufficiently carefully. This applies both sides of the Atlantic.[21]

Before describing just what ASE *was* doing to improve performance and presentation for the future, let us digress briefly to consider the fitting of existing 10cm radar. Sufficient sets were being produced: how could escorts be swiftly fitted without having to interrupt their convoy duties by having to go into dockyard hands? The answer proved to be a prefabrication scheme, conceived by Dr Landale and Commander Yates of ASE. Radar huts suitable for destroyers and for corvettes were prefabricated by Metropolitan Vickers Ltd near Manchester. The components of the radar sets were 'plan-packed' in cases by the various manufacturers (principally Allen West Ltd of Brighton).[22] Initially, these were all sent direct to the port where the fitting was to be done, the hut lifted on board the ship and bolted down, and the radar was then installed in the hut by base staff.

Early in 1942, with the appearance of the first main production design of type 271, a refinement to this scheme came into operation. Huts and plan-packed sets were both sent to the works of W.H. Smith Ltd. at Old Trafford, Manchester, where the radar with all necessary equipment was installed in the huts – down even to ash trays for the operators. The sets were run up and tested and could then be sent anywhere in the world – by road, by rail or by sea. 'The prefabricated type 271s arrived in excellent condition,' it was reported from Alexandria, 'and to everyone's joy and amazement, the sets worked perfectly on first switching on.'[23]

But the first priority in 1942 was not ships in the Levant but the 150 or so escort vessels in the Western Approaches command not so far fitted with 10cm radar. After every three months or so of steaming, a steam-driven vessel requires its boilers to be cleaned and, in small ships, this has to be done in harbour, taking about a week. Coordinated by the Command Radar Officer, type 271PF (PF = prefabricated) sets, as they were termed, were fitted in two stages, each one occupying a boiler-cleaning period between convoys. In the first period, the steel platform to support the PF hut was erected. In the second, the type 271 set in its PF hut was fixed onto the platform and the necessary power supply provided. The first type 271PF was fitted in the minesweeper *Salamander* at Rosyth in April 1942.[24]

The plan-packing scheme applied to all 10cm sets but the PF scheme applied only to type 271 in small ships. Types 272 and 273 and type 271 in other types of ship had to be fitted in a dockyard.

25. Rear-Admiral R. R. McGrigor, Assistant Chief of Naval Staff (Weapons), leaving the corvette *Stonecrop* after an inspection in 1941. An unfinished dockyard-constructed WS type 271 lantern can be seen forward of the funnel.

(IWM)

26. The first prefabricated WS type 271 leaving the works of W.H. Smith Ltd, Trafford Park, Manchester, in April 1942 for fitting in the minesweeper *Salamander* at Rosyth. Mr Smith himself is on the right standing next to Lieutenant (Sp.) J. Nathrow, RNVR, one of the two officers supervising the work. The other was Lieutenant (Sp.) F.R. Roberts RNVR, who took the photo.

(K.J. Roberts)

One of the advantages of the 10cm sets over the 1.5m type 286 or 291 in escort vessels had been the elimination of the undesirable 'back echoes' which so complicated the radar picture when a large number of targets was present, as with a convoy. The solid reflecting mirror of the 10cm aerials completely eliminated back echoes appearing 180° from the true echo, but side lobes are inherent in any directional aerial and these produce 'side echoes' at various angles on either side of the true bearing of the target in the main beam. This can greatly complicate the radar picture near large reflecting objects such as a convoy or land if these side lobes are more than a few per cent of the strength of the main lobe. Experimental measurements during development indicated that the side lobes of the type 271 aerial were well below 10 per cent of the main beam but reports from sea claimed 25 per cent and upwards. Early in 1942, the problem was investigated in some detail in several ships and the cause was traced to the lantern structure whose elegant teak pillars and flat windows were scattering and reflecting radiation to a degree that was unacceptable, varying according to the direction the aerial was pointing.

The solution was to replace the prefabricated octagonal lantern of type 271 by a six-foot-diameter moulded cylinder of perspex, the first of which was fitted in the destroyer *Hesperus* in December 1942. For technical reasons, it took a little longer to produce the larger cylinder needed for type

273, but *Duke of York* had been so fitted before her successful engagement with *Scharnhorst* in December 1943.[25]

MALTA CONVOYS

During the early part of 1942, the carriers *Indomitable, Illustrious* and *Formidable* were in the Indian Ocean, and *Victorious* and *Furious* in the Home Fleet. Those in the Indian Ocean had particularly good opportunities for 'working up' in the period of relative calm which followed the withdrawal of the Japanese Fleet into the Pacific. Numerous exercises in the coordinated defence of the Fleet by two or three carriers were carried out and much was learnt of practical fighter direction at sea while some war experience was gained during the invasion of Madagascar. Similar exercises were carried out under the more restricted conditions in the Home Fleet.

Experience in the combined operation of aircraft from two or more carriers soon showed that the facilities that had been based on single-carrier operations in the Mediterranean were inadequate. Minor adaptations in equipment were made in *Illustrious* in Ceylon (including the scrounging of a PPI from an RAF GCI station) and a more elaborate lay-out based on current RAF practice was carried out in *Victorious* in England. In those days, the Admiralty had issued no 'Staff Requirements', only the simple order: 'Full facilities for fighter direction are to be fitted.' Improvements were largely the result of private enterprise, the cooperation and enthusiasm of ships' staff, and the cautious goodwill of Commanding Officers.[26]

In the Mediterranean, the siege of Malta continued. Of the many operations to keep Malta supplied, three must be mentioned here. The first of these, the second Battle of Sirte, took place on 22 March 1942, when Rear-Admiral Philip Vian, with five light cruisers and destroyers, covering a convoy of four supply ships from Alexandria to Malta, drove off the Italian battleship *Littorio* attended by three cruisers and destroyers. Three of the four supply ships reached Malta, only to be sunk by air attack within a few days. Of the 26,000 tons of cargo loaded, only about a fifth was safely landed.

All the cruisers in the action had air warning and gunnery radar and all but one of the destroyers had type 286, mostly with the fixed aerial; a few had type 285. Two radar-related anecdotes are worthy of record:

- By Commander R. L. Fisher of the destroyer *Hero*, during a heavy gale while returning to Alexandria:

> *Hero* was, I think, the only ship present without radar. At all events we didn't have any and it was pretty well impossible to keep station on a

pitch dark night in such conditions. Having started the night as 'inside right' on the anti-submarine screen, we found ourselves at dawn 'outside left' or thereabouts. The rest of the force must have been amused to see us on their radar screens wandering across their front.[27]

- By Captain A. D. Nicholl of the cruiser *Penelope*, christened 'HMS Pepperpot' after subsequent bombing in Malta:

> We had been fitted with RDF but my experts had failed entirely to get it to work. As a matter of fact they reported triumphantly that it was at last working after our return to Malta after the battle. Half an hour later it was blown to smithereens by a near miss.[28]

In June, two Malta convoys were despatched simultaneously, 'Vigorous' from Alexandria and 'Harpoon' from Gibraltar. The 'Vigorous' convoy was covered by eight cruisers and twenty-six destroyers under Admiral Vian, containing substantial reinforcements from the Eastern Fleet and relying on air support from Malta and the Western Desert. In his report, Vian mentioned an incident where high-level bombers approached from the western sun, unseen until they were nearly overhead. 'Without blind radar-controlled fire, many unopposed attacks would have been made.'[29] After intervention by Italian heavy surface forces, the convoy turned back. The Navy lost a cruiser (*Hermione*, by a U-boat), three destroyers and two merchant ships, and the operation failed to revictual Malta. The Italians lost a cruiser and had a battleship damaged.

The 'Harpoon' operation was conducted by Vice-Admiral A. T. B. Curteis in the cruiser *Kenya* and his forces included the battleship *Malaya*, two more cruisers, and the old and slow carriers *Eagle* and *Argus*, neither of which had fighter direction arrangements, their only radar being type 286 or 290 (with 285 in *Eagle*). They could put barely a dozen fighters into the air and the AA cruiser *Cairo*, with the convoy, did what fighter direction there was with her type 279. Though overwhelmed, they managed to shoot down eleven enemy aircraft. After the main covering force had turned back before the Sicilian Narrows, the convoy was attacked by an Italian surface force and three merchant ships were sunk. Two out of six merchant ships reached Malta. The Navy lost two destroyers while a cruiser, three more destroyers and a minesweeper were seriously damaged.

OPERATION 'PEDESTAL'

The post-mortems on all these operations were unanimous on two matters – the need for more fighter cover, and the inadequacy of fighter direction arrangements. The chief change made for the August convoy, called

Operation 'Pedestal', was the inclusion of the carriers *Victorious* (flagship of Rear-Admiral A. L. St. G. Lyster, whose Staff Officer (Operations) and Fleet Fighter Direction Officer was Charles Coke, by now a Commander), *Indomitable* and *Eagle*. Between them they could put up seventy-two fighters. The forces taking part were under the command of Vice-Admiral E. N. Syfret who had, in addition to the three carriers, two battleships, six cruisers and two dozen destroyers, as well as submarines.

Before the convoy reached Gibraltar from the United Kingdom, Admiral Syfret arranged an exercise called 'Berserk' in the Atlantic in which, for the first time in the Royal Navy, five carriers operated in company – *Victorious, Indomitable, Furious, Eagle* and *Argus* – all under Admiral Lyster. This was of the utmost benefit in exercising fighter direction and cooperation between carriers. All except *Argus* were to take part in 'Pedestal', *Furious* ferrying thirty-eight Spitfires which she was to fly off to reinforce the defenders of Malta – which she successfully achieved.

The force passed the Straits of Gibraltar on the night of 9/10 August and was detected early on the 11th. In the afternoon, while *Furious* was flying off her Spitfires for Malta, *Eagle* was hit by four torpedoes fired by U73 and sank in eight minutes. Air attacks began at dusk that day and intensified all day on the 12th. The broad beams of the naval air warning sets quickly produced saturation on the plots whenever a large number of aircraft were in the sky, with the result that enemy units could slip through undetected and lost fighters could not be identified and homed, a problem which soon became familiar to every trainee FDO in the Navy, because battles fought in the mock-ups at the Fighter Direction School were all loosely based on 'Pedestal'.

Nevertheless, until the carriers and capital ships were about to be detached from the convoy at dusk on the approach to the Sicilian Narrows, losses by air attack had been remarkably light, most of the attacks being driven off by fighters, one merchant ship being hit, sinking later. *Victorious* had been hit by a heavy bomb on the flight deck but suffered only superficial damage. Then, at dusk on the 12th, came an attack by more than a hundred aircraft. The fighters were down to half their original strength and the pilots were worn out. The defence was overwhelmed. *Indomitable* suffered very severe damage putting her flight deck out of action. Nevertheless, on the 11th and 12th, the sixty British fighters remaining after *Eagle* had sunk had been able to shoot down thirty or so enemy aircraft at the cost to themselves of thirteen, besides those lost aboard *Eagle* and *Indomitable*. Ships shot down seven.

The convoy, still thirteen strong, pressed on towards Malta. At 8pm, the cruiser *Nigeria* and AA cruiser *Cairo* were hit by torpedoes from an Italian submarine, the former having to head back to Gibraltar and the latter sunk. This was disastrous because it had been arranged that fighter

27. The carriers *Victorious, Indomitable* and *Eagle* exercising in the Atlantic in August 1942 immediately before the 'Pedestal' Malta convoy. The photograph was taken from *Victorious*'s masthead by her radar officer, Lieutenant (Sp.) J. C. Maynard, RCNVR. The after aerial of WA type 279 can be seen in the foreground.

(J. C. Maynard)

cover should be provided by Beaufighters from Malta, controlled by *Nigeria* and *Cairo*, both specially fitted with VHF radio before sailing.[30] During the night, Italian MTBs sank *Manchester* and four merchant ships. Then enemy aircraft attacked at dawn, fighters from Malta being at a grave disadvantage without fighter direction ships. Five merchant ships entered harbour, at least two in a sinking condition, and 32,000 tons of stores were unloaded, as well as the precious oil from the tanker *Ohio*, kept afloat by a destroyer on either side.

'Pedestal' highlighted the deficiencies in the Fleet's fighter direction and air warning radar arrangements – the need for better communications, plotting and display arrangements to cope with the saturation problem; and the need for radar equipment which could measure heights and detect low-flying aircraft. For the first of these, some plans existed already and *Victorious* and *Formidable* had a prototype Aircraft Direction Room layout hastily installed before sailing in November to provide air cover for the invasion of North Africa, Operation 'Torch'. Providing low-cover and heightfinding radar was a much longer-term problem which will be mentioned in the next chapter.

IMPROVED SURFACE COVER: TYPES 271/3Q

In August 1941, Professor Oliphant's team at Birmingham University had made another breakthrough in valve design when J.Sayers, working with Randall and Boot, had invented the idea of 'strapping' the magnetron, allowing a five-fold increase in efficiency. As a result, GEC produced the CV56 strapped magnetron, enabling ASE to increase the transmitting power from 5kW with the NT98 to 70kW with the CV56.

When the first of these magnetrons appeared at Eastney, development of a more powerful modulator than that used in the original type 271 was already in hand, but forcing even more power out of the previous magnetron NT98 was a sledgehammer process. Could the delivery of this new magnetron short-circuit that process? The new valve was fitted in an experimental set in a trailer on the shore at Eastney and there was great interest to see just how large an echo would be returned from the standard target, the Nab tower. But no echo was seen: all that happened was that the polythene in the solid-cored coaxial cable melted. Fortunately, exploratory work on waveguides at ASE had been progressed just in time to enable the coaxial cable to be replaced by a waveguide feed capable of withstanding the higher power.

The main design features of the new Mark 4 sets (types 271Q and 273Q) were settled by December 1941 and the Signal School's workshops at Eastney immediately began to construct three prototypes. At the same time, Marconi and Allen West were commissioned to produce ten more prototypes each, based on ASE sketches and drawings.

The principal features were these:

- design so that they could directly replace type 271/3P, using the same office and aerial enclosure and similar sized panels and aerials;
- the strapped magnetron CV56, permitting 60–100 Kw peak power, with two possible pulse lengths, short for good discrimination and accurate ranging, long for greater power and longer ranges;
- a greatly simplified and more powerful modulator, using the discharge-line principle;
- waveguide instead of coaxial cable feed to the transmitting aerial, equivalent to doubling the transmitting aerial gain.

The first Eastney prototype set that was ready went to the corvette *Marigold*, to replace one of the earliest 271s. The conversion went without a hitch and sea trials on passage from Liverpool to Tobermory in May 1942 proved successful, an ocean-going submarine trimmed down being detected at 5 miles, a Focke-Wulf aircraft at 100 feet at between 10 and 12 miles.[31] Incidentally, the first trials of the new Hedgehog ahead-throwing anti-submarine mortar took place at the same time.

The trials in July of the second prototype, type 273Q, in the fleet flagship *King George V*, were equally successful. The cruiser *Cumberland* was held out to 28 miles, the destroyer *Marne* to 14 miles, a small submarine to 7.5 miles and a Sunderland aircraft at 5,000ft followed out to 27.5 miles and seen at 30 miles when turning at a lower altitude. The battle cruiser *Renown* gave an echo that was four times the amplitude of the echo of the same ship on *Prince of Wales*'s type 273P.[32]

Once again Admiral Tovey was delighted and his letter forwarding the trial results to the Admiralty was most complimentary to ASE:

> The results of these trials are highly satisfactory, those on low flying aircraft giving great promise for a future 'centimetre wave' set for the detection of low flying aircraft.
>
> Looking back on three years of war now past, it is amazing that such great strides of invention and production should have been made in so short a time. It is no exaggeration to say that RDF has revolutionised practically every aspect of sea warfare and its full effect is yet to be seen. Despite the difficulties to be expected and in fact experienced, in training the fleet to a new outlook on attack and defence and in the use of a completely novel aid to the armament, the high rate of supply and the general soundness of design has achieved a success that can be measured by the confidence placed on RDF by the fleet. This, I consider, reflects the greatest credit on those responsible for design, production and installation of Naval RDF gear.
>
> Jack C. Tovey[33]

Looked at nearly fifty years later, the Commander-in-Chief's kind words would seem to have been fully justified by events.

The wide variety of uses to which naval centimetric radar was already being put is well illustrated by the allocation of the twenty remaining Mark 4 prototypes. One went to replace type 271M in *Duke of York*, one for type 273Q for the new battleship *Howe*, two to 273s with stabilized aerials (one being *Belfast*), six to pre-fabricated type 271s, four to instructional establishments, two to the Army, one to mine warfare, one to contractors, and two to special applications.[34]

Contracts were immediately placed with industry for 1,000 sets, of which 400 were for the Army. Production commenced in December 1942.[35] General fitting of types 271Q and 273Q to replace types 271/P and 273/P was authorized in January 1943. Presumably because the next generation of 10cm sets (the Mark V series, types 276/277, described in the next chapter) was imminent, type 272P sets (with aerial remote from the office) were never upgraded to type 272Q.[36]

THE BATTLE OF THE BARENTS SEA

To wind up this chapter, let us return to Arctic waters to see the part our radar played in the convoy battle that has come to be called the Battle of the Barents Sea. On 31 December, convoy JW.51B was approaching Russia, escorted by destroyers and corvettes under Captain R. St. V. Sherbrooke in *Onslow* and covered by the cruisers *Sheffield* and *Jamaica* under Rear-Admiral R. L. Burnett. Attacked by the pocket battleship *Lützow* (ex-*Deutschland*), heavy cruiser *Admiral Hipper* and six destroyers, the convoy escort fought a delaying action (for which Captain Sherbrooke was awarded the Victoria Cross) until our cruisers made contact, forcing the German force to withdraw. The British losses were the destroyer *Achates* and minesweeper *Bramble*, while the Germans lost one destroyer and *Hipper* was damaged. The convoy escaped virtually unscathed.

Written immediately after the action, the Commander-in-Chief, Home Fleet's remarks sum up the part played by radar:

> In the close escort no surface warning of the enemy was given by either type 286 or type 271, but, with a visibility of over 7 miles, this was not surprising. *Achates* [with both 271 and 290] reports that, after being sighted, *Hipper* was held up to 8.5 miles which was probably the maximum range possible.
>
> Type 273 enabled the cruiser force to detect and plot two enemy units, probably *Lützow* and *Hipper*, and make a surprise attack on one of them. At this time, the enemy were undoubtedly concentrating their

attention on the targets to the southward (the destroyer escorts) and were thus caught unawares. The cruiser force was similarly surprised by an enemy destroyer later on.

The vital necessity for maintaining a vigilant all-round lookout by RDF cannot be over-stressed. The tendency to use type 273M as a rangefinder – made particularly strong by its excellent ranging panel – must be guarded against . . .

In ships of the cruiser force, type 284 gave entirely satisfactory results [though] it is unfortunate that neither ship made use of RDF spotting. Type 285 in destroyers gave satisfactory results.

Type 252 IFF in *Sheffield* responded to *Kent*'s interrogator at a range of 12 miles. Apart from this one success, the set appears to have been ineffective.

. . . British RDF sets stood up well to the test of battle, only one set (*Jamaica*'s 273) being reported to have broken down owing to gun blast. [It was below the bridge, abaft 'B' turret; see p. 112.] The excellent results expected from type 273 were fully borne out in practice. It is becoming increasingly evident that a set of this type is required not only in cruisers and capital ships but also in fleet destroyers.

It is clear that the personnel in *Sheffield*'s plot during the action carried out their duties with considerable skill and were successful in passing out a great deal of valuable information. In a modern capital ship or cruiser, the amount of RDF information now available for plotting is fast outstripping the ability of one officer to compete with it. When this is combined with a large number of enemy reports and RDF reports from other ships, there is bound to be some lag – if not actual confusion – in the plot. The matter is under investigation in the Home Fleet and improved internal communications will do much to assist.

Finally, the outcome of this engagement might indeed have been very different had RDF not been available or if it had been jammed. No useful results were obtained from optical rangefinders.[37]

TAILPIECES

• By Rear-Admiral S. E. Paddon, RCN, former radar officer of *Prince of Wales*, who joined *Warspite*, Eastern Fleet flagship, in January 1942.

Another story of Admiral Somerville I can give exactly as it occurred, involving a message he made in my presence. The background was this: *Warspite* was supposed to receive a new suit of radar, but much of it had been sunk en route to the West Coast of the States, and though the Americans did everything they could, by improvising here and there, I had a real bag of tricks on my hands when I finally got on board. There was difficulty with all of the sets. But this time I had some assistance. I had two New Zealand junior officer for only five radars. This was in marked contrast to my eleven radars and no help at all in the *Prince of Wales*. I was living the life of Riley.

However, we had real problems with our radar. One day, the Admiral, who was not too happy with the performance of the radar, sent for me, and said, 'Paddon, I want you to send a message to the Admiralty and tell them what our problems are.' So I prepared a very matter-of-fact, deliberate, carefully worded message on what we needed to put the equipment right, and took it up to the Admiral to sign and despatch. He took one look and said, 'That's not what I wanted.' Then, in its stead, he wrote the following message which I can quote: 'To Admiralty from Commander-in-Chief, Eastern Fleet. Herewith the state of the RDF in my Flagship *Warspite*. Type 281 minus interrogator resembles blind man without dog. Type 284 works up to 18,000 yards, provided the guns don't fire. Two types 285 are remarkable in their consistency in that they give negative results. Type 273 is the only radar I have. Admiralty, please help my flagship.'[38]

History does not relate what action the Admiralty took.

• While *Illustrious* was operating in the Indian Ocean, a section of fighters directed by her senior FDO, Lieutenant-Commander M. Hordern (now Sir Michael), RNVR, shot down a Japanese intruder which was shadowing the Fleet. To report this success through the intercom to Rear-Admiral Denis Boyd one deck above, the opening words of Hamlet's speech after discovering the dead Polonius sprang to Hordern's lips:

> Thou wretched, rash, intruding fool, farewell!
> I took thee for thy better . . .[39]

• Bob McCormick, from Ottawa, joined *Manchester* in 1941 as her first radar officer, and the ship eventually holed up in Scapa Flow. One day he went to the quarterdeck to find a heated argument going on between Guns and Torps. Both were RN, and at least one of them had two and a half stripes. A *King George V* class battleship was coming to the anchorage, but still hull-down.

Guns was saying that the battleship was *Duke of York*, and calling Torps all sorts of names because Torps thought it was *Anson*. Bob took one quick look at the far distant ship and said: 'You're both wrong. It's *KGV*.'

Thunk! The argument stopped, and then both 'professionals' started laughing. What was this *single*-striped, *green*-striped, *wavy*-striped *Colonial* doing interrupting a professional discussion between his betters?

Bob had never heard either of them laugh before, and he was nettled. 'What we should do is put in a pound apiece, and the one who's right gets the three quid,' he suggested.

'Done!!'

Bob had the last laugh, and the three quid of course. *KGV* was the only one of the class to have type 279 air warning radar, and, from almost as far as the eye could see, the masthead configuration could be distinguished from that of type 281, which was fitted in the others.[40]

8

1943: Sicily, Salerno and the Sinking of the *Scharnhorst*

Mar	Arctic convoys suspended
May	Battle of Atlantic turns in Allied favour
May 8	Adml Fraser C-in-C Home Fleet
May 13	Axis surrender in North Africa
Jul 10	Invasion of Sicily: 'Husky'
Sep	Eastern Fleet moves back to Ceylon
Sep 3	Invasion of mainland Italy: 'Baytown'
Sep 7	Italy surrenders
Sep 9	Assault on Salerno: 'Avalanche'
Nov 15	Arctic convoys resume
Dec 26	Sinking of the *Scharnhorst*

RDF BECOMES RADAR

At the beginning of 1941, the British Commonwealth had stood almost alone against the Axis powers: by the beginning of 1942, thanks to the aggression of Germany and Japan, it had two powerful allies, Russia and the United States. During 1942 and 1943, the Allied fortunes turned from almost uninterrupted disaster to almost unbroken success. This chapter tells the story of radar in the Royal Navy during the last of those years – when, thanks in no small measure to 10cm radar, airborne as well as seaborne, and to HF/DF, the Battle of the Atlantic turned in the Allies' favour; when, thanks largely to British metric radar and the development of naval fighter direction, the German air force was denied air supremacy over the waters of the Mediterranean; when the Royal Navy's Eastern Fleet in the Indian Ocean was greatly strengthened and the United States Navy in the Pacific began to push back the Japanese.

Within the compass of this chapter, radar changed its name – by Allied agreement – within the British fighting services. The order went out in June 1943 that, as from 1 July, the term 'radar' was in future to be used instead of 'RDF'; RDF Officers were to become Radar Officers; Petty Officers (RDF) were to become Petty Officers (Radar).[1]

A month or so earlier, the Navy had abandoned the 1938 inter-service abbreviations for the type of set – SS for ship-to-ship, SA for ship-to-aircraft and so on. While the Army and RAF continued to use CD, CHL, GCI, ASV, etc., to describe the purpose of radar sets, the Navy changed to the abbreviations WA, WS, GS, and so on, as defined on p. 340.[2]

RADAR DEVELOPMENT PRIORITIES

Within ASE, in parallel with the massive ship-fitting programme of the first generation of naval radar sets which demanded so much support, development teams had been progressing a great variety of new techniques in all radar areas. During 1942, pressure on resources had been building up so that agreement was necessary on future priorities for this work. On 17 December 1942, an important Admiralty meeting took place under the chairmanship of Rear Admiral R. R. McGrigor, Assistant Chief of Naval Staff (Weapons), to consider future radar priorities. It is interesting to look at the items placed in Priority A – 'items which will proceed at high priority simultaneously, and for which dated commitments can be accepted' – because these decisions were to govern most of the work at ASE almost to the end of the war, though, of course, additional items arose and a few priorities changed as the war proceeded. The dates that the projects mentioned actually became operational after their sea trials are noted in square brackets after each item :

Priority A

1. Maintaining the performance and reliability of existing sets.
2. Skiatron for *Indomitable* [1943].
3. 272/3 Mk.V. [10cm surface warning; type 276 1943, type 277 1944].
4. 274 [10cm surface gunnery; 1944].
5. 275 [10cm HA/LA gunnery; 1945].
6. Fighter direction (beam sharpening for 281 [abandoned]; transmission of height [1944 with 277]; display of range without interrupting spinning [1944]; Dr Bohm's height-finding).
7. Identification in connection with items 3 to 6 [IFF III; 1943]
8. Target indication for close range weapons.
9. Blind fire for Bofors Mk.IV and STAAG [abandoned].
10. Blind fire for MGBs without accurate ranging [abandoned 1943].
11. Shipborne beacon to replace 251M.
12. 3cm technique [types 262 and 268, 1945].

28. The radar officer of the famous Second Support Group, Lieutenant (Sp.) Ceri Fisher RNVR, transfers in a 27ft whaler under oars from *Starling* to *Woodcock* in the Bay of Biscay in 1943 to mend the latter's radar. Such transfers in mid-ocean were commonplace for radar officers of escort groups in the Battle of the Atlantic, among whom was the future President of the Royal Society, Sub-Lieutenant (Sp.) George Porter, RNVR, in the sloop *Rochester* from 1942 to 1944.

In his first convoy with Commander F. J. Walker, OG.78 early in 1942, Fisher transferred no less than six times between Liverpool and Gibraltar, keeping all the 36th Escort Group's radar working. Between 1942 and 1944 he witnessed the destruction of 23 U-boats, more than Walker himself. The photo was taken by *Starling*'s gunnery officer.

(Alan Burn)

29. Massachusetts Institute of Technology Mission visits the Admiralty Signal Establishment, Lythe Hill House, Haslemere, in July 1943. *Standing left to right*: Sir Henry Tizard; [USN]; J. D. S Rawlinson, ASE; . . . .; Sherston Professor of Chemistry, Canada; Cdr. G. F. Burghard, RN, ASE; Sir George Thomson; [Captain RN]; G. M. Wright, Chief Scientist, ASE; Commander Andrew Yates, ASE; [USN]; S.E.A.Landale, ASE; H.E.Hogben, ASE; . . . . . .; *Seated*: C. E. Horton, ASE; . . . . . .; Rear-Admiral C. S. Holland, DSD Admiralty; Rear-Admiral Fuhrer, USN; Captain B. R. Willett, Capt.Supt. ASE; . . . .; Dr Lee DuBridge, Director of Radiation Laboratory, MIT; C.S.Wright, DSR Admiralty.

(ARE Portsdown)

The Priority B list – 'items which will be progressed whenever staff and resources are available: dates are uncertain' – was as follows:

Priority B
13. Teachers [1946+].
14. Anti-jamming application to all sets [1944; S-band 1945].
15. Single-mast 281 with continuous rotation. (Note: to be further reviewed after results of Skiatron trials in *Indomitable* are known) [single-mast 1943, continuous rotation 1945].
16. Auto-following for STAAG mounting [type 262, 1945].
17. 273 Mk. VI.[Type 295 discussed on p. 186 below; 1951 as types 982/3].
18. 279M: reduction of spurious frequencies.
19. Auto-following for 275 [developed but not adopted for service]
20. RDF rangefinder sets for gyro sights [not followed up].

Then followed six items in Priority C ('intermittent at low priority') and two at Priority D ('under consideration').

Claiming that ASE was understaffed compared with TRE, Horton said that, to ensure that all Priority A items were completed in time, he must have additional staff of six experienced scientists or engineers, twelve assistants for them, six draughtsmen, and twenty-four mechanics. If Priority B were to be done as well, he would need a 50 per cent increase of both staff and space.

It was agreed that the Admiralty should put up a case for increase of staff to the Interservice Radio Board and that ASE should investigate the possibility of placing more work out to private firms.[3]

A month later, representatives of Messrs Ferranti Ltd (Mr V. Z. de Ferranti), The Gramophone Co. Ltd (Mr H. M. Bowen), and Metropolitan-Vickers Electrical Co. (Mr W. Symes) met at ASE to consider a list of sixteen items which ASE considered might be developed by Industry on the understanding that the firm undertaking development should also be responsible for production.[4] Unfortunately, no copy of the list has come to light.

ESCORT CARRIERS AND FIGHTER DIRECTION SHIPS

Before discussing the equipment for the air defence of the Fleet, let us briefly introduce two novel types of ship, the first of these being the escort carrier, the prototype for which, *Audacity*, was mentioned in Chapter 6. The second to be built in Britain was the converted merchantman *Activity* in 1942, followed by *Nairana* and *Vindex* in 1943, and *Campania* in 1944. Meanwhile US-built Lend-Lease ships joined the

30. The US-built escort carrier *Begum,* with American radar, in 1945. WA type SK aerial at the top of the lattice mast, WS type SG scanner on a spur just below, and aircraft homing beacon type YE on the top of the polemast.

(IWM)

31. The US-built escort carrier *Tracker*, with British radar. WA type 79 aerial at the top of the mast, aircraft homing beacon type YE on a spur below, lantern for WS type 272 above the bridge, and HF/DF on a polemast forward. The caption issued to the Press in 1943 says: 'A very noticeable list does not prevent physical training during the evening.'

(IWM)

escort fleet, starting with *Archer* in November 1941, followed between March 1942 and February 1944 by thirty-seven more, affectionately know as 'Woolworth' carriers. The first two British-built ships had types 79B and 272 radar, the last three, 281B and 272 (in *Campania*, 276 and 277 instead of 272). Of the US-built ships, the 'Ruler' class (*Emperor*, *Nabob*, etc.) had US radar – 1.5m SK air search, 10cm SG surface search, and US fire control radar – while the *Archer* class (*Biter*, *Pursuer*, etc.) had the British type 79B air warning and 272 surface warning. The US search sets had powered continuous aerial rotation and PPIs well before most British sets were so fitted, to the envy of many British officers. Many also had HF/DF, particularly valuable in that, when a U-boat transmission was heard, one or more aircraft could immediately be directed to search down the bearing. HF/DF was also used for homing aircraft such as Swordfish that only had HF radio.

Then there were the fighter direction ships – the converted merchantmen *Palomares* and *Ulster Queen*, originally classed as Auxiliary AA Ships for convoy escort, fitted with types 279 and 285. In 1943, they were renamed Fighter Direction Ships, joined by *Stuart Prince* and, as we shall see, given new radar.

RADAR FOR AIR DEFENCE

The principal tool of the fighter direction officer was air warning radar of which there were now two British versions available – the 7.5m type 79 or 279, and the 3.5m type 281. In 1943, a single-mast version of type 281 became available, called type 281B, to complement the type 79B and 279B which had first appeared in 1941, therefore suitable ships could now have two air warning sets, one on each mast. The first type 281B went to sea in the new maintenance carrier *Unicorn* in March 1943. Because of the general fitting of 50cm gunnery sets, type 279 no longer needed its ranging unit, so it reverted to its original name of type 79 (or 79B when with one mast).

Although type 281's shorter wavelength gave better low cover than type 79, it also lowered the tip of the first lobe (see vertical coverage diagram in Appendix C) so that it only gave cover up to about 28,000ft as compared with 40,000ft for type 79. With two-mast working, cover was reduced still further by the beam-switching facility (to give bearing accuracy of about half a degree) which, as originally designed, could not be switched off, though a modification was soon made to allow the operator to have the choice of either maximum detection or beam-switching on a target of reasonable strength. (With single-mast working, beam switching was not possible anyway.) These factors meant that aircraft could fly over the first lobe, with detection by the second lobe

32. The carrier *Indomitable* in March 1943 just after her fighter direction arrangements had been modernized as a result of experience in 'Pedestal', being fitted with two WA sets, types 281B and 79B, PPIs and skiatron (see p. 171). The aerial of the WA type 79B aerial can be seen abaft the island with that of type 281B off the top of the picture. The tripod mast is surmounted by the 'dustbin' of type 72 aircraft homing beacon, the British equivalent of the US YE. The GA type 285's 6–unit fishbone aerials can be seen mounted on the HA Director, with three 2–unit aerials of GC type 282 mounted on pom-pom directors below. The lantern of WS type 272 is mounted on the front of the bridge with the MF/DF aerial immediately above.

(Popperfoto)

roughly halving the first-detection range. This also confounded height estimation, as described in Appendix C.

In deciding whether to fit 281 or 79 where only one set was needed, the Admiralty had to weigh the virtues of the two sets one against the other:

Type 79/B	*Type 281/B*
• More reliable in operation	• Capable of continuous rotation
• Better in harbour or inshore	• Could operate PPIs and Skiatron
• Better for high cover	• Better for low and medium cover

Taking these factors into account, it was decided that, from mid-1943, the policy should be for large aircraft carriers and special fighter direction ships to have both 281B and 79B; other large ships in the Fleet to have either 281B or 79B approximately in the ratio 2:1; depot ships and others who spent much time close to land to have type 79B; existing ships with two-mast versions to convert to the one-mast version as soon as convenient.[5]

Early FDOs could not see a radar display direct but had to rely upon reports by voicepipe or telephone from the operator at the A-scan in the radar office, but from 1943 PPIs began to be fitted close to their 'intercept position', so they could see the radar information at first hand. The face of a PPI could be used for plotting the tracks of targets relative to own ship with a wax pencil, though the small diameter of the tube resulted in a rather small-scale picture which could soon become cluttered in any complicated situation.

The Skiatron display was a development of the PPI to simplify immediate and direct plotting of radar targets and of tracks and vectors, without the room having to be darkened. Echoes in PPI form, from a special CRT, were projected optically upwards onto a translucent circular ground-glass screen, and appeared as dark shadows (or 'stains') on a light background. The ground-glass screen was engraved with a circular 'spider's web' of some 24in in diameter (as against the normal British naval PPI's diameter of 9in), and could be used as a plotting surface using a wax pencil, for plotting tracks of aircraft either from actual radar echoes or by dead reckoning.

The following remarks, written by the pioneer fighter direction officer David Pollock in 1948, are of interest at this point in our story:

> In 1942 permission had been refused to *Illustrious* to fit an RAF PPI on the ground that the broad beams of type 281 would create saturation. Subsequently trials showed that while the broad beam gave a result which was not comparable with those given by the narrower beams of the GCI sets used by the RAF, yet practice enabled a fair interpretation to be made. Intercept officers reading their own PPI could obtain a better and quicker appreciation of a situation than from a plot told

from the ordinary [A-] scan. More plots could be told from a PPI and more raids plotted and intercepted.

The Navy in particular carried out extensive trials with the Skiatron and adopted it for general use as intercept plots. Many who had experience with it preferred the Skiatron, while others, and particularly those who had worked in RAF GCI stations, preferred the PPI.

Skiatron and PPI presentation brought night-fighter control within reach of HM ships. The first trials were carried out with a Skiatron in *Indomitable* in the Irish Sea in the spring of 1943. RAF aircraft were used and the results, although not up to the GCI standards to which the RAF were accustomed, were distinctly promising.

In height-finding and low detection, naval equipment was still far behind operational requirements. Another year was to pass before any equipment would be available.[6]

During the first half of 1943, two fleet carriers, *Illustrious* and *Indomitable,* were each fitted with two air warning sets, types 281B and

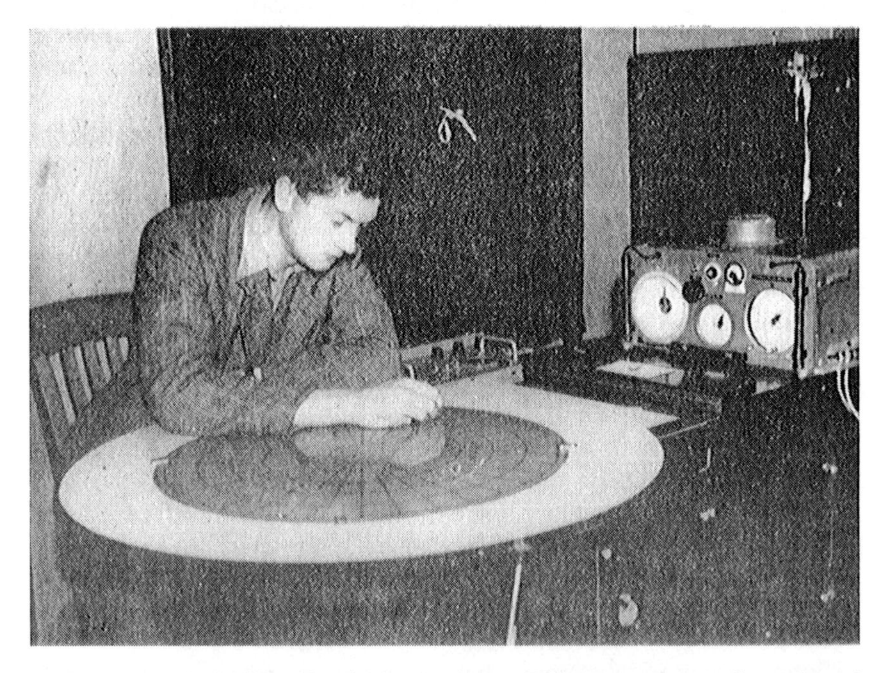

33. David Lloyd of ASE at the skiatron, where a PPI picture is optically projected onto the plotting surface from below, for use by a fighter direction officer when making aircraft interceptions. It first went to sea in March 1943 in *Indomitable.*

(Times Newspapers Ltd)

type 79B, as well as having their fighter direction arrangements modernized, taking into account the lessons learnt in Operation 'Pedestal'. PPIs were fitted in the aircraft direction rooms (ADRs) of both ships, and *Indomitable* did successful trials of the Skiatron, displaying type 281B, in the Irish Sea in June 1943 before sailing to take part in the invasion of Sicily in July.

For best presentation, the PPI requires continuous power rotation of the radar aerial, not available on type 281B. During 1943, ASE began designs to enable this to be done. Type 281BQ, as the modified set was called, came into service in 1945 and will be described in more detail later. The longer wavelength type 79B was never modified for PPI presentation because of its much wider beamwidth.

THE MARK III IFF SYSTEM

The Mark III system was devised by an RAF scientist F.C.Williams in 1940. In it, airborne and shipborne IFF would no longer respond directly to primary radar signals but only to special interrogators, in effect low-powered secondary radar sets. They operated in conjunction with main radar sets, so that the transmitted pulses were synchronized and the received signals displayed on the same tube as the radar echoes. A separate aerial, and usually separate power supplies, were also required with each interrogator. With the limited mast and office space available in ships, the introduction of IFF Mark III therefore involved heavy new commitments, particularly for the Navy at a time when all available effort was fully extended in fitting the Fleet with radar.

In spite of every effort having been made to change over to the new system as smoothly as possible, the task proved too great for the resources available. Although the introduction was carried out in stages, by changing over in one area at a time, it was quite impossible to ensure that all operational ships, aircraft and ground stations in a given area were fitted with the necessary equipment by the date agreed upon for that area. The ensuing period was generally one of great confusion, particularly in the western and central Mediterranean where the change-over coincided with a number of important operations. The uncertainty of radar identification caused particular anxiety at this time when the enemy's air effort was on a considerable scale. The AA cruiser *Delhi* reported that, during her month close inshore off Salerno, friendly aircraft displayed IFF Mark III, Mark IIN, Mark IIG, Mark II, Mark I, or nothing, apparently almost at will.[7]

As the fitting of IFF Mark III became more complete, and as operators and mechanics became more familiar with the new equipment, the situation gradually improved.[8]

THE SICILY AND SALERNO LANDINGS

The invasion of Sicily on 10 July ('Husky') and the assault on the beaches near Salerno south of Naples on 9 September ('Avalanche') were two operations in which radar played an important part. Many of the deficiencies shown up in the 'Pedestal' convoy and in the North African landings had been rectified, but many new lessons were learnt. In both operations, fleet carriers in Force H provided offensive and defensive air cover, *Formidable* in both operations, *Indomitable* in 'Husky', *Illustrious* in 'Avalanche'. The last two carriers now had two air warning sets each, PPIs, and upgraded fighter direction arrangements, all of which proved most successful, particularly the ability to use combined Vertical Polar Diagram charts for height estimation, as described in Appendix C. *Indomitable's* Skiatron (showing type 281B) was well liked though its use brought to light a dilemma: to show a proper picture on the Skiatron, the aerials had to be kept rotating continuously (in practice, 360° clockwise, then 360° anti-clockwise, and so on); but to measure echo amplitude for height determination or to display IFF responses, the aerial had to be stopped.[9] The eventual solution to this dilemma was the Sector Display, to be described in the next chapter.

During 'Husky', *Indomitable* was torpedoed by a low-flying aircraft at night, demonstrating once again the need for efficient low-seeing radar. The only ships present who had anything like it were two battleships and a cruiser with type 273, which obtained plots of the incoming unidentified aircraft but, because the radio policy in force that night did not allow silence to be broken, the attacker was not reported until it was too late.[10]

The landings on Sicily took place within easy reach of shore-based fighters but this was not the case at Salerno, where fighter cover over the beaches was mainly provided by Force V, assembled for the occasion under Rear-Admiral Vian in the light cruiser *Euryalus*, with the maintenance carrier *Unicorn* and escort carriers *Attacker, Battler, Hunter* and *Stalker*, temporarily employed as assault carriers, carrying only Seafire fighter aircraft. To assist in fighter direction over the beach-head, the fighter direction ship *Ulster Queen* with full fleet-carrier radar and a naval FD team led by Commander Coke (replacing the *Palomares* who had been torpedoed *en route*), and LST 305 with GCI radar and an RAF control team, were close inshore. (The LST had also taken part at 'Husky'.)

Reports on these operations bring out many points of interest:

- the difficulties of operating metric radar close to high land (nothing new here!);
- super-refraction, causing many spurious echoes;
- height estimation, difficult enough without being exacerbated by the above two factors;

- the success of centimetric surface radar in detecting low-flying aircraft, particularly with type 273Q;
- the enormous value of centimetric radar ranges of points of land for navigation generally and during bombardment operations in particular;
- the qualified success of the Mark III IFF system, recently brought into operation in the Mediterranean theatre supposedly to supersede Mark II system – but see *Delhi's* remarks above;
- the use by the Germans of the new radio-controlled glider bomb which severely damaged the battleship *Warspite* and the cruisers HMS *Uganda* and USS *Savannah*, as discussed on p. 176 below.

The following, written immediately after 'Avalanche', tells how radar was used in 1943; though the extracts quoted mostly give adverse criticism of the equipment, it should be said that its good points were also reported upon, though, human nature being what it is, not so often:

- By *Illustrious* operating offshore:

Out of eleven interceptions attempted, only two were successful. The first of these turned out to be Beaufighters and the second an Italian SM 79 which was escorted back to Sicily by one of our Martlets. [Italy had signed an armistice on 3 September.] Failure to intercept was largely due to errors in estimation of height.

On more than one occasion, friendly and 'bogey' [unidentified] plots actually coincided on the PPI and the angle of interception seemed favourable, but no 'Tally Ho' was forthcoming . . . The main difficulty experienced was that of giving height on any aircraft above 15,000ft. Such aircraft invariably missed the first and quite often the second lobes in the case of 281 and the first lobe in respect of type 79.

This failure does not seem to be confined to this ship. It may be explained perhaps by the peculiar Mediterranean conditions . . . which seem to be borne out by the following:

- Saturation echoes of land appeared on 'A' scan up to 80 miles. Half saturation echoes of land were also visible at 120 miles. This seems to indicate that the first lobe is diffracted downwards.
- On the night of 8/9 September, 'bogey' aircraft were first detected at 25 to 30 miles. From reports it seems that these aircraft were flying at 70 metres . . .
- Phenomenal results with 271/2/3 series.[11]

- By the AA ship *Delhi* close inshore:

Perhaps what has been most clearly demonstrated in the month's proceedings reported is the proved inadequacy of the present radar equipment for close support of beach operations. Without labouring the technicalities, warning sets suffer from land echoes to such an extent that even if the radar guard situation admits of undivided attention to specific incoming raids, these are invariably lost before they can be transferred to gunnery sets; nor does the adequacy of communication between warning and gunnery sets enable the aerials of the latter (and armament) to be laid and trained beforehand in the position of best advantage.

The gunnery set is therefore only in an approximate position (usually without any height indication) to obtain a snap target, which it can rarely do, and the problem remaining is to engage the target by visual or aural observation according to the light or darkness prevailing.

Adding to this the completely nebulous situation which still exists as regards aircraft identification (since friendly aircraft display IFF Mark III, Mark IIN, Mark IIG, Mark II, Mark I, or nothing, apparently almost at will) advantage is lost by day in engaging targets beyond recognisable distance, and by night blind barrage fire only can be resorted to, with its unwelcome defensive character . . .[12]

GERMAN DEVELOPMENTS IN ANTI-SHIP GUIDED WEAPONS

On 27 August 1943, ships of the Royal and Canadian Navies conducting an anti-submarine sweep south of Finisterre were attacked by a number of Luftwaffe aircraft using a new weapon – an air-launched, powered missile known as the Hs.293 glider bomb, which was under continuous radio guidance from its parent aircraft. The Germans achieved complete surprise and the British sloop *Egret* was sunk and the Canadian destroyer *Athabaskan* damaged. At this time the only defence was close-range AA gunfire and smoke screens – but see Tailpiece on p. 190.[13]

On 9 September 1943, the Italian government capitulated to the Allies, just before the Allied assault on the Italian mainland at Salerno. The surviving main body of the Italian Fleet, comprising three battleships, six cruisers and eight destroyers, left Spezia for Malta to surrender to the Royal Navy. A squadron of the Luftwaffe equipped with *another* type of air-launched weapon, a radio-guided armour-piercing bomb known as the FX.1400, made an unopposed attack on the Italian Fleet on 9 September 1943 en route for Malta, sinking the battleship *Roma* and damaging the *Italia*. On the 11th, the same squadron attacked Allied shipping off the Salerno beaches with FX.1400 bombs, seriously damaging the US cruiser *Savannah* and the British *Uganda*, and on the 16th, the battleship *Warspite* was also very seriously damaged.

Following these attacks, Elint teams started the search for the radio control signals which might be involved in the new weapons, as will be discussed in the next chapter.

PLANS FOR SECOND-GENERATION 10-CENTIMETRE WARNING RADAR

Let us now continue the story of the development of centimetric warning radar, which we left on p. 158 above, discussing the production of the type 271/3 'Q' series with the 'strapped' magnetron, which began general fitting in 1943. As soon as this programme was under way (so ensuring that the fleet had the best possible surface warning sets in the shortest possible time) senior officers at ASE, civilian and naval, had in 1942 started to plan the next generation of centimetric sets, to take advantage of the many technical developments which had occurred since the design of the first type 271, and to do their best to satisfy the needs of the fleet after experience in three years of war. The submarine threat had been contained, at least for the time being, but the threat from the air had intensified and our ships were particularly vulnerable to attacks by low-flying aircraft, against which our existing radar was not effective. Fighter direction techniques were being developed and there was a great need for radar, not just for early air warning, but also to assist in fighter interception of enemy aircraft when between 70 and 15 miles of the Fleet, the so-called Fighter Interception Zone. Better methods of heightfinding were urgently needed.

By October 1942, the main features of the new designs had been settled and a detailed description was circulated to the Naval Staff, the introductory paragraphs of which were as follows; in this description, Mark III refers to the original series, Mark IV to the improved 'Q' series, Mark V to the proposed new series, soon to be renumbered types 276/7:

> RDF Type 271/273 Mark V [types 276/277] is being developed to replace Type 271/273 Mark IV [271/3Q] and Type 272 Mark III [272P]. The chief features of the new set will be:-
>
> (a) Considerably increased power, even compared to Mark IV.
> (b) A single mirror (AUK) [single paraboloid aerial for type 277] or sandwich (AUJ) type mirror [single cheese aerial for type 276] with common transmitting and receiving arrangements. The aerial system will be exposed, i.e. there will be no 'lighthouse' structure.
> (c) PPI presentation.
> (d) Stabilisation in azimuth.

(e) In the case of Type 273 Mark V [277], stabilisation in elevation also.
(f) Larger office and, in general, heavier equipment.

Although the production and fitting of Mark V sets is still some way off, and Mark V should be regarded as the 1944 edition of the WS set, the expected details contained in the enclosure are circulated now as a number of decisions will shortly have to be taken.[14]

Like the Mark IV series, the basic radar of the Mark V was common to both sets, the difference being in the aerials which were both to have continuous power rotation, so important for PPI presentation. Type 277 had variable rotation speeds up to 16 rpm with the ability to stop within ½° of a selected bearing to 'hold' a target, or for heightfinding. The type 276 aerial rotated at a fixed speed of 10 rpm. Unlike previous 10cm sets, common transmit/receive working was incorporated for the exposed continuously-spinning aerials.

The Naval Staff paper continued, saying that the tactical uses of the two sets would be: (a) surface warning; (b) aircraft warning below 3° with the type 277, below 20° with the type 276; (c) fighter interception of aircraft at elevations below 3° (out to 16 miles with aircraft at 50 feet, 45 miles at 5,000 feet); and (d), with type 277 only, heightfinding by measurement of elevation (bracketing, not beam splitting) to an accuracy of 1° between elevations of 4° and 40°. For gunnery, both sets could be used for indication of surface targets and might in the future (with the cooperation of the Director of Naval Ordnance) be used for indication of aircraft targets. They could also provide accurate ranges to the main armament in an emergency. The paper also forecast what proved to be somewhat over-optimistic range performances against various types of target.[15]

Before that paper was written, the first prototype of the new transmitter and receiver had been exhaustively tried out ashore from mid-1942 in a cabin trailer with a cheese aerial fixed on the roof, with the whole cabin capable of being rotated, but very slowly. This was type 277T, first at Selsey Bill, then on the cliffs at Ventnor in the Isle of Wight, then to Great Orme in North Wales – all of which trials apparently fully justified the aspirations of the designers of the seagoing versions of the high-power radars, though it has to be noted that the trials were performed holding the aerial on the target bearing, whereas the motor-driven aerials of the seagoing types 276 and 277 would be spinning at several revolutions per minute.

A brief digression concerning another prototype 277T . . . In March 1943, there was an account in the Press of how three hundred school children at a school in Ashford missed death by seconds when their school was destroyed by twelve FW190s. Much stress was laid on the efficient organization that enabled the entire school to take cover within a minute of the alarm being given. In fact the raiders were picked up by Capel shore station type 277T near Dover, at a range of 20 miles within two minutes of their having crossed the French coast. The plot was told to the RAF's Stanmore filter room and identified as hostile. As a result the sirens were sounded within 70 seconds of the first plot and 2½ minutes before the planes crossed the English coast, which gave just sufficient warning for people to take cover. ASE said that trials had shown that the set, 550ft above sea level, was capable of detecting wave-hopping aircraft at 35 miles, 'the first real "answer" to the low-flyer problem'.[16]

The original 1942 concept was that there should be two sets with the same electronic equipment but different aerials: type 276 (with a cheese aerial fixed in elevation horizontally) for the smaller cruisers and fleet destroyers; and type 277 (with a paraboloid aerial which could be set to any elevation and was capable of measuring elevation) for capital ships, aircraft carriers, larger cruisers, and all escort vessels able to bear the topweight of the larger aerial. However the need for target indication of aircraft for gun direction (as discussed on p. 182 below) became most urgent, and this gave rise to the need to extend air cover from the 20° of type 276 to a 70° angle of elevation. To adapt developments to this new specification implied design of a new very wide fan-beam aerial and a special display unit (TIU) for preliminary target tracking. In addition, it seemed eminently reasonable at the time to assign to this target indication radar the dual role of target indication (TI) and surface warning (WS), thus greatly easing the masthead siting problems in the smaller war vessels. So the development of the type 276 aerial was dropped and the modified set, with a tipped-up aerial and renamed type 293, was substituted for type 276 in forward fitting plans.

In July 1943, the Fleet was told of the Admiralty's plans for the future fitting of centimetric warning radar. Large cruisers and above were to have type 277 for WS, and type 293 for TI. Small cruisers, destroyers, AA sloops, etc., were to have type 293 to serve for both WS and TI. To allow this to be done, all future destroyers would be fitted with a lattice foremast to accommodate also the HF/DF aerial. However, as discussed below, until the redesigned type 293 aerials were available, type 276 was to be fitted for TI as an interim measure in all classes.[17]

The first sea trials of the high-power equipment, the laboratory model type 277X with a solid paraboloid mirror aerial and rudimentary stabilization, took place in the trials ship *Saltburn* in April 1943, together with a laboratory model of EMI's 'JE' PPI. Sea trials of the target-indication type 293X with tipped-up cheese aerial (discussed on p. 182 below), together with the first pre-production 'JE' PPI, took place in the destroyer *Janus* in August, both trials being strictly experimental, as indicate by the 'X' suffix. These sea trials, especially those of type 293X, were planned to determine experimentally the shape of the air cover diagram and the surface performance.

The results for type 277X were much as had been predicted, but as regards type 293X, although high air cover had been achieved, the range of the radar fell far short of that required for either surface warning or target indication. Target indication was possible for close range weapons only and indeed operational trials of type 293X in *Janus* had to be continued until the end of 1943 to determine rates of change of bearing, ability to discriminate between multiple targets, and other factors critical to gun direction. However, at least one flotilla of fleet destroyers was equipped with the original version of type 293.

Meanwhile, ASE had realized in September 1943 that development of a higher-gain aerial for the dual-role type 293 would take some time and that some interim action was needed if the many ships building or refitting were not to be denied effective surface and low air cover radar. Although specific development of type 276 had been dropped in favour of type 293, there existed in the type 271Q transmitting aerial a design that could easily be adapted to type 276: the 293 pedestal was suitable, the mirror was already waveguide-fed and optimized for gain, so the only technical change needed was to seal the mirror aperture for exposed aerial working. After prototype trials in *Tuscan* in November 1943 had demonstrated adequate surface performance and air cover up to 10,000 feet out to a reliable 10 miles or more, this type 276 began to be fitted not only for surface and low air warning, but also as an interim TI radar pending availablility of the new aerial for type 293, a situation that was to persist until the first fitting in January 1945 of type 293M, whose larger-aperture 8ft × 7½in aerial, tilted up 15°, satisfactorily provided high air cover while retaining adequate surface capability.

General fitting of types 277 and 276 began in 1944, though *Virago* was fitted with type 276 in time to take part in the *Scharnhorst* action in December 1943. Type 276 was to become obsolescent in 1945 as soon as type 293M became available, though many ships – the carrier *Formidable* for example – still had type 276 when the war ended.

10–CENTIMETRE RADAR TO 1943: A SUMMING-UP

In November 1940, a development programme was launched with the primary objective of the detection of surfaced submarines. It was clear from the beginning that the magnetron immediately available could be made to provide more power as techniques evolved and therefore a basic element in the programme was the development of modulators ultimately to give 1MW driving power. Before the middle of 1942, the ultimate target of this programme had been clearly visualized (certainly by Landale during the *King George V* trials[18]) – type 277 with its 500kW transmitter, exposed power-driven aerial, and plan display. On the way, the programme spawned three times – first at 5kW in mid-1941; then at 70kW in mid-1942 (271/3/Q); then at 500kW in the spring of 1943 (276/277T). As we shall see in the next chapter, the seagoing version of type 277 appeared in the spring of 1944.

From the beginning, there existed a guiding philosophy that the only useful place for major improvements in performance was in ships: admiring laboratory achievement had no added value. There is no doubt that this programme gave a solid foundation to 10cm development for all Allied nations. Equipment from this programme was used by all three services and urgently sought as soon as production commenced. Nor should it be ignored that its conception was early and at a time when there could be the minimum of feed-back from operations to aid technical judgement on possible developments. The naval designers were working on the fringes of current scientific knowledge and technological capablility.

By the end of 1942, the Navy had beaten the early radar into submission and attitudes had changed dramatically. Radar was not just another burden, it was essential equipment. ASE had grown and reorganized. Acting on hard-won war experience in the Mediterranean and Atlantic, the Naval Staff became vocal and made known what was wanted from radar. For 10cm radar, this launched two wing programmes to the core programme for type 277 – first, target indication for weapon control, second, dedicated fighter direction radar, both to be considered below.

The developments in the Navy's 10cm radar in 1942 and 1943 were truly remarkable, perhaps best illustrated by contrasting actual operational results at the beginning and end of that period. In the Straits of Dover in February 1942, the flotilla leader *Campbell,* fitted with an early version of type 271 abaft the bridge, did not detect the battle cruiser *Gneisenau* until she was 9.5 miles away, when almost the whole of her hull and superstructure would have been visible from *Campbell*'s radar aerial behind the bridge. Off the North Cape in December 1943, the battleship *Duke of York,* fitted with type 273Q high up on the foremast, detected *Gneisenau*'s sister ship *Scharnhorst* at about 22 miles, when only a few square metres of her control tower were over the horizon.

TARGET INDICATION AND GUN DIRECTION REQUIREMENTS

By 1943, some larger ships had as many as eleven gunnery radar sets. However, the tactical handling of weapons in the Fleet at sea – gun direction – remained rudimentary, with slow assessment and very limited ability to make the best possible use of the weapons under coordinated air attack, developed by the German Air Force with considerable success in the Mediterranean during 1941 – and which the Japanese Navy was bound to use also. A fast response with good assessment and rapid coordination of all available weapons would be necessary to realize the full capability of the weapons now available and prevent wasteful overlapping or dangerous gaps in the assignment of defences.

An important development during 1942 was the PPI and, by 1943, plans were being made for them to be placed in various parts of the ship such as the bridge and plotting positions (later to be known as the Action Information Centre), allowing the command team to appreciate at a glance the situation as seen by the warning radar.

During 1942, it had been recognized that the characteristics of the existing metric air warning radars were not ideal for gun direction in fast-moving situations. Whilst radar coverage over the full range – from high altitude to surface – had been greatly improved, the wide beams (35° to 75°) and slow rotation rates of metric radars (2 to 4 rpm) resulted in large errors in establishing the present position of targets and made quick assessment of the overall situation difficult. These difficulties were emphasized in 'Pedestal' and, in December 1942, a Naval Staff Requirement was put forward for a specialized target indication (TI) set, based on the new generation of 10cm warning radar. It was decided that the new TI radar should be based on type 276, with its PPIs, azimuth stabilized displays, and 500kW transmitter. It would have a different aerial to provide the very wide fan beam described on p. 179 above. In June 1943, plans were made to carry out trials of this tilted cheese aerial in the fleet destroyer which had a new lattice mast capable of carrying it. It was part of the overall plan for smaller ships which suffered from topweight problems that this radar had to provide both surface and air warning, and, *faute de mieux,* even occasionally be used for fighter direction.

Extensive operational trials of *Janus*'s experimental type 293X, including comparison with type 276 later in the year, amply demonstrated the value that could be obtained from a centimetric radar with a narrow horizontal beamwidth (3.5°) and 10 rpm spin rate with 0.7 microsecond pulse length, as a basis for target indication and gun direction. However, ranges were below expectation both for surface targets and for aircraft. Two operating conditions were novel: the use of 10cm radar against aircraft which both reduced echo size and greatly increased signal fluctuation; and power-driven continuously rotating aerials which illuminated each target for only a short time with each sweep of the aerial, thus limiting the number of pulses received from each target. Both factors significantly reduced detection range performance, which could only be restored by increasing the size of the aerial aperture from the 6ft × 4in used so far to keep weight and windage as small as possible.

In view of these hitherto unknown factors, fitting plans had to be modified; cruisers and above due for type 293 were fitted with type 276 to ensure maximum surface capability as well as using it temporarily in a target-indication role. Some destroyers were fitted with type 293 and some 276, pending design by ASE of a new higher-gain aerial for the former. This became available at the end of 1944, having an 8ft × 7½in aperture, whose higher gain resulting from the narrowing of the vertical beam and reduction in the horizontal beamwidth to 2.6° gave increased range performance. Range performance thus being solved for destroyers, design began in 1945 of a yet higher-gain aerial with 12ft horizontal aperture to improve performance for cruisers and above.

The experimental trials in *Janus* were completed in October 1943 and plans were made for the first installation of a Target Indication Room in the new 'Battle' class destroyer *Barfleur* which commissioned for trials in September 1944. The next chapter will discuss the complementary device – the Target Indication Unit Mark 2 (TIU2) which enabled the targets – evaluated and selected from the 293 displays, to be passed quickly to the chosen weapons without the need for voice transmission of data, so eliminating another source of delay and error.

10-CENTIMETRE GUNNERY RADAR

The first discussions on the possibility of a 10cm surface gunnery set for the main armament had taken place as early as June 1941, even before the first big-ship 10cm surface warning set had been fitted in *Prince of Wales*.[19] On 16 October 1941, at an Admiralty meeting chaired by Captain G.M. Langley, director of the new Gunnery Division of the Naval Staff, the projected new set was designated type 274, and outline staff requirements were laid down, demanding 'equipment which will permit accurate blind firing to be carried out at extreme range against a selected ship of a number of ships in close order',[20] extreme range being defined as 40,000 yards (20 miles) battleship to battleship, 32,000 yards (16 miles) cruiser to cruiser. The set was to be fully integrated into the fire control systems with the aerial mounted on the main armament directors; able to be applied to existing ships as well as to future construction; and be in production by 1942.[21] All of this was achieved, except that, as we shall see, the first sets did not go to sea until 1944.

The existing main armament directors were suitable for carrying the cheese aerials needed for type 274, which was designed to ensure adequate signals from destroyer targets at maximum gun range. It was also hoped that shell splashes at maximum gun range would give adequate signals so that spotting corrections could be applied in blind fire. To obtain accurate 'blind' bearings, beam-switching in azimuth was incorporated. In the event, the bearing accuracy against small targets (though not on larger targets) was better than ±3 arc-minutes (the width of a destroyer end-on at about 13,000 yards). The peak power output of the transmitter was 500kW and, because of the difficulties of switching this high power, separate aerials were used for transmission and reception, beam-switching being applied to the receiving aerial only. The range displayed was 48,000 yards (24 miles) and a precision ranging system was incorporated, as in type 284P, to give a range accuracy of about 25 yards, varying cyclically over each thousand yards but not increasing with range. The half-microsecond transmitter pulse improved range discrimination to about 80 yards. The 14ft cheese aerials gave a beam a little over 1° wide horizontally, compared with the 5° of type 284, which greatly improved angular discrimination as required by the 1941 Naval Staff Requirements.

To spot fall of shot, type 274 was fitted with a special expanded range display showing only 2,000 yards of the range scale. From it, the operator could estimate fairly accurately the difference in range between the target and the mean point of impact (MPI) of a salvo. This device did not, however, permit the operator to quantify the MPI's 'error in line', it being possible only to estimate roughly whether the MPI was to the left, to the right, or 'on' the line of sight. The proposed solution to this problem was an additional small scanning radar working at 1.25cm – type 931 – which will be discussed in a later chapter.

BLIND FIRE REQUIREMENTS

Blind fire against aircraft posed an additional problem – the need for precise measurement of elevation as well as of bearing. This demanded a centimetric wavelength but it was not possible to design a 10cm equipment with the size of aerial that could be fitted on *existing* HA directors and which would at the same time give adequate range performance. Accordingly, the design of a new 10cm radar set, type 275, and the design of a *new* HA director specially for it, were put in hand in 1942. Opportunity was taken to introduce into the new Mark VI director remote power control and stabilization against ship motion, as well as to provide space for two 4ft diameter aerials with elevation motion, and for the transmitter and centimetric parts of the radar receiver. All this was a major engineering task and required more than two years for development and production, so that it was not until late in 1944 that type 275 was first fitted for trials in the destroyer *Barfleur*.

Types 275 and 262 (described below) were the first British naval radars to use conical scanning for tracking the target in bearing and elevation. The requirement was that it should be capable of *picking up* a target at 36,000 yards (18 miles) and be able to *track it accurately* from 30,000 yards (15 miles). Owing to the limitation on the size of the aerial mirrors, under certain conditions tracking was possible only under 25,000 yards (12.5 miles). As time was insufficient to develop a conical scanning system for the very high voltages involved in transmission (400kW peak power output), separate circular parabolic aerials had to be used for transmission and reception. The basic radar set was almost identical with type 274. Training and elevation operators below decks in the HACP (or TS in destroyers) were provided with separate displays and meters showing the deviation of the target from the aerial axis in bearing and elevation, permitting tracking in both planes by remote control of the director from below.

With conical scanning, the inherent accuracy of this system should have been about 2 arc-minutes, but the aiming of the complete director through its power-control system, together with stabilization problems, degraded this to 8 arc-minutes at best. This was to have a profound effect on the design of later radar-controlled systems, but, despite this degradation, type 275 was the first accurate radar system for HA gunnery at longer ranges capable of full blind target-tracking in bearing and elevation.

As with type 285 which it was designed to supersede, type 275 performed both surface and anti-aircraft functions when fitted in destroyers without a special surface gunnery radar.

CLOSE-RANGE GUNNERY

In 1940, the 40mm Bofors gun had been introduced into the Navy and it was planned to introduce an advanced short-range gun system, based on a self-contained mounting with a gyro-controlled rate-measuring sight and hydraulic power control with on-board fire control, also hydraulic. The prototype of this was completed about the time that ASE was developing a 'search-and-lock-on' 3cm radar, to be compact and able to find and track the target automatically in all three coordinates, blind. It was to use a 2ft-diameter paraboloid aerial – spinning off-axis at 30 cycles per second to generate the auto-aiming signals – and to have an auto-ranging system.

In 1942, it had been decided to modify the new Bofors gun system then being designed – known as the STAAG (Stabilized Tachymetric Anti-Aircraft Gun) – to carry two Bofors guns, and the new radar type 262 on the mounting to provide blind fire. The whole would be fully automatic and entirely self-contained, with all the equipment – guns, predictor, radar – on the mounting itself. Such a system was particularly suitable for smaller ships which could not carry the type 284 or 285 systems, and which did not have the space for separate guns and fire-control directors. It could also supplement the AA systems of larger ships by providing close-range systems with a better performance than simple pom-poms with type 282P. Another version of type 262 was to go into service with the radar mounted on a small director known as CRBFD (Close Range Blind Fire Director) to control separate guns in barrage fire at short ranges.

Meetings took place with EMI Research Ltd. in December 1942, and ASE provided a brief specification for an experimental radar to be developed by the company in close cooperation with ASE – type 262X. The work progressed rapidly during 1943 and the system was ready for non-firing trials early in 1944.

PLANS FOR FIGHTER DIRECTION RADAR

Bitter experience in the 'Pedestal' convoy in August 1942 prompted the Admiralty to take the most urgent steps to improve the ability of the Fleet to defend itself against air attack. For the longer term, the Admiralty issued Staff Requirements for three new radar systems:

- *type 960* – a 3.5m air warning set giving ranges up to 160 miles, to replace types 79 and 281: the firm of Marconi was awarded a development contract in August 1943;
- *type 990* – a 10cm low air cover set to give horizon ranges to 150 miles or so;[22]
- type 294 or 295 – a special fighter direction radar for carriers and fighter direction ships, to provide gapless cover to the limit of the Fighter Interception Zone, 30,000ft out to 70 miles, specifying a heightfinding accuracy of 500ft at 50 miles.[23]

By the end of 1942, ASE had outlined a project to meet as far as possible the Staff Requirements for type 294/5 which were (a) the continuous plan display of all aircraft to an elevation of 20°, and (b) accurate heightfinding on any aircraft below that elevation without interrupting the plan-display picture, which was not easy to arrange.

Initially, the basic set (called type 294) was to be based on the 10cm type 277 with a 500kW transmitter, outlined on p. 177 above. The number of sites available for radar aerials in a ship is very restricted, particularly for those as heavy as this one was likely to be, because of the need for a fully stabilized platform. To make the maximum possible use of any one site, especially where that site was required to have an all-round view, ASE decided to develop a scheme where the plan-display and heightfinding aerials were on the same mounting. If two sites were available, so much the better: the ship could have two sets. It was to have a new display system integrated with types 960 and 990.

Right from the outset, ASE realized that the range obtainable with 294 with the 277 transmitter would not meet the Staff Requirements. To do so, it would be necessary to replace the transmitter with one of higher power, when it would be renamed type 295. But to approach the Staff Requirement for range even with the 295 would demand not only a transmitter of 2MW peak power output but also the use of a longer pulse, so increasing the mean power, though this would make it more difficult to discriminate between targets close to each other. However, there seemed a good prospect of obtaining a transmitting valve of the required power, as one giving something approaching the required output, though on a slightly different wavelength, had already been experimentally made in Birmingham. It was therefore decided that both types 294 and 295 should march forward together,[24] although, as we shall see on p. 199 below, it was not long before both 294 and 295 were dropped in favour of the new fighter direction radar types 980 and 981.

Internal ASE scientific manpower resources being stretched to the limit, early research and development of types 294/5 was placed to outside contract – to Howard Grubb Parsons for the aerials, to Metropolitan-Vickers and Vickers Armstrong, Crayford, for the stabilization, to BTH for the transmitter and modulator, to mention a few.[25]

Such was the situation at the end of 1943. However, by about 1945, it became clear that ASE had underrated the problem of good fighter direction radar and that a much more complex system was needed. This crystallized in 1947 into what was to become type 984. Types 294 and 295 were splendid attempts at solving a difficult problem but with little hope of achieving a practical solution.

THE SINKING OF THE *SCHARNHORST*

On 26 December 1943, units of the Home Fleet off the North Cape of Norway were covering the passage of two Russian convoys, the eastbound JW.55B and the westbound RA.55A, against attack by German surface ships. In 72° north latitude on that day, there might be some glimmer of arctic twilight around noon but the stormy weather made certain there would never be much. For the rest of the day, virtual total darkness prevailed.

At 4am, the westbound convoy was well to the west of Bear Island and does not come into the story. The eastbound convoy was 50 miles south of Bear Island steering ENE at eight knots, escorted by fourteen destroyers and three smaller vessels. The Home Fleet ships were disposed in two groups, the cruisers *Belfast, Norfolk,* and *Sheffield,* under Vice-Admiral R. L. Burnett, about 150 miles to the east of the convoy; and the battleship *Duke of York,* cruiser *Jamaica* and four destroyers under the Commander-in-Chief, Admiral Sir Bruce Fraser, about 210 miles to the south west.

The British large ships all had types 281, 284, 285 and 273Q (*Duke of York* with a cylindrical perspex lantern and a pre-production model 'JE' PPI), except *Jamaica* who had 272P with the aerial abaft the bridge in place of the 273 she had had forward of the bridge in the Battle of the Barents Sea a year earlier. The German battle cruiser *Scharnhorst* (with a single 80cm FuMO 27 Seetakt radar on the foremast and possibly a 55cm Hohentwiel ASV on the mainmast[26]) left Altenfjord with five destroyers on Christmas Day to attack convoy JW.55B which had been shadowed by German aircraft since Christmas Eve. Admiral Fraser received an 'Ultra' signal at 2.16am on the 26th saying that *Scharnhorst* had probably sailed the night before. The Germans were not aware of the presence of *Duke of York.*

First contact was made at 8.40am when *Belfast*'s type 273Q detected *Scharnhorst* at 12.5 miles when the latter was only about 30 miles from the convoy. The enemy was sighted visually at 6.5 miles and there was a brief exchange of fire with *Norfolk* before touch was lost. That exchange of fire was to have an important influence on the outcome because one of *Norfolk*'s 8in shells hit *Scharnhorst*'s foremast, completely carrying away the mattress aerial of her Seetakt radar.

Shortly after noon, *Belfast* regained contact and a short gun action ensued when *Norfolk* had a turret and all radar sets knocked out, though it was not long before her type 273 was back in operation. Meanwhile, Admiral Fraser was approaching from the south-west and *Duke of York*'s type 273Q detected *Scharnhorst* at 22.75 miles at 4.17. Her gunnery set, type 284M(3) gained contact at 35,000 yards (17.5 miles) but the echo amplitude was insufficient to hold bearing firmly until about 26,000 yards (13 miles).[27]

At 4.50, *Scharnhorst* suddenly found herself flooded with light from a starshell – without warning from look-outs or radar and with her turrets trained fore and aft. *Duke of York* and *Jamaica* immediately opened fire at a range of 12,000 yards (6 miles), thereafter using blind fire until the range had opened to 21,000 yards when fire was checked at 6.24 because – and only because – shell splash observation by radar became difficult.[28] It was during this blind fire phase that Duke of York managed to slow down the enemy, so leading eventually to her destruction.

Scharnhorst's gunnery was at first erratic but soon settled down, frequently straddling *Duke of York* at ranges between 17,000 and 20,000 yards so that the latter was extremely lucky not to incur serious damage in a duel that lasted for ninety minutes until *Scharnhorst* ceased fire at 6.20. Eleven-inch shells passed through both of *Duke of York*'s masts – of which more anon – but luckily neither exploded. *Scharnhorst* fought most gallantly against overwhelmingly superior forces and finally sank at 7.45pm.[29]

One 11in shell passed just below the deck of *Duke of York*'s type 273 radar office which was mounted on a platform above the junction of the two tubular struts altogether forming a tripod mast. The resulting shock was to the considerable alarm of the occupants, the ship's radar officer, Lieutenant H. R. K. Bates RNVR, and the operators, Able Seamen (Radar) J. Badkin and G. Whitton. When they picked themselves up, they discovered to their amazement that they were unharmed and that the radar was still operating, though there were no echoes on the scan. Switching off the office lights, Bates climbed up into the aerial enclosure to discover with the aid of a torch that the stabilizing gyro had toppled and the two circular aerials were pointing skyward. He got them horizontal and stabilized again, so that type 273 was again fully operational, but it was lucky that the set's electrical cables passed along the front of the mast and were undamaged.

Soon after, an 11in shell passed through the mainmast, severing all the lead-cased cables running up the mainmast to type 281's receiving aerial, depriving her of air warning capability. When the ship was in Kola Inlet later, the damaged mast was repaired by a Russian welder and the cables were repaired by Lieutenant Bates and his team; the cables above and below the shell damage had to be separated, identified core by core, and bridged across the damage temporarily – all in temperatures of around 16°F until rewired on return to Scapa.[30]

Bates received the immediate award of the Distinguished Service Cross, and Badkin and Whitton of the Distinguished Service Medal, the Commander-in-Chief's own citation (which was not communicated to the Press) reading as follows:

> For great skill, gallantry and devotion to duty in operating in an exposed position and under heavy fire special equipment which enabled me [the Commander-in-Chief] to bring the enemy to action so that the forces under my command were able to destroy him.[31]

AB (Radar) E. McKay at the range panel in the gunnery Transmitting Station of *Jamaica* and AB (Radar) S. A. Stevens at type 291 in *Scorpion* were also awarded the DSM and more than a dozen radio mechanics and radar operators were mentioned in despatches, a higher proportion than was received by any other category of rating.[32]

The aim of this night action – the safe and timely arrival of the convoy – was achieved. But this was only made possible by radar and its associated plotting arrangements, and was largely managed not from ships' open bridges but from the plot below. Admiral Burnett conducted the whole action from *Belfast*'s Chart House, one deck below the bridge.[33] Admiral Fraser had this to say, writing about his flagship:

> In general the speed of wireless communication and the exceptional performance of radar reflects the greatest credit on the personnel concerned, and in this night battle contributed in great measure to its success.
>
> Plotting arrangements in the Fleet flagship worked well, and were of great assistance both to me and to the ship. I myself alternated between the plot and the Admiral's bridge, the Chief of Staff remaining in the plot. I feel very strongly that the officers in the plot must always be in the closest contact with the Admiral, who should obviously be on the bridge.[34]

He concluded the radar section of his official report thus:

> There is no doubt that, despite its shortcomings, British radar is still far superior to any yet encountered in German ships, and that this technical superiority and the correct employment of the gear enabled the Home Fleet to find, fix, fight, and finish off the *Scharnhorst*.[35]

TAILPIECES

- By Fleet Radar Officer, Mediterranean, after 'Husky':

  At least four of *Indomitable*'s radar offices were continuously at a temperature of over 110°. It is suggested that radar offices should not be placed next to the funnel.[36]

- By the battleship *Valiant* after 'Avalanche':

  Raids detected by types 273 or 285 were regarded as low and those by type 279 as high, no other means of height estimation being successful.[37]

- From Terence Robertson's *Walker, R.N.* (London, 1958), p. 123:

  Against the 'Chase-me-Charlies' [HS.293 glider bombs] there was no defence until, one day in the Bay [of Biscay], an escort was attacked by an aircraft which launched its 'glider bomb' just as a scientist aboard switched on his electric razor to test out a theory. To the amazement of the ship and of the enemy aircraft, the new weapon gyrated about the sky in a fantistic exhibition of aerobatics finally giving chase to its own 'parent' . . . In Liverpool there was a sudden run on shops selling electric razors.

9

1944: Normandy, Before and After

Jan 22	Allied landings at Anzio: 'Shingle'
Apr 3	FAA attacks *Tirpitz* in Altenfjord: 'Tungsten'
June 4	Allies enter Rome
June 6	Allied landings in Normandy: 'Overlord'/'Neptune'
June 19	Battle of the Philippine Sea
July	First U-boats with schnorkels
July 5	Eastern Fleet attacks Sabang
Aug 15	Allied landings in southern France – 'Dragoon'
Oct 24	Battle of Leyte Gulf: first Japanese suicide air attacks on Allied ships
Nov 22	Formation of British Pacific Fleet (BPF) under Admiral Fraser in Ceylon
Nov 28	First Allied convoy arrives Antwerp after clearance of River Scheldt

THE ANZIO LANDINGS

On 22 January 1944, two Allied divisions, one British, one American, were landed over beaches near Anzio about 30 miles south of Rome – operation 'Shingle' – to attack the right flank of the German army who were facing the main British 8th Army and US 5th Army north of Naples. This amphibious assault differed from those on Sicily and Salerno in that no aircraft carriers were present, but two fighter direction ships took part: *Palomares,* who had the misfortune to strike a mine off the beach during the early stages of the assault, and *Ulster Queen,* who controlled day fighters with marked success, besides issuing air raid warnings to the beaches and anchorage based on German radio intercepts and radar. Both ships returned to the United Kingdom after the operation and, as we shall see, *Ulster Queen* returned to the Mediterranean in time for the South of France invasion in July, fitted with Air Force GCI radar for the control of night fighters.

GERMAN ANTI-SHIP GUIDED WEAPONS

After the trouble experienced at Salerno with radio-controlled glider-bombs, three ships fitted with apparatus for listening on the enemy wavelengths and jamming the transmissions of controlling aircraft were included in the Anzio assault force. However, according to Roskill, their crews were not yet adequately experienced in their highly specialized task[1] and glider-bombs did a great deal of damage, including the sinking of the cruiser *Spartan*.

Nevertheless, during this last attack, Lieutenant Field, ASE's RCM representative with the Mediterranean Fleet, was able to intercept several bomb radio-control signals from the parent aircraft, in the 6m waveband. He successfully deduced the format of these signals for both azimuth and elevation steering of the glider bomb. He then set up a workshop in Naples to produce a simple jammer, and fitted some twelve of these in Mediterranean Fleet ships. However, effective jamming was not easy because of the difficulty of tuning the transmitter rapidly enough once the particular missile control wavelength had been established during the bomb's flight.

There was now considerable alarm in the Mediterranean Fleet – and not a little in ASE – concerning the threat posed by the Hs.293 missile. Luckily, a complete one with an unexploded warhead was found on one of the Anzio beaches, which Field salvaged and despatched to RAE Farnborough for detailed examination. RAE's confirmatory analysis of the control system ultimately led ASE to solve the twin problems of jamming several missiles simultaneously, without the need for the rapid tuning of several individual jammers, and of the rapid determination of the wavelength of the glider bomb control system during the few seconds between its launch from the parent aircraft and its arrival at the target.

The FX.1400 bomb seems to have been phased before the Anzio operation. As a short-term expedient on ASE's advice, an order was issued to the Fleet that, when ships were expecting attack from glider bombs, anyone possessing an electric razor – in those days, probably not very many – should go on deck as quickly as possible, activate the razor from the nearest power point, and hold it aloft at arm's length. As we saw on p. 190, these measures had worked in the Bay of Biscay but whether they had any practical effect on subsequent operations is not known, though they probably had some positive psychological value. What is certain, however, is that they caused not a little mirth in the Fleet at the time and, after nearly fifty years, this is one of the very few things about wartime RCM still remembered by the present author.

A new jammer known as type 651 was developed, which effectively jammed all the missiles in flight at one time, so that they would all

34. The destroyer *Scourge* in June 1944 escorting the *Arethusa* carrying King George VI to visit the beach-head in Normandy, with a lattice mast and 'modern' warning radar – the new 10cm WS type 276 (later replaced by WC type 293) on the front of the lattice mast, replacing the older type 271 or 272 in destroyers, as in the *Tumult* in Fig.24. *Scourge* also had 50cm GA type 285 on the rangefinder–director, and 1.5m WC type 291 aft, with HF/DF at the top of the foremast.

(IWM)

35. The AA sloop *Lark* in 1944 with a lattice mast and the same 'modern' radar as *Scourge* in Fig.34. Compare with *Erne* in Fig.23.

(IWM)

become uncontrollable and, during the next few months, this was fitted in nearly every large ship in the Royal Navy.

NEW-GENERATION 10-CENTIMETRE WARNING SETS: TYPES 276/7, 293

The fitting in ships of the Fleet of the new generation of 10cm warning sets – with their PPIs and continuous powered rotation – proceeded apace in 1944, particularly with many new destroyers and sloops joining from builders' yards. The most widely fitted at this stage was the surface warning type 276 with its 4ft horizontal cheese aerial in small cruisers, destroyers and AA sloops. On the whole, the users' reaction from sea was most favourable, the maintenance staff rather less so. How marvellous to have a PPI on the bridge, said the Senior Officer of the 39th Escort Group in the sloop *Black Swan* (with type 276) in April, the whole value of radar being enormously increased thereby, particularly in night operations – and no less welcome in daytime! He was now able to appreciate the tactical situation instantly, while blind navigation became reliable and accurate. But side echoes were tiresome in convoy work.[2]

On the other hand, owing to pressures of staff limits, accelerated development and the paramount need for rapid introduction into service without the usual preparations that would be made in peacetime, there were many technical teething troubles. ASE had to send an officer to Alexandria in March to help solve type 276 problems in *Tuscan* and *Black Swan*;[3] and Captain, Radar Training, in HMS *Collingwood* found it necessary to call a special meeting in September, attended by officers from sea, to discuss the many seagoing maintenance problems.[4] Nor, apparently, was ship-fitting straightforward. In February, the Port Radar Officer at Belfast told ASE that he had had great difficulty in fitting the waveguide for type 276 into the frigate *Erne*. It took the dockyard, he said, three times as long to fit type 276 into a British ship as it did to fit the American type SL (which performed exactly the same function) in the Lend-Lease 'Captain' class frigates.[5]

As we saw in Chapter 8, the decision had been made back in August 1943 that type 276 (designed for surface warning) should only be a stopgap until it could be replaced by the type 293 giving both surface and air warning as well as being suitable for target indication and gun direction. (The only difference between the two sets was in the aerial and its mounting.) Trials of type 293 with its first aerial, tilted and 6ft wide (AUR), in August 1943, proved far from satisfactory as regards range on both surface and air targets. Nonetheless, a fair number of destroyers were so fitted in 1944 and, in November, the Commodore (Destroyers), Home Fleet, was able to send ASE tables of the performance of type 293 in seven modern destroyers compared with the performance of type 276 in six other destroyers. In fact, the results of these somewhat unscientific trials seemed to show that there was little to choose between them.[6]

However, although this first 293 aerial was wider than the 276's (6ft as against 4ft), it was significantly narrower in the vertical (4in against 10in) so the aerial gain was only around half that of type 276 aerial. One might have expected this difference to appear clearly in the destroyer comparison above, but it was probably masked by maintenance and tuning difficulties experienced in small ships faced with handling the early models of new and advanced equipment, more especially as fleet destroyers had not previously been equipped with 10cm radar.

Meanwhile, trials of a new eight-foot-wide aerial (AQR) for type 293 had proved successful. From the beginning of 1945, the present aerials of all existing type 276 and 293 sets began to be replaced by the new 8ft aerial, the resulting radar being known as type 293M.

The other version of the new-generation 10cm warning sets, type 277 with its paraboloid aerial and heavy mounting, took a few months longer to develop than type 276 with its simpler aerial. Designed primarily for surface and low-air warning, it was also classed as a heightfinder. Its circular paraboloid aerial, stabilized in elevation, gave a pencil-shaped beam which could be pointed in any direction; aerial elevation could be measured with reasonable precision, from which the height of the target could be derived. It was fitted in large cruisers and above (and later in smaller cruisers as well), in 'River' class frigates and, with pre-fabricated huts, in 'Loch' class frigates and 'Castle' class corvettes.

The first operational fittings took place early in 1944 – in the new carriers *Indefatigable*, *Implacable* and *Campania*, in *Victorious* refitting at Liverpool and, soon after, in the battleship *Howe*. In its primary role, that of surface warning, it was highly successful. For low air cover, it proved only fair in the Pacific with reliable detection at 25–30 miles, fairly frequent detection (with anomalous propagation) at 45–50 miles – but this was not enough to counter raids by the fast Japanese aircraft.[7] The only radar in the British fleet which could provide adequate low air cover was the American SM-1 with its 8ft-diameter aerial (described below) – 50 miles normal, 80 with anomalous propagation – but *Indomitable* was the only British ship so fitted at the time.

For measuring height for interception purposes, type 277 did not prove a success, though, to be fair to the scientists, this function had been presented to the Naval Staff as something of an afterthought, a bonus resulting from the elevation stabilization needed for optimum surface cover – but a bonus seized upon as being highly desirable by the Naval Staff in general and by the Air Warfare Division in particular. The difficulty for the operator was this: because of the narrow pencil-shaped beam, it had to be 'put on' to the target by other radar information and,

36. The carrier *Victorious* in May 1944 with 'modern' warning radar – two WA sets, 7.5m type 79B on the foremast, 3.5m 281B on the main; 10cm WS type 277 with the circular 'dish' on the top of the island, for surface and low-air cover and (a function which was to prove none too successful) heightfinding; the 10cm WC type 293 scanner on the mainmast for target indication; and the normal type 285 and 282 gunnery sets. The YE homing beacon aerial was mounted on a pole mast between the two tripods. It was with this equipment that she fought the war in the Far East.

(IWM)

37. The carrier *Indomitable* in April 1944, immediately after her refit and damage repairs in Norfolk Navy Yard, Virginia, where she was modernized to the same standard as *Victorious* in Fig.36, except that she had two American sets – the 10cm scanning fighter control radar SM-1 on the front of the island in place of type 277 and, instead of type 276, the 10cm surface search SG with its scanner on the foremast. Both these sets proved highly successful in the war in the Far East, particularly the SM-1 with its heightfinding capabilities.

(IWM)

when the aerial wandered off, there was no obvious way of knowing in which direction the aerial should be moved to regain the target, a defect which was overcome in the American SM-1 by conical scanning. But there was another factor, probably more important, which the Command had to consider. To measure a height, the aerial had to be stopped, so all-round surface and low air cover was lost, seldom acceptable particularly in view of its known heightfinding difficulties. The answer was, of course, a second set, but no operational ship had this during the war years.

RADAR FOR COMBINED OPERATIONS

The RAF's 9cm H_2S Mark II radar was modified for naval purposes in 1943 (by mounting the scanner upside down), dubbed type 970, and used in the Sicily and all later landings, principally to provide rocket landing craft with the correct range from the beach at which to fire their rockets. In the spring of 1944, trials of the RAF's 3cm H_2S Mk.III radar were carried out to produce a 3cm navigational radar which was known as type 971, used in small craft in conjunction with Outfit QM (later Decca Navigator) during the Scheldt Estuary clearance described on p. 221 below. In June 1945, the order went out that all existing type 970 sets (10cm) were to be converted to type 971 (3cm).[8]

FIGHTER DIRECTION RADAR DEVELOPMENT

As we saw in the last chapter (p. 186), ASE started designs early in 1943 for a new radar system for air defence and fighter interception consisting of the metric type 960 for long-range high air cover, centimetric type 990 for long-range low air cover, and type 294/5 for fighter interception, with two aerials on the same mounting spinning together to give continuous gapless plan display of all aircraft out to 80 miles with heightfinding to the same range without stopping the aerials spinning.

By the spring of 1944, it was obvious the requirements would not be met in any time-scale of use to the Navy in the Pacific War. The hoped-for 2MW transmitting valve had not materialized, so range requirements could not be met; and there was still no immediate prospect of being able to measure heights to the required range and precision without stopping the aerial. ASE therefore proposed that the two functions of plan position and heightfinding in type 294/5 should be split and, using the 500 kW transmitter of type 277, make two entirely independent sets, one of which – type 980 – would spin continuously for plan position, and the other – type 981 – could be stopped on the target for heightfinding. Granted this

38. LCT (Rocket) 334. These rocket landing craft took part in the Sicily and all subsequent major assaults. They were fitted with type 970 radar – modified RAF 9cm H_2S Mark II sets with the lantern on a lattice structure behind the bridge – which was used to range on the beach to give the moment of firing, aided in the Normandy landings by readings of the Decca Navigator. *(IWM)*

would mean finding two aerial sites in a ship instead of one, but each function could be better performed because there would now be room for novel aerial designs to increase aerial gain and to allow sweeping in elevation.[9]

This proposal was agreed: type 294/5 was replaced in the project list by types 980 and 981. Type 990 was dropped but the metric type 960 went ahead. Further progress will be described in Chapter 11.

US RADAR IN BRITISH SHIPS

This is a good moment to list the American 'search' radar sets (the American name for British 'warning' radar) fitted in certain British ships during the war, particularly those transferred to the Royal Navy under Lend-Lease, notably the *Smiter* class escort carriers (CVEs), and 'Captain' and 'Colony' class frigates. The twenty-three CVEs of the *Smiter* class had the 1.5m SK air search (with a large mattress aerial), and the 10cm SG surface search radars. (The other fifteen CVEs of the *Archer* and *Tracker* classes had British radar.) Most of the frigates had the 1.5m SA air search and 10cm SL surface search radars, though some had HF/DF at the masthead instead of SA. All these sets had continuous powered aerial rotation and PPIs. Compared with the British metric sets, the American SK and SA air search radars gave shorter ranges and had no height estimation capability; they scored, however, in giving better low air cover and having a much better PPI picture.

Because, at that time, there were no British equivalents, two other American radars were fitted in British ships for special purposes. The first of these was the 10cm SM-1 heightfinder and fighter direction set, giving both plan position out to 80 miles and heights out to 50 miles or so. When searching, the 8ft aerial could sweep continuously, scanning any desired vertical segment of 12° up to 75° elevation. Alternatively, sweeping could be at a fixed elevation. High accuracies for heightfinding were achieved by triaxial stabilization and conical scanning. SM-1 was fitted in the carrier *Indomitable* at Norfolk, Virginia, in February 1944. (At the same time she was fitted with surface search radar SG, as was her sister ship *Victorious*, then refitting in Liverpool.) So successful did the SM-1 prove (not only for heightfinding and fighter control, but also for the detection of weather systems so important in aircraft operation in the tropics) – and so slowly was the development of the equivalent British set, type 981, proceeding – that another twenty-six sets were ordered for the Royal Navy. However, only the light fleet carrier *Ocean* and the fighter direction ships *Boxer* and *Palomares* had been fitted when the war came to an end, and none of these saw any action.

39. The US-built frigate *Antigua* in 1944, with American radar – 1.4m SA air search with antenna at the masthead, and 10cm SL surface search with its antenna in a radome just below. The 'Captain' class frigates, also US-built, had the same radar. In both classes, some of the ships had HF/DF instead of SA.

(IWM)

The second of these special American sets was the 10cm SJ surface search sets for submarines, fitted in *Tiptoe* and *Trump* in mid-1944 to give them a surface search capability pending the availability of the British type 267W which (as we shall see in Appendix D) did not go to sea until March 1945.

RADAR DISPLAYS AND THE ACTION INFORMATION ORGANIZATION

In warning radar, as with gunnery radar, the very multiplicity of sets had by 1943 given rise to the concept that the Command – of a ship, of a squadron, of a fleet – should no longer think of radar sets as individual weapons, but should treat them as being part of a system, a weapon system of the fighting ship, of the fighting squadron, of the fighting fleet. Thus was born the Action Information Organization (AIO) whose function in a ship was to collect information from all available sources (radar being the largest contributor), to relate and 'filter' it, and to serve it up to the Command in a clear and digestible form.

What made all this possible was the development of displays which could be sited remotely from the radar set itself. In particular, this applied to the PPI and Skiatron which had the tremendous advantage of requiring almost no specialist knowledge to interpret. Furthermore, it was sometimes possible to make a single plan display capable of being switched from set to set.

For satisfactory plan display, it is necessary to have continuously rotating power-driven aerials, which began to be introduced for all warning sets (except type 79/279) in 1944. However, there was certain information which could only be obtained from a range-amplitude type of display, such as estimation of the number of aircraft in a group, precise measurement of range, some forms of interrogation of IFF, and echo-amplitudes for height estimation. To study such a display, it had in the past been necessary for the aerial to be stopped or to be moving very slowly, to the detriment of all-round lookout and the quality of the picture on any plan displays that might be fitted – bad enough when the aerials were hand-rotated, but with continuous power-rotation, quite unacceptable. This gave rise to the development of the sector-selected display (generally known at the time as a sector display) where a range-amplitude trace was automatically switched onto the tube as the aerial passed through a selected bearing. The CRT screen was of the after-glow type, so that the picture persisted from one excitation to the next, each excitation occurring as the aerial passed the selected bearing.

For height estimation with metric sets (see Appendix C), special sector display panels for types 281 and 79 were fitted in radar offices or at the Height Filtering Position (in the Radar Display Room [RDR] if there was one). These sector displays enabled echoes over 6° sectors to be kept under accurate observation without interfering with the sweep rate of the set.

Similar sector displays were introduced to display those IFF responses which could not be displayed on a PPI, without having to stop the sweep of the main set. They were fitted in the appropriate radar offices (or at the

40. The 6in cruiser *Swiftsure* in 1944, the first cruiser with 'modern' radar and Action Information arrangements and equipment. For radar, she had WA type 281B on the mainmast, the 10cm WC type 293 aerial for target indication on top of the foremast and the circular mesh 'dish' of the 10cm WA type 277 just above the DCT, on which is mounted the double 'cheese' of the new 10cm GS type 274. The centimetric GA type 275 was not ready for her so three 50cm GA type 285s were mounted on the HA Directors either side of the forward superstructure and aft.

(IWM)

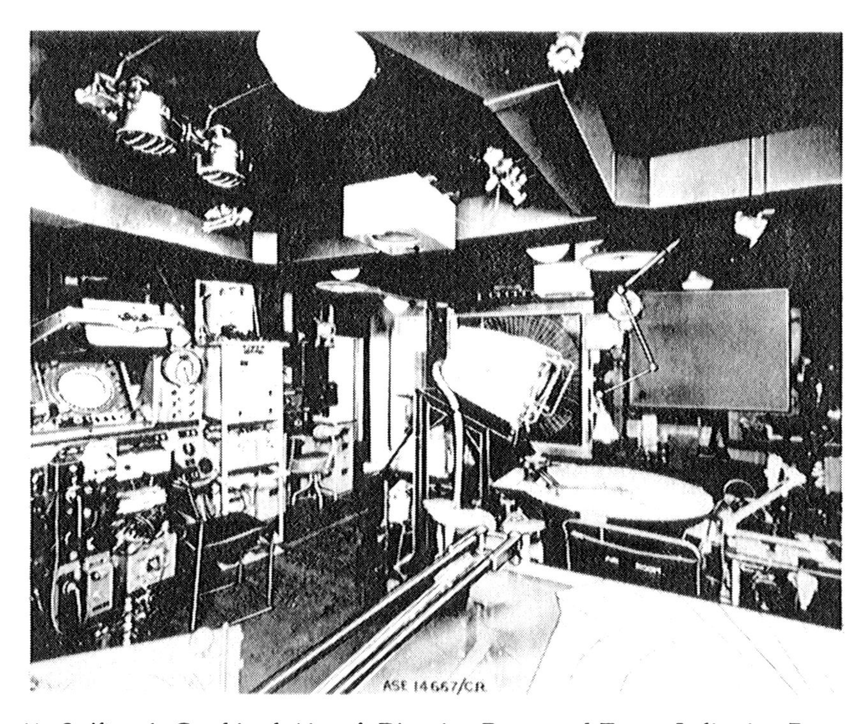

41. *Swiftsure*'s Combined Aircraft Direction Room and Target Indication Room, 1944. On the left is an early target indication unit with ranging outfits and a PPI displaying signals from either type 293 or 277. In the foreground is the automatic plotting table with a PPI displaying types 281 or 277. At right centre is the FDO's intercept position with skiatron and PPI capable of displaying signals from any one of the three sets. The Main Air Display Plot is here obscured by that same PPI.

(IWM)

Central Interrogation Position in the RDR if there was one) for use with types 281 and 277; and at the Target Indication Position in the Aircraft Direction Room (ADR) or RDR for use with type 293.

Two other displays used with type 277 when heightfinding must be mentioned briefly: the height position indicator used with type 277 – a sort of PPI in the vertical, displaying range and angle of sight, from which the target's height could be read off on the mask on the face of the tube; and the Azicator ('azimuth indicator') used in heightfinding to put the 277 aerial onto the bearing (azimuth) of the target indicated by 281. Both of these displays are discussed in Appendix C.

Fitting the full AIO set-up involved significant structural changes in ships, difficult enough to arrange in ships still under construction, even more difficult in existing ships which commanders could ill afford to spare from current operations. But it had to be done: in the naval war in the Pacific, the main threat was expected to come from the air, and it was here that the AIO would come into its own, for fighter and gun direction; so, during 1944 and 1945, as many ships as possible were at least partly modernized in this respect, priority being given to those going to the Pacific.

ADMINISTRATIVE AND PERSONNEL CHANGES

Before returning to operational matters, some important administrative and personnel changes affecting radar which took place in 1944 must be mentioned. From its inception, radar in the Royal Navy – both the equipment and the men to design, develop, maintain and operate it – had been the responsibility of the Signal Branch. On 21 January 1944, responsibility within the Admiralty for research, development, production and fitting of all communication and radio equipment, including radar, was transferred from the Director of the Signal Department (DSD) to the newly-established Director of Radio Equipment (DRE). His department, which became responsible for ASE, was largely manned by the former *matériel* sections of DSD, the Deputy Director of the new Department being Captain G. F. Burghard who had been Head of the Application Staff at ASE before becoming Deputy Director of the Signal Department. This reorganization was a continuation of the Admiralty's policy of separating the functions of the policy-making Naval Staff *Divisions* from the *matériel* Admiralty *Departments*. Thus, the new Radio Equipment Department reacted to the policies of the Gunnery Division (for gunnery radar), the Navigation Division (for warning radar), the Air Warfare and Training Division (for fighter direction), and the Signal Division (for communications).

At the same time, responsibility for all radar maintenance and operating personnel was transferred to the Director of Training and Staff Duties and a new radar school was set up in HMS *Collingwood*, a 'stone frigate' near Fareham in Hampshire, under the Captain, Radar Training, who was also responsible for the training establishment *Valkyrie* in the Isle of Man, and for the Radar Training Flotilla.[10]

So, to the chagrin of many specialist signal officers, the Signal Branch relinquished a responsibility they had held since 1935: the Director of the Signal Division (as the Signal Department now became) and the Captain of HM Signal School were no longer specifically concerned with naval radar, which successive incumbents of these posts – and their staffs, naval and civilian – had done so much to promote during the previous eight years.

42. The 5.25in cruiser *Black Prince* escorting a Russian convoy. She had WA type 281, WS type 272 with a lantern on the foremast, and the usual cruiser gunnery radar.

(IWM)

Just before these wider organizational changes took effect, important staff changes were made at ASE. On 31 December 1943, Captain B. R. Willett, who had been appointed Experimental Commander at the Signal School in November 1937, Experimental Captain in October 1939, and Captain Superintendent of the ASE when it came into being in August 1941, was succeeded as Captain Superintendent by Captain Patrick W. B. Brooking. He had been Deputy Director of the Signal Department at the beginning of the war and had commanded the new cruiser *Sirius* in 1942 and 1943, taking part in many actions, including the 'Pedestal' Malta convoy in August 1942, which had proved so important in the history of naval fighter direction.

On the scientific side, C. E. Horton, who had been head of the Radar Department since its inception in 1938, moved to the Admiralty, to be succeeded at ASE by S. E. A. Landale who was at that time head of ASE's equipment division dealing with tactical, anti-submarine, fighter interception and shore station naval radar. Landale assumed the title of Superintending Scientist (Radar), J. D. S. Rawlinson remaining deputy head of department.

While these changes were being planned, the whole question of radar operators at sea was under review. As we saw on p. 112 above, it had been decided in 1940 that all radar operators should be 'Hostilities Only' ratings, a situation which obviously could not continue. In February 1944, therefore, a new scheme was introduced, splitting radar operators into two non-substantive branches of the seaman branch:

- *Radar Control Ratings (RC Ratings).* To man all gunnery and target indication sets and to carry out certain important fire-control duties in the TS and/or calculating positions.
- *Radar Plot Ratings (RP Ratings).* To man all warning sets and to carry out plotting duties in the Action Information Centre and related positions.

In a ship, the RC ratings came under the control of the Gunnery Officer and the RP ratings under the Navigating Officer, who had been made responsible for the Action Information Organization. On 1 November 1945, the whole of the training of RC and RP ratings was assumed by the respective training schools – HMS *Excellent* and HMS *Dryad*. Training of Radio Mechanics remained at HMS *Collingwood*.[11]

THE INVASION OF NORMANDY

'THE OBJECT OF OPERATION "NEPTUNE" IS TO SECURE . . . A LODGEMENT ON THE CONTINENT FROM WHICH FURTHER OPERATIONS CAN BE DEVELOPED.'[12] So ran the opening words of the operation orders of Admiral Sir Bertram Ramsay, Allied Naval Commander, Expeditionary Force (ANCXF) – orders for operation 'Neptune', the assault phase of the Allied invasion of Europe, operation 'Overlord'.

The functions of the naval forces taking part in the Normandy assault on 6 June 1944 were to provide for 'the safe and timely arrival of the assault forces at the beaches, the cover for their landings, and subsequently the support and maintenance and rapid build-up of our forces ashore . . . The operation,' Admiral Ramsay continued, 'is a combined British and American undertaking by all services of both nations.'[13]

For a most readable and authoritative summary of all aspects of what is acknowledged to have been the greatest amphibious operation in history – the planning, the assault, the build-up – the reader is referred to volume iii of Stephen Roskill's magisterial *The War at Sea*.[14] In this operation, radar – and naval radar in particular – played a vital part, and its full story, both planning and execution, could fill a book in itself. But here, space permits us to do no more than to summarize the specifically British naval aspects connected with radar – in navigation, in air defence, in shore bombardment, in radar countermeasures and deception, and in surface plotting and interception.

The radar aspects of the planning of 'Neptune' came under Commander R.T. Paul, Chief Signals and Radar Officer to ANCXF, whose senior colleagues on the naval radar side were Lt-Cdr F.J. Emuss RNVR, a specialist radar officer, and Mr A.W.Ross, a senior scientist from ASE, both of whom also played an important part in the post-assault phase. Two other staff officers should be mentioned: Lt-Cdr Ian Steel and Lt-Cdr W.J. Borthwick RNVR, responsible for navigation and fighter direction respectively.

One of the more difficult aspects of the planning was caused by the very high levels of secrecy that had to be maintained before D-day, for which a special code-word, 'Bigot', was coined. The limited number of people 'in the know' were said to be 'bigoted'. And, of course, there were various levels of bigoting, some being allowed to know more than others. All of which made it difficult for ANCXF planning staff when communicating with people outside their own organization.

In the operation, radar, radio countermeasures (RCM) and radio communications were interlocked to an unprecedented degree. To avoid saturation of the ether and complete loss of all efficiency – not to mention denying the enemy intelligence of our operations – drastic restrictions were imposed. Enemy listening stations must not be alerted during the pre-assault phase, so radar silence was maintained by the assault forces on passage except that 10cm surface warning radar was allowed in some cases to deal with enemy attack or for essential navigational purposes.

After the assault, there were to be no restrictions on radar, but in the event, because of the proximity of land and the presence of large numbers of aircraft, ships and balloons, the 'clutter' on radar displays reached unprecedented proportions. Furthermore, there were complications due to mutual interference arising from VHF radio communications and RCM, and between the main cross-channel radio link and type 281 which, owing to a misguided piece of 1943 planning, shared the same frequency band.[15] Many of these problems had been foreseen and steps taken before sailing to fit equipment to ships to minimize such mutual interference. However, there still remained many problems after D-day with which Ross in particular had to deal. In contrast, interference from enemy sources was in the event minimal, except for 'Window' dropped at night by German minelaying aircraft.

RADAR COUNTERMEASURES AND DECEPTION

Within the overall 'Fortitude' deception plan,[16] two operations in particular affected our story. Two small RAF and RN forces were to simulate large Allied assault forces just prior to, but well away from the area of, the real assault. The RN contribution was to total thirty-four small craft towing convoy-type barrage balloons fitted with radar corner-reflectors (which give large echoes) and RCM jammers (capable of producing spurious targets). The RAF contribution was to be provided by 617 and 218 Squadrons, flying in carefully controlled orbit patterns in relays over each spoof naval force, dropping controlled clouds of Window – reflecting strips of metallized foil – to increase the apparent size of each force to look like a convoy 14 miles wide by 16 miles long. On reaching Area Z south of the Isle of Wight, the two spoof forces were to turn east towards the Dover Straits, while the real assault forces, following slightly later, would turn south towards Normandy.

Finally, there was to be a third diversionary force associated with the heavy ships of the naval bombarding force to be stationed off Le Havre. This force was to comprise fifteen convoy-type balloons anchored in groups of three, plus four minesweepers each flying two balloons at different heights, an idea of A. W. Ross, so that radar echoes would fade independently. Each balloon was to be fitted with a corner reflector to enhance the radar echo to the size of that of a battleship, in order to draw the fire of German coastal batteries.

In about April, the RAF had started bombing attacks on most of the German radars on the Channel coast,[17] sites outside the intended invasion zone receiving more attacks than those within it, to confuse the Germans as to the zone of assault. Quite deliberately, care was taken not to damage some radars which might be expected to detect and report the spoof assault forces.

In the initial approach of the real – as opposed to spoof – assault forces, it was of the utmost importance to avoid early detection and to distract the enemy. RCM by both naval and air forces had a vital part to play in this – in neutralizing German coast radar by jamming (where the station had not already been neutralized by air or ground attack), and in diversionary operations mentioned above. And the enemy, once roused, had to be prevented from using radar to control his coastal batteries against allied ships and craft.

On the naval side, an immense amount of planning was needed and a vast programme was undertaken to fit ships and craft with special jamming and other electronic equipment (mostly RCM, but some for navigation), all coordinated by ASE in collaboration with the US Commander of Landing Craft Bases in Europe. Many items were not in production and had to be specially manufactured in the USA. Because of this, fitting did not start until April 1944 and had to be continued right up to D minus 2 days.[18]

THE OUTCOME

After dark on 5 June, the movements of the two deception forces up-Channel east of Area Z were detected by the Germans, as planned. They were reported by the few German radars still operational there, and an alert was issued for *that* zone – but not for the Normandy area. The American invasion force U, heading for Utah beach, had cut south of Area Z because of the longer distance that had to be travelled from south-west UK ports, and they were detected briefly by radars on the Cotentin Peninsula. These soon reported to German HQ that strong and effective jamming was being experienced. As for the Normandy area, no special German sea or air patrols were ordered initially, since it was considered that the prevailing weather was totally unsuitable for a large-scale invasion. However, detection of the British and American airborne forces on the Eastern and Western flanks of the actual assault zone to secure its boundaries led to an eventual German alert.

Apart from those German radars following the deception forces, no worthwhile intelligence concerning the real invasion was obtained from the few remaining German coast radars covering the assault beaches, owing to the effects of naval jamming.[19] Tactical surprise was complete.

As far as the third diversionary force was concerned, whilst the evidence is not conclusive, it seems that the decoy 'battleship' targets were successful in drawing German coast battery fire onto themselves rather than on the bombarding ships. In the event, none of the real or artificial targets were hit.

A remarkable event took place in the Dover Straits at 5pm on D-day – a convoy of eleven large merchant ships carrying reinforcement troops from the Thames to the Normandy beaches passed through the straits *in broad daylight* in safety – the first convoy to do so in the face of the German long-range coastal guns for over four years. Undoubtedly, the convoy was protected from the German long-range guns in the Pas de Calais area by the jamming of the associated fire-control radars by the Dover Z-station jamming barrage.

MINESWEEPING AND NAVIGATION

The minesweeping organization was, perhaps, the most intricate of all the many measures taken to ensure the safety of the invasion convoys. From a swept area five miles in radius south of the Isle of Wight – officially Area Z, familiarly 'Piccadilly Circus' – eight channels (collectively known as 'the Spout') led due south towards the German mine barrier in mid-Channel. From there, ten channels through the mine barrier – two for each assault group, each channel 400 to 1,200 yards wide, marked by lighted buoys – had to be cleared by fleet minesweepers steaming ahead of the invasion convoys. Radiating out from the ends of these approach channels were a number of lanes for bombarding warships. Finally, the waters where the assault ships had to anchor had to be cleared of all types of mine that the enemy might lay.

RADIO AIDS TO NAVIGATION IN THE ASSAULT

The need for precise navigation was paramount – for the minesweepers to establish the correct channels, for the ships and craft to follow them, for the landing craft to beach at the correct places – all to be done at night approaching a low-lying coast. It was very largely achieved by the help of radio aids to navigation as well as radar.

Though these radio aids are not strictly radar, they use many of the same electronic techniques and are so important to the story that their operational use deserves mention. Of these, the most widely fitted was the RAF 'Gee' system (outfit QH) first used by the Navy in the Dieppe raid in August 1942 and subsequently established as a standard system for surface navigation. For 'Neptune', the initial legs of the swept channels were planned to coincide with Gee lattice lines. Some 860 ships taking part were fitted and it proved invaluable both to craft and bombarding ships.

Nevertheless, there was a need for a stand-by navigational system to guard against any possibility of Gee being jammed, so a new, and at the time very secret, radio position-finding system was brought into use although still under development. This was the 'Decca Navigator' system (outfit QM), conceived in 1939 by an American, W. J. O'Brien, which had aroused Admiralty interest in 1941 and done its first sea trials under the supervision of the Decca Company and ASE in September 1942. Though the two systems were both 'hyperbolic' and similar in broad principles, Decca was potentially more accurate than Gee and, in modern parlance, more 'user-friendly' because the results were presented direct on clock dials ('decometers') instead of on a cathode ray tube. A disadvantage of the early models was that the indicator in the ship had to be set up initially in an accurately known position and if there was a break in reception for any reason, this had to be repeated, in case the decometer had 'slipped a lane'.

For 'Neptune', four Decca transmitting stations were set up with great secrecy, at Chichester, Poole and Beachy Head to cover the area of the assault, and on the Isle of Sheppey as a 'decoy' in case the Germans discovered any part of the plan. The ladies responsible for calculating the Decca lattice lines – which gave a good indication of where the landing beaches might be – worked in pairs in a hut at ASE. So secret was their work that an armed guard was provided.[20]

Nineteen pre-production Decca receivers were made by the Decca Company at very short notice and these were fitted in the leaders of twelve minesweeping flotillas, in five headquarters landing craft (LCH), and in two navigational motor launches.[21] There were also monitoring receivers ashore.

Perhaps because it was so secret, perhaps because so few sets were fitted compared with Gee, the use of Decca Navigator in 'Neptune' seems not to have been mentioned in any of the published accounts, though the system worked extremely well, proving more accurate than Gee. if less reliable because of lane-slipping.

Thanks largely to Gee and Decca, the 18th Minesweeping Flotilla, for example, was only about four minutes late and 400 yards out of position on reaching the destination point off the beach, despite strong winds and tides during the passage. A reasonable margin of error! To the chagrin of the sweepers, who could have made good use of it, the Decca chain was switched off on D+1, presumably because the system was then so secret, though there were also transmitter troubles.[22]

RADAR FOR NAVIGATION

And, of course, once radar silence was lifted, radar proper was used extensively for navigation, particularly by bombarding ships. To help

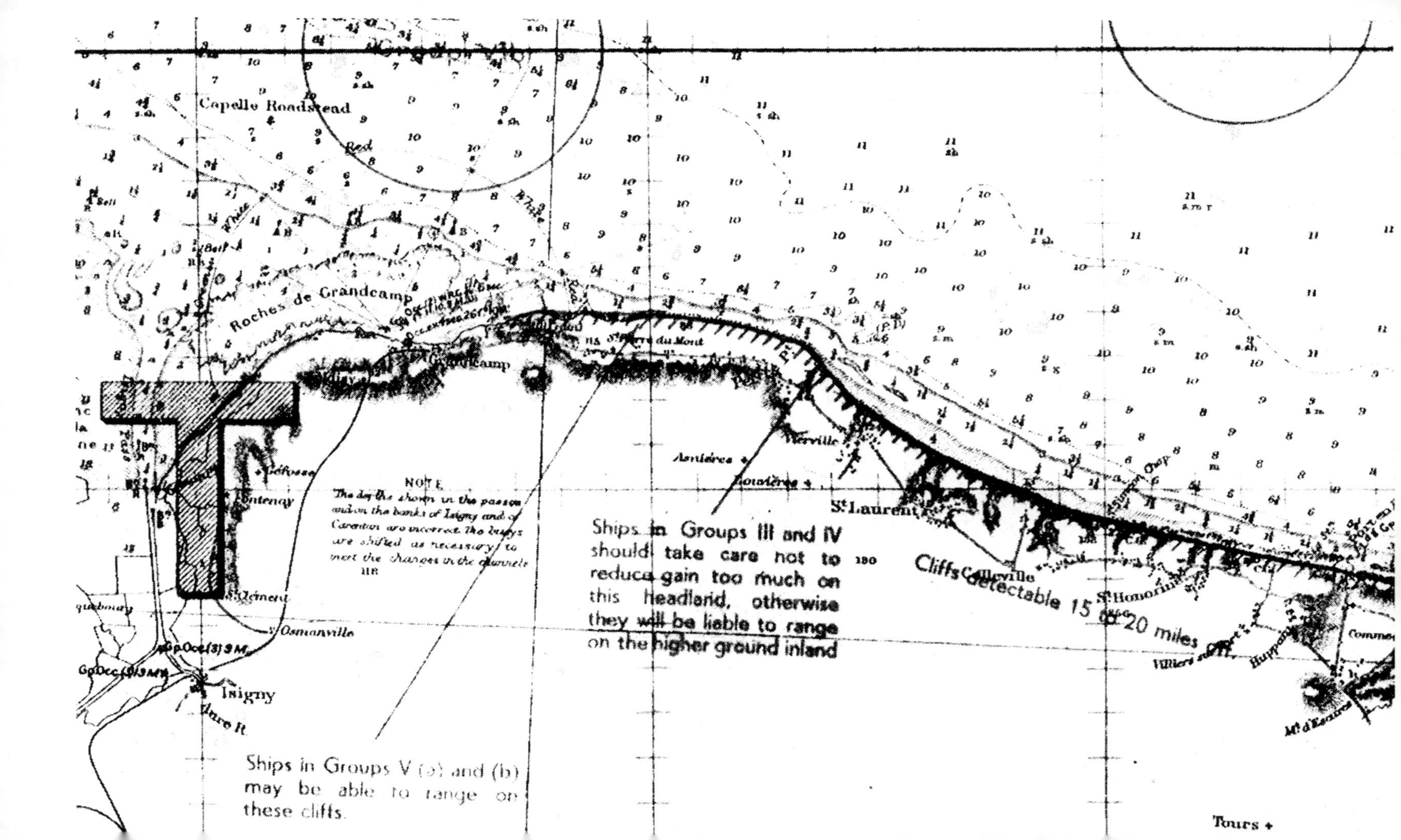

Capelle Roadstead
Roches de Grandcamp
Grandcamp
Vierville
St. Laurent
Colleville
St. Honorine
Isigny
Osmanville
Fontenay
Tours
NOTE
The depths shown in the passage and on the banks of Isigny and of Carentan are incorrect. The buoys are shifted as necessary to meet the changes in the channels
Ships in Groups III and IV should take care not to reduce gain too much on this headland, otherwise they will be liable to range on the higher ground inland
Cliffs detectable 15 to 20 miles
Ships in Groups V (a) and (b) may be able to range on these cliffs.

43. Radar chart for the assault on the Normandy coast, detail from Admiralty chart GR 2613, published 15 March 1944, marked 'Bigot. Top Secret'. The portion reproduced shows the centre part of the assault area. The following Explanation was printed on the chart:

> This chart has been prepared to facilitate the use of Surface Radar (WS) as a Navigational Aid. It can be used with Radar Types 271Q, 273Q, 276, 277 and SL. It is intended for use with high power sets fitted with PPI, but may be of some assistance with low power sets (e.g. Type 272) and where only trace presentation ('A' Scan) is available.
> Physical features which may be distinguished and ranged upon are overprinted in GREEN. The leading edges of the features are outlined.
> CAUTION – *To distinguish marked from adjacent features it is essential to reduce the Radar Gain* UNTIL THE ECHOES CLEARLY CORRESPOND TO THE FEATURES INDICATED ON THE CHART. *This may prove difficult particularly when ranging on inland slopes.*
> *NOTE – Group Areas indicated are roughly those which may be occupied by ships when bombarding. Other ships in the vicinity of these areas should use the features suggested for the groups.*
>
> *(Hydrographer of the Navy)*

---

them, a chart was prepared, overprinted to indicate which land features could be safely ranged upon to fix position in low visibility. This chart (GR.2613), part of which is reproduced in Fig. 43, was reported to have proved most helpful.

Some smaller craft were fitted with type 970, which were modified RAF type H_2S 9cm power-rotated radars with PPIs, which had been successfully tried out in the Sicily landings. Some seventy sets were fitted in rocket landing craft, headquarters landing craft and motor launch navigational leaders. Apart from the usual ranging and navigational assistance derived from a surface warning set with PPI, the value of these sets was increased by the attachment of a 'reflectoscope' to facilitate accurate beach-finding by comparing the land echoes seen on the PPI with so-called PPI predictions for each beach, prepared as transparencies that could be viewed simultaneously with the PPI picture.

These predictions were made by a team of about six WRNS officers who were mathematicians, assembled hurriedly as late as 16 April, working under the guidance of J. Hamilton of the Naval Operational Research Department and R. F. Hansford of ASE. They had to be officers for security reasons; very, very few people were allowed to know the target beaches so far ahead.

The predictions shown on the transparencies represented what should be seen on the PPI of a radar set of known characteristics with its aerial at a given height, at a certain position off the coast, taking into account all known topographic and hydrographic information, including the results of clandestine visits to the beaches. Navigational information such as

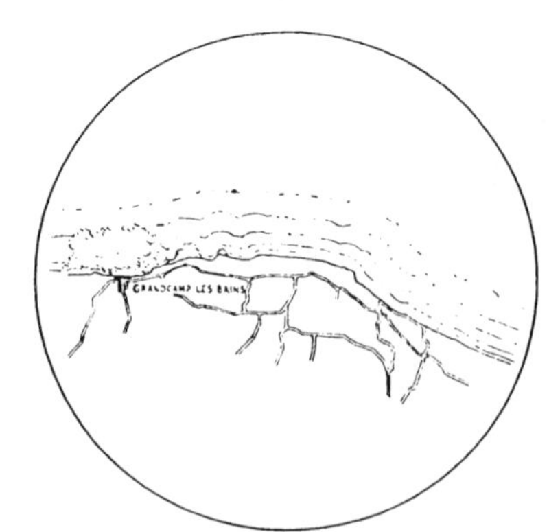

Plate 1. Chart of coastline

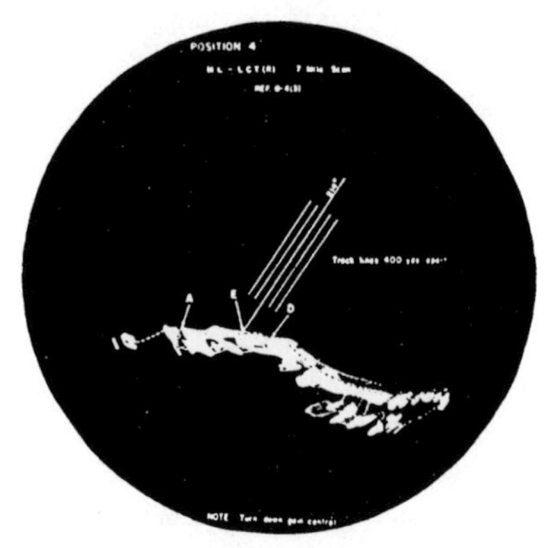

Plate 2. Prediction of P.P.I. picture

Plate 3. Photograph of P.P.I.

Plates 1–3. Radar predictions used for the Normandy landings

44. PPI predictions for type 970, as described on p. 215. From *Journal of the Institute of Navigation*, 1, 2 (April 1948), 133.

track lines and the position of radar landmarks useful for fixing was added, as can be seen in Fig. 44.[23]

Two other minor radio aids deserve mention: metallized flags on the dan buoys marking the swept channels to increase the echo response on radar; and radar beacons (type 78T) triggered by type 291 carried in the Harbour Defence motor launches marking the entrances to the swept channels. The former worked well, the latter not so well.

AIR DEFENCE OF THE BEACHES

The Normandy assault differed from the earlier Mediterranean operations in that no carriers took part, all air cover being provided by shore-based aircraft. To the great surprise of the Allied command, the German Air Force remained comparatively inactive during 'Neptune', despite large concentrations of shipping. During the initial stages of the assault, a few attacks by bombers and fighter-bombers were experienced, but day attack caused little damage and was soon given up in favour of night attack (including a proportion at dusk and dawn), almost exclusively by aircraft laying ground mines. Though these attacks were on a small scale, their persistence caused a toll of shipping which caused concern, particularly in the Eastern (British) Task Force Area. Enemy aircraft were able to lay mines without the casualty rate inflicted on them with other forms of attack.

During the initial stages of the operation, the Navy provided three fighter direction tenders (FDTs), converted LST(2), each fitted with two RAF 1.5m type 15 GCI and one 50cm type 11 early warning radar,[24] with RAF teams manning the control equipment. Two FDTs operated in the assault area and one in mid-Channel for convoy protection and it was the RAF intention that the FDTs should be relieved of their air warning duties (but not for air interception) within two days by shore GCIs to be landed on D-day, one through the US sector, one through the British sector. In the event, the former was badly damaged on the beach and the latter was unable to relieve FDT 217 until D+6.

In fact, through no fault of the RAF teams, the FDTs failed to make any appreciable difference to the main threat to the naval forces – mining – owing to the inability of RAF equipment to see low-flying aircraft. This failure is illustrated by the near-miss by torpedo on FDT 13 in mid-Channel and the sinking of FDT 216. In the latter case, the first warning came from the sound of aircraft engines and from the 271 radar in the corvette *Burdock*. Throughout the operation, the vulnerability of the FDTs was a continual source of worry and prevented them being used to best advantage.

45. Fighter Direction Tender 13 with special radar for assault operations: 1.5m RAF GCI on the forecastle, 50cm GCI amidships and AI radar beacon right aft. The other aerials were for R/T communications and DF.

(IWM)

Air warning radar of the big ships in the assault area proved relatively ineffective in detecting and tracking hostile aircraft. Permanent echoes were normally present on most bearings out to 10 miles – and on some out to 25 miles – due to land, back echoes and shipping. Consequently, it was virtually impossible to maintain a continuous track of more than one aircraft at a time when within 20 miles, and even this could not be guaranteed if the aircraft passed within solid land echoes. Although many 'friendlies' showed IFF, their presence slowed down the rate of search (because of the need to stop and interrogate) and confused the tracking of unidentified plots generally. On some occasions there were as many as fifty aircraft within detection range, often difficult to distinguish from shipping.

Actually, land echoes were not as bad as in similar amphibious operations in the Mediterranean, but as expected the longer-wavelength type 279 proved itself more useful than type 281, particularly for long-range detection when close inshore. The three British headquarters ships *Largs*, *Bulolo* and *Hilary* only had low-power 1.5m combined warning sets type 291 but these proved to be of greater value than expected (except when the enemy dropped 'Window' during minelaying sorties) and the cruiser *Argonaut* was able to hand over fighter direction control to *Largs* where RAF information was available to assist.

With few exceptions, gunnery sets were severely handicapped in 'Neptune' owing to the inability of the air warning sets to 'put them on' to targets.

GLIDER BOMBS

Since the devastation caused by Hs.293 glider bombs during the Anzio assault in January, the type 651 jammer has been fitted in nearly every large ship in the Royal Navy. For the forthcoming Normandy invasion, Field was responsible for installation afloat; in the event, there was no airborne missile attack against ships in the invasion forces until 22 June when a glider bomb apparently directed against the headquarters ship *Hilary* appeared to swerve at the end of its run, probably deflected by the earlier type of jammer, Type 650, which had been kept in a state of immediate readiness.[25] The Germans apparently realized that the RN now possessed suitable countermeasures, and only a few further attacks developed. During one such attack against Allied escort groups off the Brittany coast on 20 July, the Canadian frigate *Matane* was damaged. Otherwise, Allied air superiority in the English Channel and North Sea prevented further attacks.

SURFACE RADAR COVER

It was planned that surface radar cover in the assault area should in the initial stages be provided by ships, particularly by Admiral Vian's flagship, the cruiser *Scylla* in the Eastern Task Force Area. However, as it was desirable to supplement this by introducing unsinkable sets with better range as soon as possible, three mobile naval radar stations (MONRADS, type 277T), with accompanying 'Y' units (MONWHYS), were landed, starting on D+1. In the event, night patrols and interception were controlled by *Scylla* with such success that little help was needed from the shore. *Scylla* used her own surface warning radar with PPI (type 276) supplemented by a filtered plot, and a Coastal Forces officer in her bridge plotting room vectored Coastal Force craft by R/T with consistent good effect.[26]

At the same time, the Commander-in-Chief, Portsmouth, faced with the problem of intercepting E-boats outside normal UK radar cover off the Cherbourg Peninsula and Cap d'Antifer (and later at other critical points), decided to extend the radar cover by using US-built 'Captain' class frigates fitted with American SL search radar (with PPIs and automatic plotting tables) to control units of MTBs attached to them. Each frigate had an experienced Coastal Forces officer embarked, who acted as Control Officer, vectoring the MTBs in much the same way as aircraft were vectored by FDOs. Four frigates were allocated for this duty and proved very successful in controlling interceptions in over thirty actions. In effect, a close blockade of the German Navy's principal E-boat bases was instituted by *Scylla* and the frigates with their attached Coastal Forces craft, so much so that German craft seldom made the open sea without being brought to action. Thus, Allied losses by E-boat action, which might have been serious, were reduced to negligible proportions. This technique of surface force direction – using much the same techniques as aircraft direction – was not only successful in anti-E-boat operations, but also in offensive actions against ships and craft other than E-boats, as discussed in more detail in Appendix E.

Meanwhile, the MONRADS settled themselves in but were not fully exploited until after *Scylla* had departed (she was mined on 23 June). The British Assault Area plot ashore at Courseulles took control on 1 July, fed by two MONRADS for surface watching and one for warning against low-fliers. The latter reported directly to RAF No.24 Sector and, in company with the radar reporting ships, frequently obtained warning of low-fliers before the Air Force radars. A British MONRAD was landed on one of the US beaches and later moved to Cap de la Hague where it provided excellent cover over the Channel Islands.

Lieutenant-Commander Peter Scott, RNVR, artist and ornithologist, acted as Coastal Forces control and liaison officer at Courseulles for a time. He later worked with the US Navy acting as controller of US PT boats from Cherbourg, embarking in American destroyer escorts.

OPERATIONS IN THE SCHELDT

The Allies captured the great port of Antwerp virtually intact on 4 September but, for reasons which need not concern us here, it did not become available as an entry point for military supplies and reinforcements until nearly three months later. The key to the control of the 80 miles of the River Scheldt was the fortified island of Walcheren on the north bank. The Allies began their assault on that island on 1 November, the sweeping of the estuary and river began on 4 November, and the first convoy reached Antwerp on 28 November.

The navigation of the tortuous channels of the Scheldt, with their very strong tides and currents, was a far more difficult problem than the navigation of coastal waters for the Normandy landings, particularly in the fogs and low visibility expected in winter. Firstly, a Decca chain was set up to cover the area, of which more anon. Secondly, four landing craft and a motor launch were fitted with RAF H_2S radar, converted to work on 3cm (designated type 971), which gave a far better PPI picture for pilotage purposes than the 10cm sets used in 'Neptune'. Thirdly, three radar beacons were made available for use as navigational marks.

The first experimental craft arrived in November and two months were spent in navigational trials, particularly in making a radar survey of the river and estuary, by taking PPI photographs at about one mile intervals, from which transparencies were prepared, to be used with the 'reflectoscope' principle, as had been done by the LCHs in the Normandy assault. A number of dummy runs were carried out in clear weather and, on 19 February 1945, a passage was made from Antwerp to Terneuzen in conditions of dense fog, pilotage being by radar throughout, only one object being sighted visually (an anchored barge) during the last 25 miles of the passage – probably the first successful blind pilotage in such difficult waters.[27]

The idea had been that these craft should be used as navigational leaders for convoys at night and in low visibility. As it happened, the weather turned out to be exceptionally good that winter and the swift Allied advance eastward allowed daytime river traffic to be resumed, so the scheme was never put into practice although valuable experience in the use of radar for blind pilotage was gained.

To assist navigation in the Scheldt, a Decca chain was set up with stations at or near Bruges, Ghent and Antwerp. The Decca company quickly developed a battery-portable receiver QM2 for use with this chain, suitable for carrying on board by a pilot. However, for the reasons given above, the chain was never extensively used for its primary purpose but the Admiralty took the opportunity to use it to evaluate the Decca navigator system as an aid to hydrographic surveying. The trial was carried out by the survey ship *Franklin* and Commander E.G. Irving's report on the results was a key factor in the Admiralty's later support for a Decca chain to cover the Thames estuary.

CENTIMETRIC GUNNERY RADAR

Continuing the story of the development of gunnery radar which we left on p. 186 above, the fitting of the type 282/3/4/5 series of 50cm gunnery radar in the Fleet was virtually complete by the end of 1943 and the latest modifications – greater transmitter power, more accurate ranges and bearings, displays in director and TS, spotting tubes – were being made as fast as possible. Action reports, some of which are quoted in this book, testify to the success of the equipment, produced so quickly with nothing to build on. In surface actions, blind fire – well, nearly blind fire! – had become almost commonplace. Short-range and barrage fire against aircraft was much more effective, but the performance of medium- and long-range AA gun systems was still limited by the mechanical problems of the large directors needed to carry the radars, and the resulting fire control problems.

The greatest advance came late in 1944: the introduction of the proximity fuze, known as the variable time (VT) fuze, developed for the allied armies and navies under the control of the US Navy's Bureau of Ordnance in a crash programme started in 1942 with British particpation. In essence, this was a miniature radar set mounted in the nose of the AA projectile with a power supply inside the fuze, designed to meet the shock of being fired from a gun. The radar was energized on firing and continued to transmit until a return signal was received from some target. When the strength of signal had reached some predetermined level, the projectile's ignition train was started and the bursting charge of high explosive would scatter lethal fragments around the line of flight. At a stroke, the VT fuze removed the long train of events from the AA fire-control problem – fuze prediction; fuze setting; gun loading to a predetermined firing rate; and fuze 'launching' from the gun at the correct time for the prediction already made – a train of events subject to a mass of errors, for which the addition to existing directors of remote power control and stabilization of the line of sight against ship motions had not made the expected contribution, though using gyro rate-measuring sights had given some increase in the probability of hitting.[28]

Since the war was expected to continue for several more years, the highest priority for the gunnery team at ASE was now to get the centimetric radars to sea – the 10cm types 274 surface and 275 long-range air sets (the latter to serve both purposes in destroyers) and the 3cm type 262 for the 40mm short-range armament. Unlike their 50cm predecessors, which had to be grafted onto various existing directors and fire-control systems and had to be accommodated below decks in separate offices, the new centimetric equipments could be developed with, and integrated into, the modified or new design of weapon systems evolving at the same time. The main radar panels could be in the TS or HACP alongside the fire control equipment or, with type 262, on the gun mounting itself. Also, the centimetric transmitting and receiving equipment could be mounted

46. The destroyer *Barfleur* which in late 1944 and early 1945 carried out the sea trials of the new 10cm GA type 275 (the nacelles on the DCT) and the Target Indication Room with WC type 293 (with the aerial at the top of the lattice mast). She also had HF/DF forward and WC type 291 aft.

(IWM)

close to the aerials on director or gun mounting respectively. In some cases, radar control panels were also provided in the HA director to allow local control of the director using radar information. These changes resulted in more efficient working of the system and higher performance from the radar, but, for reasons discussed later, considerable gunnery problems remained.

Because it could be added to existing directors, the low-angle type 274 was ready first. Of the first four sets, two went to seagoing ships – the battleship *Howe*, and the new cruiser *Swiftsure* – and two to training establishments – the gunnery school's Fraser Battery at Eastney, and the radar school's HMS *Collingwood* near Fareham on the other side of Portsmouth harbour – all being complete by April 1944. Thereafter, all capital ships and cruisers building or coming in for refit were also fitted, replacing type 284 in existing ships. The new type proved highly successful in the few surface actions that took place before the war's end, particularly with blind fire, as, for example, in *Norfolk* in Operation 'Spellbinder' off Norway in January 1945, described in Chapter 10.

Type 275 took a bit longer. The radar was ready, being virtually electronically identical with type 274, but the production and fitting of the new Mark 6 and US Mark 37 directors (on which the type 275 and other equipment was to be fitted) was delayed and full operational trials of the radar did not start until the new 'Battle' class destroyer *Barfleur* commissioned in September 1944. She carried out sea trials of type 275 and of the first target indication room with type 293. As for type 262, the first shore trials of the radar system early in 1944 led to non-firing trials with the set mounted on a STAAG ashore early in 1945.

SPLASH-SPOTTING RADAR

Visual shell-splash spotting has always been used in naval surface gunnery to assist in 'finding the target', and doing the same by radar was an obvious corollary. With type 274, range spotting of fall of shot was possible at greater range and with better discrimination than with type 284, but line spotting was possible only to the extent that it could be seen whether they were falling left or right of the target (provided that they were within the beam) but not by how much. In December 1943, the Admiralty asked the National Research Council of Canada to develop a splash-spotting radar capable of accurately measuring the position of shell splashes relative to the target in both range and line. To achieve the necessary discrimination, a wavelength of 1.25cm (K-band) was chosen.

No attempt was made at NRC to work everything out *ab initio*; whatever K-band components could be bought in the United States were thankfully incorporated. The scanning aerial, however, was a distinctly

Canadian development. By September 1944, sufficient progress had been made for the Admiralty to ask NRC to produce twelve sets, with spares, for installation in large naval vessels, the new set being designated type 931. Following successful user trials in Canada, the first sets reached Britain in August 1945, just as the Pacific War was ending.[29]

RADAR AS PART OF A WEAPON SYSTEM

During the early years of the war, glaring deficiencies in all parts of the weapon systems were shown up by battle experience, and *ad hoc* improvements were made to the fire control systems, directors and guns, under pressure of circumstances which prevented systematic study, design or trials of complete weapon systems. Even radar was, to begin with, only an 'add-on' substitute for the optical rangefinder. Gradually however, over the period 1940 to 1943, it was becoming clear that the overall improvements in performance resulting from this work were well short of expectations, and the accumulation of experience and limited test results were beginning to indicate that, for further real progress, a weapon system would have to be designed as an integrated whole. Even this was not thought sufficient by some and, as early as November 1943, the Naval Staff had issued an initial Staff Requirement for 'directed projectiles' with the object of avoiding fire-control prediction problems by using missiles guided in flight to their targets, rather than unguided shells.

These system aspects of weapons were particularly emphasized by the development work on type 262 early in 1944 when some systematic basic trials against aircraft showed that radar was inherently capable of an aiming r.m.s. error of 1.3 arc-minutes with a smoothing time of 0.3 seconds (the reference standard), whereas, when mounted on STAAG, the best achievable was only 6 arc-minutes, nearly five times worse. It was clear that imperfections in the mounting and aiming servo-mechanisms degraded the blind performance, even at the best settings possible. Trials with type 275 showed that the inherent accuracy of this radar, which was about 2 arc-minutes, was degraded to 8 arc-min. even with a well set up associated director system, not always achieved at sea.

These results and other deficiencies led to the issue in 1944 of a Naval Staff Requirement for an up-to-date (*sic*) system to control guns up to 5.25in calibre in all types of ship from destroyers upwards, and capable of defence against aircraft and ships. The Staff Requirement admitted that, beyond 7,000 yards against aircraft, guns would probably be superseded by 'controlled projectiles' – the earliest official reference in British naval circles to the possibility of guided missiles, although there had been informal discussions between the Admiralty and ASE on possible means of guiding projectiles using an existing radar suitably modified, as, for example, type 275. Some of the earliest of these informal discussions took place between J.F. Coales of ASE and Captain R.F. Elkins of the Naval Ordnance Department.

This Staff Requirement set in motion in 1944 a major programme of work at ASE and AGE (Admiralty Gunnery Establishment), covering long- and medium-range gunnery systems ('LRS 1' and 'MRS 3, 4 and 5'), which will be discussed briefly in the postwar Chapter 11.

TARGET INDICATION

During the early years of the war, when enemy attacks were usually light and uncoordinated, tight control of a ship's gunnery defences was not essential to its survival. The wide beams of the 50cm gunnery sets enabled them to be used for searching, provided a rough lookout bearing was given from the appropriate warning set through the tactical or air plots. The gun directors concerned were then trained to and fro until the target was picked up. In the case of type 285, it was usual to search with an elevation of 10°, which ensured detection of reasonably high aircraft without a marked loss in range against low aircraft or surface targets.

By 1942, however, the Germans were employing coordinated and concentrated attacks in the Mediterranean and there was no doubt that the Japanese would exploit these techniques in the Pacific. Furthermore, the improved 50cm sets had considerably narrower beams for better resolution, and this was taken further with the 10cm and 3cm sets. It thus took much longer to search for the targets and the probability of their escaping detection until it was too late to engage them was greatly increased, particularly since there was no assurance that targets would be engaged in the best order if they approached from several directions.

To achieve the most effective defence, it was obvious by 1943 that a high-resolution system was required, one which enabled targets to be detected separately and their movements established in time for the target indication officer to plan the best way of using his various weapons to meet the threat, and then to tell the weapons where to look – all with the least delay. Furthermore, the increase in night air attacks on the Fleet made it necessary to make the fullest use of whatever blind-fire capability there was – only possible with target indication by radar. These requirements could not be met satisfactorily by existing warning radars or by existing internal communication arrangements where accuracy and time was lost at every stage.

Thus, the requirements for target indication were, first, a radar set of high definition and data rate, and second, an indicating system allowing the controlling officer to make a rapid assessment of the overall situation and permit a swift response in pointing out to the weapons the targets to be engaged. For the radar, the full requirements were that it should provide accurate bearing, range and elevation of all air and surface targets within a radius of 15 miles from the ship, and that the information should be brought up to date at least once a second – to compete with the very high rate of change of bearing of aircraft that are close to.

As we saw on p. 179 above, a start was made in satisfying some of those requirements with type 293, based on the Staff Requirement of December 1942, but in the first set in *Janus* in July 1943, the aerial rotation speed (15 rpm) and bearing accuracy were inadequate and the range of detection was insufficient. Though these deficiencies were later alleviated by the larger aerial of type 293M, it was evident that this could only be a stopgap. Furthermore, the Target Indication Unit Mark 2 (TIU2), although superior to the earlier *ad hoc* systems, was too limited in its target-handling capacity and too slow in its response to the tactical situation. Therefore a new Staff Requirement was raised to cover the longer-term needs, and ASE prepared two guidance specifications, one for a new radar, type 992, and one for the Target Indication Unit Mark 3 (TIU3), to form the basis of a draft contract with EMI. These specifications were issued in October 1943 and work under this contract continued for the remainder of the war and after.

The TIU2 and its associated Range Transmission Units, first fitted in 1944, were used for passing to gunnery control positions – five in large ships, three in small – the bearings and ranges of targets seen on type 293, as selected by the target indicating officer.

Although the first fully-fitted Target Indication Room did not become operational until 1945 – in the 'Battle' class destroyer *Barfleur* – several large ships had interim systems fitted, often in the Aircraft Direction Room, during 1944.

THE LANDINGS IN SOUTHERN FRANCE

Whereas no aircraft carriers took part in the landings at Anzio or Normandy, those in southern France in July (operation 'Dragoon') involved seven British and two American escort carriers with some 220 fighters embarked, commanded by Rear-Admiral T. H. Troubridge in the light cruiser *Royalist*, which was fitted with the latest radar and communications. Also taking part were two fighter direction ships: FDT 13, fresh from 'Neptune', which controlled all aircraft over the Assault Area, whether for beach cover or for army support;[30] and *Ulster Queen*, fitted since Anzio with types 79B and 281B air warning, 277 surface and low air, and an RAF GCI set which offered long range, high accuracy and discrimination and a device to reduce land echoes, though its mattress aerial was far too large to be fitted in any but a specialist ship. But she was now fully equipped to direct both night and day fighters.

In September, a naval striking force of cruisers, destroyers and escort carriers under Admiral Troubridge moved to the eastern Mediterranean to discomfit the German evacuation of Crete and the Aegean islands. In these operations, the *Ulster Queen* played an important part. To quote Roskill:

> Nonetheless we found it very difficult to stop the evacuation of the German garrison from Crete by transport aircraft, generally by night; and it was not until the specially equipped Fighter Direction Ship *Ulster Queen* began to work off the coast in co-operation with the RAF Beaufighters from Egypt that our counter-measures became effective. By early October we had intercepted and destroyed nearly a score of the enemy's transport aircraft, and the evacuation was seriously checked . . .[31]

The following is from a report written immediately after the operations:

> During the eight days she [*Ulster Queen*] was actively engaged, an average of six Beaufighters [fitted with AI Mk.VIII radar] were available each night and a total of twenty-four interceptions were achieved, resulting in nineteen destroyed and five damaged. Of those destroyed, eleven were Ju.52s [the troop carriers] . . .
>
> For the most part, the enemy aircraft flew low and the 277 played as large a part as the GCI in effecting interceptions. It is noteworthy that the old naval set type 79 was responsible for one interception of an aircraft flying over an island 2,000ft high.
>
> The radar conditions in the Aegean are notoriously difficult having regard to the numerous islands and consequent land echoes. In these circumstances, the results obtained are particularly satisfactory.
>
> The interceptions were carried out by two naval FDOs who had received specialist night fighter control training at RAF Northstead. An RAF controller was present in an advisory capacity.[32]

NIGHT ACTION OFF NORWAY

In the autumn of 1944, German forces in northern Norway were being supplied by convoys proceeding northwards up the coast of Norway, wherever possible making use of the Norwegian Leads, those channels inside the island fringe which allow the voyage northward from Stavanger to be carried out in sheltered waters almost throughout.

Accordingly, it was decided that plans should be made for British forces of cruisers and destroyers to intercept and destroy such enemy convoys before they reached the protection of the Leads.

Accounts of the night actions which follow here and in Chapter 10 are given in some detail because they illustrate so well what had been achieved by the end of the war with British naval radar as far as surface actions were concerned.

For the first such operation, 'Counterblast', Rear-Admiral R. R. McGrigor sailed from Scapa late on 11 December 1944 with a force of two cruisers, *Kent* and *Bellona,* and four destroyers, *Myngs, Verulam, Zambesi* and the Canadian *Algonquin*. *Kent*'s surface warning radar was type 273Q, *Bellona*'s the older type 272P. Except for *Verulam* which had 272, the destroyers all had the newer type 293 or 276 as well as the 1.5m type 291. All but *Verulam* had PPIs in the plot and all had 50cm gunnery sets.

After darkness on the 12 December, the force crossed a suspected German mine barrier at 25 knots. Soon, the loom of Stavanger light was seen, indicating that an enemy coastal convoy was at sea. (At other times, the Germans extinguished shore lights.) Altering course parallel to the coast, the force was in line ahead, cruisers leading, a formation which the Admiral had stated he intended to maintain in any night action that might ensue. Only two orders would be expected, he said – 'starshell commence' and 'engage'. The night was pitch dark with no moon, sky overcast, with a cloud base of two or three thousand feet. The sea was smooth and there was a light offshore breeze from the north-east.

At 10.55pm on the 12th, *Kent* detected a group of ship echoes between 10 and 15 miles range, *Bellona* detecting it a few minutes later when the range had closed to 7.5 miles. The plots showed that the targets were about 5 miles offshore between Egersund and Lister, steaming north. At 11.14, the Admiral ordered 'starshell commence' and *Bellona*'s splendid 5.25in starshell illuminated the target area, revealing a northbound convoy of four merchant vessels with 7 escorts. *Kent* opened fire with 8in broadsides, straddling at just under 2 miles range a merchant vessel which her radar had been tracking for some time.

Part of *Bellona's* gunnery report has radar interest:

> Initial information of the enemy was passed from the tactical plot to the TS at 2302. Type 284 had the target 4 minutes later and 'Ready to open blind fire' was passed at 2307 at a range of 9,000 yards. Visual fire was in fact opened after the starshell had illuminated the target at a range of 4,800 yards [2.4 miles] at 2314.
>
> The target was shifted six times with no difficulty . . . Blind fire was not employed at all, all ship targets being clearly silhouetted by starshell [but radar ranges were used].[33]

The attack took the enemy completely by surprise and, in the course of a brisk action, two of the four merchant ships and all but one of the six escorts were sunk. Shore batteries joined in, but their fire, though spectacular, was ineffective and the British ships sustained no damage.

Reporting the action a day or so later, the *Daily Mail* somewhat naïvely dubbed it 'The Navy's revenge for 1940!'. More interesting if less spectacular is the summing-up by Captain Charles Norris of *Bellona*:

> This action proved to be one of those happy occasions when everything went well. The weather was perfect, the visibility good. We knew our position accurately to within a mile [largely because of radar], the action developed according to plan, the radar worked perfectly and, most important of all, the Gunnery reached the highest standards, and no breakdown of any material whatsoever occurred.
>
> This satisfactory state was not achieved by chance. It was founded on the fact that everybody knew exactly what the Admiral required in all circumstances. The time spent on exercises and conferences beforehand, with resulting teamwork, paid a handsome dividend. The rather overworked, but underfed, slogan of 'teamwork' has a very real meaning when it comes to fighting as a force against a multiplicity of targets at night, close in on the enemy coasts. It is essential.
>
> Constant practice, intelligence and forethought on the control and maintenance sides of all branches reaped their just reward.[34]

TAILPIECES

- By Admiral of the Fleet Lord Lewin:

> *Blind fire* – In May 1944 HMS *Ashanti* and the Polish destroyer *Piorun* were engaging two German minesweepers off the Minquier rocks near St Malo. The night was dark and the range was too short for starshell to be effective. The gunnery radar type 285 was working well, giving both range and bearing, but no datum for elevation. Having practised much bombardment recently in preparation for the coming invasion, CPO Docwra, the director layer, seized upon the idea of using the bombardment mercury switch, which provides an artificial horizon. Interception of German communications confirmed instant hits. Translated, the report was: 'All is lost! We have been hit in the engines and the officers' pantry.'[35]

- By Captain E. M. B. Hoare, flotilla navigator of the 18th Minesweeping Flotilla in *Ready* during the Normandy assault:

The jammer – A very large radar jammer [type 91] was fitted in the Captain's cabin. This looked like an outsize upright piano and was switched on during the cross-channel passage. It was very noisy and I believe very powerful. We had no orders about switching off when the operation was under way and I think we left ours running continuously for about a week. I believe the transmitter valve life was only about 24 hours.[36]

10
1945: The End of the War

Jan 24–29	British carrier raid on Palembang, Sumatra: 'Meridian'
Feb 10	Main body of BPF arrive Sydney
Mar 15	BPF join US Pacific Fleet under Adml Nimitz
Mar 26	BPF in assault on Okinawa: 'Iceberg'
May 8	V-E Day
May 15	Japanese cruiser *Haguro* sunk by 26th Destroyer Flotilla
May 28	First Allied air attacks on Japanese mainland
Aug 6	First atomic bomb: Hiroshima
Aug 9	Second atomic bomb: Nagasaki
Aug 15	V-J Day

ANOTHER NIGHT ACTION OFF NORWAY

> The outstanding point of gunnery interest is the performance of *Norfolk*'s GS radar type 274. The ship was never in doubt of the accurate range and bearing of the individual targets, despite the closeness of the land, while, of sixty-four broadsides fired, only four were not spotted by radar. Results with types 284 and 285 were meagre in comparison.
>
> Although the fitting of beam switching to *Bellona*'s type 284 has given greatly increased bearing accuracy, it has caused a 15% reduction in range and in this action radar spotting on four-gun salvoes was almost impossible beyond 5,000 yards.
>
> The value of radar calibration on shell splashes prior to meeting the enemy was fully demonstrated in *Norfolk* who carried out a calibration in the forenoon and obtained hits with the opening broadsides that night.[1]

So reads part of an Admiralty staff minute, commenting on Operation 'Spellbinder' off the Norwegian coast in January 1945, in which Admiral McGrigor led two cruisers, *Norfolk* (*Kent*'s sister ship) and *Bellona*, and three destroyers, *Onslow*, *Orwell* and *Onslaught* (Force 1) on an operation with precisely similar intentions to 'Counterblast' a month earlier (see pp.

232–30 above) – to intercept and destroy a northbound German convoy off Egersund. This was coordinated with 'Mitre' and 'Gratis', operations by the fast minelayer *Apollo* supported by two destroyers (Force 2) and aircraft from the escort carriers *Premier* and *Trumpeter* supported by the cruiser *Dido* (Force 3), both to lay mines across the southern entrance to the Norwegian Leads. By lucky chance, the coordination of 'Spellbinder' and 'Mitre' could hardly have been improved even though the timing of *Apollo*'s minelaying had to be decided several days ahead, whereas the timing of the convoy operation depended upon enemy intelligence which had to be acted upon immediately.[2]

Norfolk had barely completed working up after a long refit during which she had had the latest radar fitted (surface gunnery type 274 and warning sets 277, 293 and 281B) together with the latest action information and target indication equipment and layout. Of the destroyers, *Orwell* had been fitted with type 293M and a proper plot a week earlier and was full of praise. The other two only had 271 or 272, Captain Hugh Browning reporting, 'The lack of a PPI [in *Onslow*] was keenly felt, the target situation remaining obscure until *Bellona*'s excellent illumination brightened the scene.'[3]

Forces 1 and 2 left Scapa together after dark on 10 January and the *Apollo* and supporting destroyers were detached for the former's minelaying operation early on the 11th. Approaching the Norwegian coast, Force 1 got a good radar fix at 7.40pm and, shortly after, *Bellona*'s radar direction-finding equipment FV1 picked up transmissions from a German Coastwatcher radar which gave no indication of having detected the British force. Then, at 10.12, the FV1 detected a Giant Würzburg shore radar which almost immediately stopped normal sweeping and appeared to concentrate the beam on the British force. At 10.50, all shore lights were extinguished, confirming that surprise had been lost, so that, when at 11.34 a second Coastwatcher station was detected, *Bellona* started jamming it.

What the plot quickly proved to be a group of some eight ships steaming towards Egersund was detected by *Norfolk*'s 277 at 17 miles distance at 11.15, and the admiral in *Norfolk*, leading a line of five ships, altered course to intercept. Six minutes later, her 274 gunnery set detected the target at 14 miles and blind fire could have been opened any time after the range had closed to 9 miles.

Then, just as the admiral was considering giving the order to fire starshell to illuminate the convoy on the port bow, several ships obtained a small echo 4 miles away on the starboard bow. The asdic (sonar) reported 'Slow-running turbines'. A possible – no, a probable U-boat! To counter the threat from Gnats (German naval acoustic torpedoes), the force turned together towards the new echo and reduced speed. At 11.41 while still turning, *Norfolk* opened fire with both starshell batteries at once, to illuminate the unknown contact to starboard and the convoy to port, by now 5 miles away. Of the unknown echo, nothing was seen and, immediately the guns fired, the echo faded – presumably the U-boat diving.

Norfolk and *Bellona* both opened fire with main armament on the convoy ships already held by radar. 'The target was seen to be a medium sized merchant ship,' reported *Norfolk*, 'the first broadside was a direct hit and this target was seen to sink after *Norfolk*'s fifth broadside was fired.'[4] Of the convoy of eight ships, two deeply laden merchant vessels and one of the six escorts were sunk in the action which followed, in which the destroyers played a full part. Then, at 12.25, an explosion was heard which could have been a Gnat detonating in someone's wake. Force 1 retired westward, turning north when clear of German shore radar to rendezvous with the carriers in Force 3, which were to give air cover during Force 1's voyage home before proceeding with their own mining operation.

Meanwhile, *Apollo* had laid her mines off Karmoy Island, 60 miles or so to the north – undetected and as close as half a mile from the shore. Once again it was radio and radar which made this possible. She made her approach by running down the beam of the German radio navigational aid Sonne (called Consol in Britain) transmitting from Stavanger. The lay was controlled by the navigating officer, Lieutenant John Noble, assisted by the radar officer Sub-Lieutenant Colin Biggs RNVR, navigating from the chart house almost entirely by radar, never seeing the Norwegian coast themselves.[5]

3-CENTIMETRE SURFACE WARNING RADAR: TYPE 268

In 1942, at the request of the Admiralty, the National Research Council of Canada had started development of a high-resolution radar on the 3cm waveband (X-band) for use in small craft, dubbed type 268.[6] Development and production were beset by difficulties and the first production models did not reach the United Kingdom until January 1945, when they began to be fitted in MTBs and other coastal craft. (See Appendix E.)

Wartime decisions were taken that this set should be fitted in 'Hunt' class destroyers and *Algerine* class minesweepers, but, so far as is known, none had been so fitted before the war ended. However, 1,600 sets were produced and were extensively fitted in the Navy after the war for navigational purposes, and several hundred in merchant ships. The set's design had a great influence on the ASE's postwar specification for the UK marine radar for the merchant navy, discussed in Chapter 11, as did the work at ASE on X-band in connection with the projected long-range gunnery set type 901.

THE 'SCHNORKEL' THREAT

Late in 1944, a new and potentially most damaging threat to Allied shipping became evident – the U-boat fitted with the 'Schnorkel', permitting it to operate submerged for long periods, showing only a very small radar target above water, not easy to detect with 10cm sets then fitted in anti-submarine vessels. In December 1944, Marconi started to develop at the highest priority a new stopgap radar set, type 972, particularly for the 'Loch' and 'Castle' class A/S frigates then building and due to be fitted with type 277. This new set was a modified 3cm type 268, the panels of which were to go in what was to have been the 277 office of the frigates, and the specially designed 8ft cheese aerial was to go on the type 277 pedestal already stabilized in both azimuth and elevation. Trials of a prototype set took place in the radar training ship *Pollux* in May 1945 – just as the German war was ending and the 'Schnorkel' threat had disappeared.

Although type 972 was never fitted in frigates, it was nevertheless put to good use in surveying ships, with an accurate ranging panel added.[7]

Meanwhile, with the end of the war in sight – or at least the end of the German war – thought had been given to how modern radio navigational aids could best be applied to the Merchant Navy. In the summer of 1944, a Government body known as the United Kingdom Conference on Radio for Marine Transport was set up to investigate the matter. One of the most important of these aids was radar and the committee asked ASE to collaborate in preparing a specification for the guidance of commercial radio manufacturers for a radar set suitable for the average merchant ship.

Talks with representatives of the Merchant Navy showed that the first need was for navigation and pilotage rather than for a simpler set for collision avoidance only. The set must have a high reliability and be simple to operate, with pictures easy for masters and watchkeeping officers to interpret direct. With the knowledge gained from these talks and with experience with sets of various characteristics in over a year of trials and operations, particularly in the Scheldt with type 971, ASE was able to produce the first draft specification early in 1945, specifying that the set should operate on the 3cm wavelength band, and suggesting a peak power of 40Kw, a pulse length of a quarter of a microsecond, and a high aerial rotation rate.

After discussion with various shipping representatives, the specification was issued by the Ministry of Transport to ship owners and to the radio industry. At the same time, ASE began to make an experimental radar that would conform fairly closely to the major electrical features of the specification, to achieve high definition and good minimum range, with a beamwidth of 2°, range discrimination of 40 yards, and a

47. The frigate *Loch Killisport* in 1945. Principally for A/S work, she had WS type 277 and HF/DF. Compare with *Erne* in Fig.23. *(IWM)*

minimum range rather better than 50 yards, in a set which would be easy to handle.[8] Meanwhile, a little before the end of the Japanese war, representatives of the shipping industry were given a demonstration at sea in *Pollux* of the capabilities of type 972, whose characteristics were not too far removed from those in the specification.[9]

RADAR FOR THE FAR EAST

As soon as the success of the Normandy invasion seemed assured, the Admiralty had turned its attention to plans whereby Britain should play a full part in the war against Japan – in South East Asia, and alongside the US forces in the Pacific theatre. The Eastern Fleet in the Indian Ocean was successively reinforced during 1944 and, on 22 November 1944, Admiral Sir Bruce Fraser became Commander-in-Chief of the new British Pacific Fleet in Ceylon, initially allocated five fleet carriers, two battleships, seven cruisers and twenty-two fleet destroyers, though not all ships had yet arrived on the station. The main body of the Fleet reached Sydney in mid-February 1945.

For the Navy, this was to be primarily an air war and in the second half of 1944 the highest priority had been given to modernizing the fighter direction and anti-aircraft capabilities of ships destined for the Pacific. Projects 'Bubbly' and 'Knobbly' were ASE crash programmes of the highest priority to help achieve this.

'Bubbly' set out to upgrade existing warning radar equipment in ships. To give some flavour of the situation at the time, let 'Bubbly' be described in the words of Captain A.V.S. Yates, Head of the Application Department of ASE in a lecture he gave to officers of the British Pacific Fleet in July 1945, first in the hangar of the escort carrier *Arbiter* at anchor off Manus Island, north of New Guinea – 'ghastly acoustics, 100°F, strapped six inches from the mike, and I spoke for 1 hour 20 minutes which was too long. All the cruiser captains came and were pleasantly entertaining afterwards'[10] – and a second time in the Shell Mex cinema in Sydney, NSW, to an audience of 140, happily for posterity, recorded in shorthand:

> . . . I want to refer to a programme which we started some six months ago. It was then appreciated [January 1945] that the war was taking on a different form – we had turned the corner and it did not seem that there was much chance of losing it. The Controller [of the Navy] ruled that ships refitting must absorb a minimum of electrical labour and it therefore appeared that only about one third of the Fleet would be able to fit our newest sets.

> While continuing to progress our long-term developments, we framed a programme which enabled the existing sets 277, 293 and 281 and their displays to be modernized by ships' staff, without help of a dockyard. Fleet Train [see p. 247 below] help may be needed in smaller ships.
>
> We divided this programme into two parts. First we produced by 'crash' methods sets for fifteen ships that we knew would be in dockyard hands and coming out this spring. Those sets have been fitted in ships of which the first out here is the *Duke of York*. She is the forerunner of the first part of what we call the 'Bubbly' programme. *Anson, Belfast, Ontario* are other ships on the way here. *Jamaica, Glasgow, Superb, Mauritius, Vindex, Dido* and up to fifteen ships will arrive on this and the East Indies station with these 'Bubbly' modifications, so those of you whose job it is, if you can find time to go and see one of those fifteen, I hope you will be impressed with what we are doing for this short-term programme, and be ready to do likewise yourselves when the stores arrive.
>
> The second part of the programme was the bulk production of these modernization kits. The object of giving the programme a name was to ensure it would not be tampered with piecemeal, in priority, when something new cropped up . . .[11]

The principal modifications in the 'Bubbly' programme improved the performance of the receivers of types 281, 277 and 293; gave types 277 and 293 redesigned panels, removing external leads which had previously given such a 'Christmas-tree effect' in radar offices; gave type 281 continuous aerial rotation (so important with PPI and Skiatron displays); and improved the displays themselves. The modified sets then became types 281BQ, 277P and 293P. Meanwhile, a yet wider aerial for type 293 had been designed for cruisers and above and this, 12 ft wide, was approved for fitting just before the end of the war, though none was actually fitted before 1946, to make the set type 293Q. Somewhat later still, type 277 also got a larger aerial, to become type 277Q.

The ships most completely equipped from an AIO and warning radar point of view were probably the refitted battleships and cruisers mentioned by Captain Yates above and the newly-built fleet carriers *Indefatigable* and *Implacable* and light fleet carriers *Venerable, Vengeance, Colossus* and *Glory*. The two heavy carriers had joined the Home Fleet in mid-1944, *Indefatigable* joining the BPF before the end of 1944, *Implacable* in June 1945. The light fleet carriers, having left Malta for the Far East on VE Day, later sailed from Sydney to join the main Pacific Fleet just as the sirens were sounding to announce the Japanese surrender. But even they did not have the full radar treatment: *Colossus*, for example, did not have continuous rotation on her type 281B though she did have a fully fitted

48. The battleship *Duke of York* in the Pacific in 1945. She had modern radar – WA type 281B, WS type 277, WC type 293, GS type 274, four GA type 285, and seven GC type 282.

(IWM)

ADR with Skiatrons, RDR, etc. She had a five-channel target indication unit for use with her 40mm guns, but not a fully-fitted Target Indication Room as such. The heavy carriers were as well equipped as the light fleets except that they had no RDR and were somewhat more cramped. Incidentally, the full treatment included the air conditioning of AIO compartments, the only parts of the ship to be so treated, making visits to the Ops. Room and ADR most popular in the tropics for those who

49. The Aircraft Direction Room of the light fleet carrier *Venerable* in April 1945. One of the intercept positions with skiatron and PPI is in the foreground, facing the Main Air Display Plot and stateboards.

(HMS Dryad)

normally worked on the bridge – or even more so, those who worked in the engine room.

These various modifications were often made piecemeal, particularly with operational ships, as with Project 'Bubbly'. The first ship to have the whole package – when the metric set came to be known as type 281BQ – was *Anson*, who completed her refit in March 1945. Several other ships were fitted before the war ended and at least one even managed to hoist up and mount the new type 281BQ aerial assembly at the top of the mast by ship's staff alone.[12]

Big ships in the Eastern and British Pacific Fleets were given American portable 'after action' type SQ surface search sets. Working at 10cm, these were designed also for use in boats with a single 3in CRT which could be switched at will to show type A, type B or PPI display. They were stowed down below for use if the main sets were destroyed.

FIGHTER DIRECTION SHIPS

Another project which was given a name and very high priority by ASE was known as 'Knobbly' – the fitting out of long-range fighter direction landing ships (LSF) for assault operations in the Pacific. What were needed were Fighter Direction Ships, equipped to direct RAF as well as naval aircraft both by day and by night, which could operate with assault escort carriers with a speed of at least 16 knots,[13] which none of the existing fighter direction ships – *Ulster Queen, Palomares* or *Stuart Prince,* or the FDTs used at 'Neptune' – could achieve. The Admiralty had therefore decided in mid-1944 that three large tank landing ships (LST 1), *Boxer, Bruiser* and *Thruster,* capable of 18 knots, should be converted.[14] Built in 1942, these ships had served with much distinction in all the Mediterranean assaults up to Anzio.

50. Fighter direction ship *Boxer* in July 1945, viewed from her port quarter. Aerials, *right to left*: WA type SK on quarterdeck; after fighter direction set SM-1 on after deck house; WA type 281BQ on No.4 mast, and WA type 79B on No.3 mast; homing beacon YE abaft funnel; VHF/DF, HF/DF and communications aerials on Nos. 1 and 2 masts; IFF interrogator type 245 at back of bridge; WS type 277 either side of bridge; forward fighter direction set SM-1 on No.1 mast at bridge level; RAF GCI for night interception in dome on forecastle.

(NMM)

In the event, the only one of the three LSTs to complete the conversion to LSF was the *Boxer*, taken in hand about October 1944 and completed in time to be entering the Suez Canal on her way to join the BPF on the very day hostilities with Japan ceased. For long-range air warning, she had British types 79 and 281 and US type SK; for heightfinding and fighter interception, she had two US type SM-1; for surface and low air warning, two type 277; and for night direction, an RAF GCI set.[15] She carried RAF controllers as well as naval fighter direction officers. Because of the multiplicity of sets and their variety of origins, the design of suitable display systems and ensuring that the sets did not interfere with each other posed many problems for ASE.

Of the earlier fighter direction ships equipped originally for convoy escort duties, we have seen how successful was the *Ulster Queen* in the Mediterranean. *Stuart Prince* commissioned in December 1943, was in the Mediterranean for much of 1944, and was in Sydney by May 1945. *Palomares* had been allocated to the Eastern Fleet when she was mined off Anzio in January 1944. She recommissioned after damage repairs in March 1945 just before the end of the war with Germany, fitted additionally, not with GCI as *Ulster Queen* had been, but with the American SM-1 and SK, as well as types 79B and 281B. She had reached Massawa by V-J Day. FDT 13, with types 277 and 291 fitted instead of the RAF type 11, reached Malta a week after V-J Day, before returning to pay off.

CENTIMETRIC GUNNERY RADAR

There were equally high priorities for fitting BPF ships with centimetric gunnery radar. Of the ships mentioned by Captain Yates, the low-angle type 274 was fitted in the battleships *Duke of York* and *Anson* and cruisers *Belfast*, *Glasgow* and *Ontario* (Canadian, ex-*Minotaur*). The high-angle type 275 was fitted in *Anson*, *Ontario* and new 'Battle' class destroyers *Barfleur*, *Armada*, *Camperdown*, *Hogue* and *Trafalgar*. Only *Anson* and *Barfleur* reached the Pacific before the war's end, and neither fired a shot in anger. *Ontario* had got as far as Malta for working up but the other destroyers had not yet left the United Kingdom. Testing type 275 under battle conditions had to await the Korean war.

As for the 3cm close-range type 262, the first trials of the radar system early in 1944 led to non-firing trials with the set mounted on a STAAG early in 1945, and firing trials at Eastney AA Range on a 'rolling platform' in June 1945. These culminated in fully realistic firings against practice targets, and a demonstration for the Admiralty and for those concerned from industry, at Eastney on 4 July, when a towed target was shot down

51. GC type 262 on a 40mm STAAG mounting, showing (1) scanning aerial mirror; (2) radar operator's control panel; (3) control officer's remote radar display. This 3cm search-and-lock-on radar first became operational in the light fleet carriers *Ocean* and *Triumph* in 1945.

(Radar Manual, 1945)

with the first salvo. The first CRBFD was also tested ashore in 1944, but the first sea trial took place rather earlier than STAAG, in the light fleet carrier *Ocean* early in 1945, closely followed by a fully operational fitting in her sister ship *Triumph*.

The results of these trials led to the prediction by Naval Staff Division closely concerned (but unnamed) that the project 'will be comparable with the advent of the breech loader in the age of the muzzle-loading gun'.[16] Sadly, postwar sea experience proved this prediction to be somewhat over-optimistic, largely owing to accessibility and maintenance difficulties.

Both ships sailed for the Far East soon afterwards but, like type 275, type 262 had to await the Korean war for battle experience. The first seagoing STAAG was fitted on the top of a forward main armament gun turret in the battleship *Vanguard* late in 1945.

THE SINKING OF THE *HAGURO*

On 9 May 1945 (the day after V-E Day) HM submarine *Subtle* in the Malacca Straits sighted the heavy Japanese cruiser *Haguro* and a destroyer on a north-westerly course (they were bound for the Andaman Islands), which she and her sister submarine *Statesman* attacked without success. As soon as *Subtle*'s report was received, the main East Indies fleet under Vice-Admiral H. T. C. Walker in the battleship *Queen Elizabeth* sailed from Ceylon to intercept. On 11th May, a Japanese reconnaissance aircraft sighted the British force, whereupon *Haguro* reversed course back into the Malacca Straits. Walker, however, expected the enemy to make a second attempt, so took his force well to the south to avoid being resighted. On the night of the 14th/15th, he detached his escort carriers and the 26th Destroyer Flotilla (Captain M. L. Power) to search the waters north of Sumatra.

At 11.50 am on 15 May, a dramatic signal was received from an Avenger aircraft of the carrier *Emperor*, 'One cruiser, one destroyer, course 140°, speed 10 knots.' The position given was just inside the northern entrance to the Malacca Straits.

Captain Power immediately ordered his five destroyers – *Saumarez, Venus, Virago, Verulam* and *Vigilant*, all with 10cm radar, PPIs and modern plotting arrangements – to increase speed to 27 knots, steering ESE. The flotilla was in line abreast, 4 miles apart, with the *Venus* (who had reported her radar was working exceptionally well) on the port wing, 8 miles from Captain (D).

Through the afternoon and early evening, nothing significant was seen or detected by radar except for the fuzzy echoes of many rain squalls on the PPI. In *Venus*, the ship's company was at relaxed action stations and Ordinary Seaman Norman Poole was on watch in *Venus*'s plot immediately below the bridge. About 10.40pm, when the flotilla was nearing the limit of the search, he was studying a rain squall echo some 35 miles away when he saw embedded in it what seemed to be a solid target. From everything he had been taught, type 293 could not possibly detect a ship at such long range, so Poole asked the bridge whether there was any land on that bearing. No, they said. But to Poole it looked like a surface target, so he spent a few more minutes studying the echo and plotting it before he reported to the bridge – a ship echo, bearing 045°, 34 miles.

We now know that this echo was indeed *Haguro* and that the unusual range of detection was the result of anomalous propagation, the effects of which – and the weather conditions that cause it – were at that time not always properly appreciated at sea. Commander H.G.D. de Chair, *Venus*'s captain, was therefore justifiably sceptical and told Poole it was probably a cloud, but to plot it anyway. Five minutes later, Poole still

52. The destroyers (*left to right*) *Virago*, *Venus* and *Vigilant* seen from *Verulam*, taken in May 1945 just before the night action in the Malacca Straits in which the Japanese cruiser *Haguro* was sunk, when WC type 293 radar played such an important part.

(IWM)

insisted that it was a surface target, and gave a course and speed – 125°, 25 knots. With that, the full team was summoned – the navigating and radar officers, the schoolmaster (the plotting expert), and the radio mechanic to the plot, the Leading Seaman (Radar) to the radar office. The radio mechanic fiddled with the PPI knobs, upsetting Poole's settings, whereupon the echo (which only Poole had so far seen) disappeared. When Poole was allowed back on the PPI, he readjusted it. He was sure this was a ship echo even though no one else could see it. Once again, he insisted he was right – and said so forcibly.

After the echo had been plotted for another quarter of an hour, all in *Venus*'s plot became convinced that it was real and that it was a ship. At 11.22, de Chair reported the contact to Captain Power in *Saumarez* – bearing 040°, 23 miles, course 135°, 25 knots. Power was as sceptical as de Chair had been, but *Venus* insisted, reporting at 11.32 that the target was altering course to starboard. Six minutes later, it steadied on a southerly course, heading straight for Singapore, with the British destroyers in an extended line some 20 miles on her starboard bow.

By 11.45 Captain Power was persuaded that this was the real thing and sent an enemy report to Colombo, at the same time ordering his destroyers to turn north to close the range, allocating each one a different sector for a coordinated 'star' torpedo attack. At 11.47, *Venus* lost contact but *Virago* got contact ten minutes later and *Saumarez* herself three minutes after midnight, *Venus* regaining contact soon after that.

For a blow-by-blow account of the subsequent action, the reader should look elsewhere.[17] Suffice it to say here that Captain Power ordered the destroyers to get into their attack sectors at 12.39, saying that he intended to fire his torpedoes at 1am; that the echo was seen to split into two at 12.48, the second echo proving to be the destroyer *Kamikaze*; that in *Saumarez,* the type 285 gunnery radar was 'put on' to the target by type 293 warning radar at a range of 15,000 yards (7.5 miles) at 12.50; and that, after firing starshell, she engaged *Haguro* at 5,000 yards (2.5 miles) at 1.05, her gunnery radar showing her first salvo to be some 400 yards short, subsequent salvos hitting after an 'up 400' order. *Haguro* returned the fire and *Saumarez* was hit in a boiler room but managed to fire her torpedoes at 1.13, followed a minute later by *Verulam,* three torpedo hits being observed. *Haguro* sank after receiving further torpedo hits from the other destroyers. *Kamikaze* escaped.

In his subsequent report on the action, Captain Power made several remarks under the heading of 'Personal impressions' which show very well how radar was used in this action where it played such an important part:

> At first contact, I went into the plot and looked over the shoulder of the PPI interpreter (the Anti-submarine Officer), watching the situation

develop and handling the Flotilla from there by TBS ['Talk Between Ships' radio] direct through the Signal Officer who is stationed . . . in the Plot. The PPI was clear-cut and covered the whole area of the action . . .

The timing of the move from plot to bridge [of the senior officer or captain] is always a subject of controversy. This one turned out right because although in the plot I did miss personal observation of important tactical developments, I had time to settle down comfortably on the bridge before the action developed. Information from the plot and PPI came clearly and adequately over the Action Information loudspeaker, but – and I think this will always be so at close ranges in fast-moving ships – appreciation of change of enemy course and speed in the plot is too slow to prevent surprises on the bridge.

The starboard wing ship (*Vigilant*) was having radar trouble and had difficulty in sizing up the situation for lack of information. It is dangerous to assume that all plots are as clear as one's own.[18]

Ordinary Seaman Norman Poole received the immediate award of the Distinguished Service Medal for 'outstanding service as radar operator in detecting an echo on the plot PPI at a phenomenal range. The courage of his convictions almost to the point of insubordination convinced the plot that his echo was an enemy cruiser which was presently surrounded and destroyed.'[19]

THE BRITISH PACIFIC FLEET

The culmination of the development of British naval radar in World War 2 may be seen in the operations of the British Pacific Fleet (BPF), formed in Ceylon on 22 November 1944. Nearly all its ships were fitted with the very latest radar and action information arrangements – plotting and communications.

Because the likely area of operations was more than 3,500 miles from the main British base, Sydney, an essential part of the BPF was floating and mobile support in the form of the Fleet Train of tankers, store ships, repair and maintenance ships, floating docks – and even a radar/radio maintenance ship, HMNZS *Arbutus*, a converted corvette capable of minor repairs, carrying a limited stock of spares in her magazines and shell rooms. The replenishment forces were protected by escort vessels and aircraft from escort carriers, the latter also carrying replacement aircraft (and removing damaged duds).

For the BPF, the Pacific war was almost entirely an air war in which seaborne radar played a central role in the air defence of the fleet – in warning, in fighter direction, in fire control. Except for submarine operations and a few shore bombardments by battleships and cruisers, offensive action was limited to air strikes from carriers. In brief, the BPF's operations can be divided into three phases: (1) January 1945, air attacks on oil refineries in Sumatra while on passage between Ceylon and Australia (operation 'Meridian'); (2) March, April and May 1945 as Task Force 57, the neutralizing of Japanese airfields in the Sakashima Gunto (and a few attacks in Formosa) to cover the southern flank of the American assault on Okinawa (operation 'Iceberg'); and (3) July and August 1945, operating as part of the US 3rd Fleet as Task Force 37 in the final assault on the Japanese mainland.

In all these operations, the central core was the carrier force, generally four fleet carriers (which would have been augmented by four newly-built light fleet carriers had the war lasted only a few weeks longer), supported by one or more battleships, six or seven cruisers, principally for gun defence, and up to eighteen destroyers. And always a few hundred miles in the rear, units of the Fleet Train ready to replenish the fighting ships' fuel, food, ammunition and stores in intervals between active operations.

This meant very long periods at sea during which radar was continuously manned: when the fleet reached Sydney for repair and replenishment after 'Iceberg', it had kept the sea for 62 days, 23 of which were 'strike' days, broken only by 5 days in Leyte. Air attacks were frequent, made the more dangerous by the Japanese use of the Kamikaze suicide bomber, first experienced by the BPF on 1 April when *Indefatigable* was hit on the base of the superstructure; her armoured deck saved her from serious damage and within a remarkably short time she was able to continue operating aircraft. Her sister carriers *Formidable, Indomitable* and *Victorious* were all subsequently hit but remained operational despite damage.

At first, the Kamikaze attacks were by small numbers of expert volunteers, but the Japanese then decided to expend any pilot not required for special duties, in mass attacks. On 6 April, over 700 aircraft attacked the US Task Forces taking part in the Okinawa assault, and 400 got past the defending fighters. Heavy attacks continued and the Americans suffered such losses and damage that the whole landing operation was in jeopardy. It was clear that many of the attacks were coming from Formosa where the aircrew were all trained. General MacArthur refused to use his air force to assist, so Admiral Spruance, commanding the US Fifth Fleet, with his fingers crossed (his own words) sent Task Force 57 – the BPF, under Vice-Admiral Sir Bernard Rawlings, with Rear-Admiral Sir Philip Vian commanding the Aircraft Carrier Squadron in *Indomitable* (fitted with the American heightfinding radar SM-1, described on p. 201), with Commander E. D. G. Lewin as Force FDO – to neutralize the two main airfields in north Formosa. This was done with sufficient success to divert much of the Japanese effort to attacking Task Force 57.

53. The carrier *Formidable* just after a Japanese suicide attack in June 1945. Except for type 276 instead of type 293, she had the same radar as *Victorious* in Fig.36.

(IWM)

The Kamikaze attacked the British task force at both high and low level, seldom in groups of more than six aircraft. The defence would normally have sixteen to twenty fighters on patrol at three different levels, and another sixteen to twenty-four at readiness on deck. Height estimation at first detection was good enough to get the right combat air patrol (CAP) started on its way. For the high attacks, all would have been well if the attack flew at a constant height, for the fleet had become remarkably proficient at height estimation by the echo-amplitude method (see Appendix C): by the time the target had come in to 50 miles, the result of pooling the fleet's estimates, with a bias in favour of the most expert ships, was generally within 1,000ft of the true height and always within 2,000ft. Unfortunately, any aircraft flying level turned out to be a passing Liberator while the attacking aircraft made drastic alterations in height as they approached. Occasionally, it was possible to detect the change of height by plotting echo amplitude, but in general the fleet relied upon *Indomitable*'s SM-1 for heights, type 277 having insufficient range to give useful information in time; the raids split into individual attacks as soon as they sighted the fleet, so it was highly desirable to intercept well before they did so.

The Japanese were fully aware of the deficiencies of Allied radar for low air cover, so low attacks presented an even worse problem. The US Navy had developed a system of early warning by having destroyer 'pickets' fitted with SM-1 and SP radar and a YE homing beacon, manned by hand-picked fighter direction teams and accompanied by an AA destroyer. These were stationed 50 miles apart and 50 miles towards the threat, and, in addition to defence against low attack for which they had their own CAP, they were required to 'de-louse' returning friendly strikes which were routed over the pickets, and which might be closely followed by Kamikaze intruders. The Japanese soon realized the pickets' importance and attacked them heavily.

No British destroyer had suitable radar, so the BPF tried at first to do without pickets. The result was that the raids would be detected by type 277 at 20 to 27 miles before splitting into individual tracks and climbing to 2,000ft before dive bombing. In the words of the Fleet FDO (Commander David Pollock, RNVR), the PPIs then looked like the bursting of a small bag of peas.[20] The fighters could intercept at 12 miles at best and would often follow their targets well into the gun zone, leaving the gunners wondering whether or not to hold their fire. Normal radar-controlled fighter direction was often impossible and a reversion to visual control was unavoidable. Fortunately, visual fighter direction had never been abandoned and every carrier had a visual FDO in the Air Defence Position, wearing a pilot's helmet to keep out the noise of aircraft and gunfire.

The difficulties proved unacceptable and the cruiser *Argonaut*, fitted with types 281 and 277 and accompanied by a destroyer, was stationed as the BPF's first picket 30 miles towards the threat. Fortunately, she survived.

For Phase 3, the final attack on the Japanese mainland, when the BPF formed part of a combined US/British Task Force, Admiral Vian had to transfer his flag to *Formidable* (and Lewin went with him) because *Indomitable* had defects which had to be remedied, so the BPF now had no SM-1. It was fortunate that the Japanese ability to attack was virtually exhausted. They mounted a last desperate sortie immediately before the surrender but none reached the main force, though they did attack the pickets.

Generally speaking, the British ships with their metric sets were superior to the US Navy in high air cover, but inferior in low cover and fighter direction because of the latter's 1.5m SK and 10cm SM-1 and SP radars. Commander Lewin had tried to make up for his lack of a heightfinder by agreeing with his American colleagues that all heights should be pooled. This was less one-sided than it may sound: the combined US/British Task Force encountered exceptionally severe anomalous propagation, so that the SM-1 beams were trapped just above the sea surface, providing wonderful low cover but no heightfinding; and the USN had neglected height estimation for which the SK was in any case not suitable because of the many lobes at 1.5m wavelength. Type 281 and the elderly type 79 came into their own.

Lewin's main problem was coping with the enormous number of aircraft on the PPI and the many communication nets. Admiral Vian in his memoirs paints a vivid picture:

> In the Aircraft Direction Room, Lewin gave an astonishing performance. On his head would be a pair of earphones, each phone listening to a different frequency. By means of a throat microphone, he could communicate with his opposite numbers in the American groups. In front of him was a radar screen. With this equipment, he directed our aircraft in the air with confidence and with the utmost skill.[21]

Radar in the Royal Navy had come a long way since *Illustrious* had sailed to the Mediterranean with a single type 279 in 1940. And five years earlier there had been no radar at all. All the radar in the BPF represented a tremendous four-year technical advance and a great tribute to those who designed and developed the equipment ashore and to those who maintained and used it afloat.

TAILPIECE

• From the *Daily Sketch,* 15 August 1945 (V-J Day). The ship was the radar training ship *Pollux*. The first lieutenant was Lieutenant T. B. McAulay, RNVR:

> *The stranger* – The small quiet man stood by himself in the wardroom of one of HM ships recently. The Navy was taking representatives of big British shipping interests to sea to show them the wartime progress made in developing Radar and how, post-war, it would make navigation almost foolproof in any weather.
>
> The first lieutenant of the ship, which was on experimental trips for all three Services, noticed the quiet man by himself and called for a gin for him. He asked (making conversation) – 'Your first acquaintance with Radar, I suppose, sir?' The stranger coughed – 'Well . . . not exactly. My name is Watson-Watt.'
>
> Sir Robert Watson-Watt is the Radar pioneer.

11
Wartime Projects Postwar

During six years of war, shipborne radar had made such strides that, by 1945, a modern battleship or cruiser had up to three warning and as many as fifteen gunnery radars. However, owing to the rapid rate of development under the impetus of war, much of the equipment fitted was already obsolescent. Furthermore, the advent of jet-propelled aircraft and guided weapons meant that very different design targets would have to be met. All this was fully appreciated in the Admiralty and at ASE, and work on new systems and techniques was proceeding rapidly in 1945. Then came the surrender of Japan and peace, which removed much of the impetus, as well as leading to the exodus of many first-rate people from direct government service at ASE. In Britain, the bleak economic prospect forced a policy of making do with what was already fitted, with some limited research and development, one of the results of which was that the British ships which took part in the Korean War of 1950–3 had largely to rely upon World War 2 radar, though the war itself did result in a measure of rearmament.

The purpose of this chapter is to round off the main wartime radar narrative by describing the postwar fate of those radar projects left unfinished when war ended. Projects initiated after 1945 will be mentioned only where they form part of the story of projects started during wartime.[1]

MERCHANT NAVY RADAR

About a year before the end of the war, the Admiralty and ASE began to take an active interest in the future of radar for the Merchant Navy and started what was to become an intensive programme of research and development on PPI-fitted high-definition centimetric radar specifically for navigation and collision avoidance. By the spring of 1945, ASE had drafted a suggested specification for a Merchant Navy radar, making use of the experience in using radar for navigation gained with type 971 off Normandy and in the Scheldt, as well as drawing upon the design features of the new Canadian-built 3cm type 268. Before V-E Day, a

Merchant Navy-oriented radar survey of the Thames estuary had been made, using an existing 3cm type 972 in *Pollux* (see p. 235 above and demonstrations at sea had been given to representatives of the Merchant Navy and the radio industry. Following this, extensive navigational trials were carried out.

In October 1945, the radio industry was invited to send representatives to ASE for discussions on operational and technical requirements for navigational and collision-avoidance radar to meet Merchant Navy needs. The Royal Navy's navigational wartime experience with centimetric radar was discussed and demonstrations given. This gave the United Kingdom industry a good and early start in the Merchant Navy field.

By January 1946, ASE had produced a 'cobbled-up' experimental model roughly conforming to the draft specification and had given further demonstrations in *Pollux* to shipping and radio industry representatives. A few months later, ASE had produced a further experimental model fully meeting the draft specification, fitted it in the sloop *Fleetwood*, and carried out trials in the English Channel. In May 1946, this radar was demonstrated to interested participants in the first International Meeting on Radio Aids to Navigation, then in session in London. At that meeting, an international specification close to that proposed by ASE was formally agreed.

From 1946, as an interim measure pending the production of commercial navigational sets, a considerable number of merchant ships – 260 by April 1948 – were fitted with ex-naval 3cm Canadian type 268 sets. Though they did not quite measure up to the *Fleetwood* specification, they were a magnificent stopgap much praised by all who used them.

By April 1948, six manufacturers in the UK were engaged in the development of shipborne radar fully meeting the agreed Merchant Navy specification, three being in quantity production. Some forty British ships and eighty foreign ships already had commercial equipment by this date.[2]

As for the Royal Navy, not only was type 268 fitted as the main surface warning set in 'Hunt' class destroyers and in minesweepers and coastal craft, but it was also fitted as an interim measure in cruisers and above until a dedicated navigational set became available – type 974, a naval version of the Decca 12, which began to be fitted in the early 1950s.

GUNNERY RADAR

When the war ended, large numbers of ships were still equipped with the 50cm gunnery sets developed in 1940, but a start had been made in replacing them with the new centimetric gunnery sets, the furthest ahead being the 10cm type 274 for controlling main armament surface fire in large ships (see pp. 183–4). Before the war ended, six capital ships and six cruisers had one or more type 274s in place of 284 and, during the first ten months of peace, the new battleship *Vanguard* and six more cruisers had also been so fitted.[3]

As mentioned on p. 224, a 1.25cm splash-spotting set, type 931, had been developed in Canada for use as an adjunct to type 274. Though its range-spotting capabilities were an improvement on those of type 274 itself, the important advance was that it provided quantitative spotting for line, which type 274 could not do. With type 931, errors in both range and line could be read off the display directly and spotting corrections made accordingly.

Twelve development models were produced in Canada for the BPF but, before they could be completed, the war was inconsiderately brought to a close. The completed sets were sent to the United Kingdom where they were installed in a few type 274–fitted ships, and Sperry's of Brentford was given a contract to design and produce a British version (thereby saving precious dollars), to be designated type 932. Meanwhile, *Vanguard* had been fitted with a splash-spotting set, type 930, based on an Army coast artillery set. This seems not to have been successful and was probably replaced by type 931/2.

The 274/931 combination was the last specifically surface gunnery system to be developed by the Navy. It remained in service until *Belfast,* last of the wartime cruisers, went into reserve in 1963.

The second 10cm gunnery set was the dual-purpose HA/LA type 275 described on pp. 184–5. Wartime fitting had been restricted to the battleship *Anson* (four sets), Canadian cruiser *Ontario* (three sets), and the first four 'Battle' class destroyers (one each). In the immediate postwar period, priority of fitting was given to the thirty-six destroyers of the 'Battle' and 'C' classes then leaving builders' yards, the only large ships to receive type 275 at this time being the battleship *Vanguard* (four sets) and cruiser *Superb* (three).

Type 275 performed well and continued to be fitted in modernized cruisers, destroyers and new-construction carriers until the late 1950s. In the early postwar period, there were insufficient funds to equip all ships with type 275, particularly where a change of director was involved, so many of the older cruisers, destroyers and frigates retained the old 50cm type 285 to the end of their days – with spares harder and harder to get. As a stopgap measure, some of the wartime *Fiji* class cruisers still active in the early 1950s had their type 285s replaced by the US 3cm AN/SPG-34 radar fitted on the side of the 4in AA gun mountings. Type 275 was eventually replaced by the MRS 3 fire control system produced by Sperry, its radar type 903 being an Anglicized version of the 3cm US Mark 35 radar. Type 903 was tried out in *Cumberland* in 1954 and fitted in the

Navy's last large cruisers, *Blake* and *Tiger*, in 1959, as well as in the 'County' class guided missile destroyers, where it was used also for tracking and controlling the Seacat close-range surface-to-air missile.

The last member of the Navy's second-generation gunnery radar family was the 3cm fully automatic 'search-and-lock-on' type 262 described briefly on pp. 185–6 above, a self-contained gunnery system known postwar as Medium Range System 1 (MRS 1). Initially, it came in two versions, type 262(1) with the radar and computer carried on the twin 40mm Bofors STAAG gun mounting; and 262(2), a basically similar system but with the radar on its own small CRBFD, separated from the guns.

Despite the highest priority being given to its development, type 262 missed the war, the first ship to be fitted being the new light fleet carrier *Ocean* when she commissioned in June 1945 with two type 262(2)s (out of the four intended) mounted on CRBFDs, followed over the next few months by similar fittings in her sister ships *Triumph*, *Theseus* and *Warrior*. The first seaborne STAAG system (262(1)) was fitted on top of the forward 15in turret of *Vanguard*, which commissioned in August 1946. Thereafter, type 262 was fitted in all classes of the postwar fleet.

Early STAAG fittings were badly affected by vibration from the guns. Also, there was little space on a gun mounting for maintenance and performance monitoring, and the removal of heavy units to the Radar Maintenance Room below was difficult when at sea. Furthermore, a peacetime trouble was the tendency to shift target by locking on to the strongest echo nearby – very disconcerting for target-towing pilots. This led to various modifications culminating in the 262Q fitted in 1952 to the new *Daring* class destroyers where the aerial was fitted immediately below the mountings' twin gun barrels, in an attempt to reduce vibration – by which time the whole system had been renamed MRS 1.

WEAPON SYSTEMS

In 1944 – with the Fleet already fully equipped with the first-generation 50cm gunnery radar and with the second-generation 10cm and 3cm sets well in hand – the Admiralty laid down requirements for a third generation of gunnery radar, to form an integral part of new fire control systems, to be designed jointly by the Admiralty Gunnery Establishment at Teddington, and by ASE. These requirements led to subsequent work on long-, medium- and short-range systems (LRS, MRS, SRS), and the much wider use of Industry in weapon development after the war.

After considerable discussion with ASE, the Director of the Gunnery Division (DGD) proposed that LRS 1 should cater for both surface and air targets, providing fully automatic search, lock-on, and control for gunfire against surface targets out to 38,000 yards (19 miles), ready for an 'open-fire' range of 28,000 yards (14 miles), and against aircraft targets out to 22,000 yards (11 miles) for an 'open-fire' range of 12,000 yards (6 miles).[4] Research and development work started immediately at ASE and it was decided that the radar should operate on the 3cm band, to be designated type 901, eventually to supersede both types 274 and 275.

While no one had any doubt that a radar could be designed which would achieve the desired performance, there were many in the Navy who had doubts whether conventional AA guns could ever deal with future aircraft attacks, even with proximity fuzes. During the late war, they had been quite ineffective at long range. How much less effective would they now be, given the greater speed of jet aircraft, however sophisticated the fire control system? Even before the LRS 1 specifications had been promulgated, the Admiralty had produced an initial Staff Requirement for a naval guided missile – a 'directed projectile' using jet or rocket propulsion, to achieve speeds of 600 mph and fly at heights of 30–40,000ft. In August 1945, when a more definitive Staff Requirement was issued,[5] the new type 901 already proposed for LRS 1 was, on the advice of ASE, selected both to control the missile by the beam-riding technique (see Glossary) and to track the target. This application of type 901 to Guided Missile System Mk.1 (GMS 1) was at first given priority over LRS 1 until, in 1949, the latter was abandoned altogether and all work on type 901 concentrated on the guided missile application, the missile concerned having the codename Seaslug.

Type 901 was the last major radar project to be developed almost entirely in-house at ASE. From the beginning of the work in August 1944, through the sea trials of Seaslug in the trials ship *Girdle Ness* in 1954, to its final operational fitting in the eight 'County' class guided missile destroyers starting with *Devonshire* in 1960, the design went through many changes, none more spectacular than those of the aerial, which started by being a 6ft dish reflector, then a 9ft dish, soon replaced by a 9ft diameter radar lens. A radar dish aerial operates on the same principle as an optical *reflecting* telescope, except that the former has both to transmit and to receive electro-magnetic radiation, whereas the latter has to receive only. A radar lens may be compared with an optical *refracting* telescope, though the refracting medium was at this time an eggbox-like structure of metal plates rather than glass.

To achieve a power of one megawatt at a wavelength of 3cm was no mean feat, but undoubtedly the greatest advance made in this radar was the ultimate achievement of a tracking accuracy giving an rms aiming error of only 20 arc-seconds – the equivalent of 2.5 yards at a range of 25,000 yards (12.5 miles). And this from a 4-axis-stabilized aerial with transmitter and receiver weighing some 12 tons in a ship that rolls, pitches and heaves.

However, as feared by DGD, early work on LRS 1 was clearly leading to a system too large and heavy for destroyers which was likely to take a long time to develop. As a result, smaller, simpler systems were also studied in 1945 and subsequent years. Of the three considered, MRS 3 finally went into production, MRS 4 was abandoned, and MRS 5 became an important research project but was not carried through to production as guided missiles increasingly took over the AA role for all applications other than very close range.

The studies for MRS 5 started at ASE in 1945 were based on a projected radar type 905 on the 3cm band, eventually to have a 6ft diameter lens system and to be capable of operating out to about 14,000 yards (7 miles). This was to be an entirely new gunnery system, untrammelled by the constraints of guided missile control. In 1947, a contract for the development of a fully stabilized digital fire control system incorporating a 3cm wavelength radar similar to the type 905 proposed by ASE was placed with Elliott Research Laboratories of Borehamwood, Herts, who were also concerned with the design of the new Comprehensive Display System (CDS) for fighter direction. Although type 905 never went into production, the study was of particular interest in that it incorporated electronic scanning for the first time in a British naval weapon radar.

TARGET INDICATION

In mid-1946, several important ships in the postwar fleet – *Indomitable, Ceylon, Dido, Black Prince,* and no less than nine other cruisers for example – still had no target indication radar at all, but many ships were fitted with type 293M with the 12ft cheese aerial, while a few had type 293P with panels cleaned up to banish the previous Christmas-tree effect. Meanwhile, a new 10ft aerial was under development for a set to be designated type 293Q, reducing the beamwidth from 2.6° to 2.0° and thereby increasing bearing accuracy and resolution, and range performance. The first type 293Q was fitted in the fleet carrier *Indomitable* which joined the Home Fleet in mid-1950 after a partial modernization, soon followed by the cruisers *Newcastle* and *Birmingham* which proceeded to the Far East and the Korean War. Meanwhile, back in 1943, ASE had issued specifications for a new radar type 992, to be used with Target Indication Unit Mark 3 (TIU 3), to form the basis of a contract with EMI to develop what came to be called Gun Direction System Mark 3 (GDS 3), to supersede type 293 and the TIU 2.

Important requirements were a guaranteed range of 30,000 yards (15 miles) against small targets at 30,000ft altitude, with good resolution and accuracy; data to be not more than one second 'stale'; and the capacity to handle eight targets simultaneously and track them automatically to ease the load on operators. In addition, it was important that the system should be designed to handle data coming from sources other than the 992 itself – both radar and electronic warfare systems of detection – and be able to deal with IFF signals. Working on the 10cm band, the aerial was a 12ft cheese. However, the system was still unable to provide elevation/height data without drawing on the support of other radars such as 277. Like type 293, it was to be dual-purpose, combining surface and low air search with target indication.

Sixteen long years after its inception, GDS 3 first went to sea operationally in 1959 in the *Tiger* class cruisers (by which time the radar had become 992Q, with a fully stabilized slotted-waveguide aerial), followed in 1960 by the early ships of the 'County' class guided missile destroyers. In 1962, the 10cm type 993, with its distinctive 'quarter-cheese' aerial, began to be fitted for the same purposes in frigates and smaller craft.

AIR WARNING AND HEIGHTFINDING

At the end of the war, type 960 was under development by Marconi as a long-range air warning set to replace types 79 and 281. Operating on the same 3.5m waveband as type 281 and essentially a cleaned-up version of that set – 960 was engineered as a complete set whereas 281 had been designed as a collection of individual units – various improvements in 960 increased the pick-up ranges to 130–150 miles and a shorter pulselength improved the PPI and skiatron pictures. For most ships, the old 281 aerial was to be used for type 960 but a mattress array was designed for fleet carriers, to reduce the beamwidth to 17° and increase the range to 200 miles.[6] One only of these mattresses was made and erected ashore for trials,[7] but that side of the project was taken no further, presumably to save costs, so the old 281 aerial was used for the 960 throughout its operational life.

The first ship to be fitted with type 960 was the new battleship *Vanguard*, commissioning in August 1946. For existing ships, only those being extensively modernized – such as *Indomitable*, *Newcastle* and *Birmingham* – had their types 79 and 281 replaced by type 960. For new ships, however, all those for which it was appropriate were fitted, including *Eagle*, *Ark Royal* and *Albion* class carriers, *Tiger* class cruisers and the new AA and air direction frigates. Type 960 provided long-range air warning for the Fleet throughout the 1950s and early 1960s, when it was gradually replaced by the 1.5m type 965 (a 'navalized' version of the RAF's Type 15 GCI radar) with its large mattress aerial (later, two such aerials stacked one above the other in the larger ships).

For over twenty years, the old four-dipole array of types 281 and 960 had served the Navy well, a great tribute to A. W. Ross and his team who designed it in 1939–40. Almost as long-lived was the aerial designed by the same team in 1940 for the 1.5m type 286P, and later used with types 290 and 291, the latter set remaining in service in destroyers and below until the early 1950s. From 1945, the aerial was given power rotation to feed PPIs, making the set type 291M.

While the development of the dedicated fighter direction radars types 980/1 was proceeding in the immediate postwar years, steps were already in hand to improve the performance of the existing centimetric warning sets by fitting larger aerials, thereby converting them to types 293Q and 277Q. Type 293Q, with its 12ft cheese first fitted in *Indomitable* in 1950, has already been mentioned above in connection with its target-indication role. Type 277Q, intended for cruisers and above which needed heightfinding capability, had a new aerial designed specially to reduce the vertical beamwidth – effectively an 8ft paraboloid with the sides cut off to a width of 5ft (Outfit ANU) – in place of the 4ft 6in circular paraboloid (Outfit AUK) of 277 and 277P. Whereas the earlier AUK dish projected a 4.5° conical beam and could not measure angles of elevations above 40°, the new ANU projected a beam that was 4.5° wide but only 2.5° high, significantly improving elevation measurement for heightfinding at all angles of elevation between 3.5° and 70°. It was the first British 10cm radar to use vertical polarization.

Type 277Q was first fitted operationally in 1947 in the carrier *Illustrious*, followed by the *Indomitable* and all other existing carriers in the early 1950s, most having two sets to ensure all-round cover. As we shall see, new carriers from 1951 were to have type 983 for heightfinding. By this time, all battleships had been placed in reserve, but modernized wartime cruisers still active were given the standard outfit of types 277Q and 293Q on the foremast, and type 960 on the mainmast, ending with *Belfast* in 1957. The new *Tiger* class cruisers had types 277Q, 992Q and 960 in 1959, while the first few 'County' guided missile destroyers had 277Q (soon replaced by the solid-state version, type 278), 992Q and 965 in 1960. 'Castle' and 'Loch' class frigates, using type 277Q principally for A/S purposes, retained the smaller 4ft 6in aerial.

FIGHTER DIRECTION RADAR

As we saw in previous chapters, design of radar type 294/5 specifically for fighter direction (FD) started in the spring of 1943, the requirements being, first, plan position set for interception control providing gapless cover out to 80 miles; with, second, accurate height out to the same range without stopping aerial rotation. The FD set was to be used in conjunction with the metric long-range type 960 mentioned above to give cover (but with gaps) to 160 miles, and a new centimetric type 990 to give cover for aircraft above the horizon, it was hoped to about 150 miles.

The Admiralty had hoped that the new FD set – initially type 294, to be followed by type 295 when higher-powered transmitting valves became available – could have the aerials for both plan-display and heightfinding functions rotating together on the same mounting, thereby saving precious space and weight. However, all attempts to devise a heightfinding method which did not involve stopping the aerials – thereby losing plan display – failed, so in April 1944 the Admiralty was forced to reconsider the whole FD equipment plan. Type 294/5 was dropped and work started on two separate sets – type 980 to give fadeless plan position for interception out to 80 miles, and type 981 to give the intercept officer accurate heights to the same range on whatever target he chose, the two sets to have identical 10cm transmitters and receivers but very different aerials. The metric type 960P went ahead but the centimetric type 990 had to be dropped because the development of its hoped-for 2MW magnetron transmitting valve did not succeed during the war.

Development of 980 and 981 had the highest priority while the Pacific war lasted (ready for the prototype to be fitted in the new carrier *Eagle* which should have been ready late in 1945) but slowed down markedly when the war ended. Nevertheless, sufficient progress had been made by mid-1946 for trials to begin at the ASE trials station at Tantallon on the cliffs on the southern shore of the Firth of Forth. Development models of types 960P, 980 and 981 were set up together to represent as far as possible a ship installation, using a 'rolling platform' to simulate ship motion in a seaway.

The transmitting and receiving panels of types 980 and 981 were not very different from those used in type 277P but were designed to be more 'user friendly'. New Universal Display Units (UDUs) for showing plan positions from any of the three sets, either separately or in combination, were tried out, each having two CRTs, the upper being a 9in PPI, the lower a 6in sector display tube. Type 981 heights were displayed on a new range-height display. These special displays were only to be used in the RDR, conventional remote PPIs being used elsewhere in the ship.

As for aerials, type 960P employed a 'mattress' of six dipoles with screen reflector (outfit ANB), stabilized in azimuth only, weighing with its pedestal a little less than 2 tons and intended to be mounted on top of a lattice mast. The plan position type 980 had an aerial comprising three 12ft horizontal cheeses mounted one above the other, together producing a fan-shaped beam which by electrical means was made to sweep vertically to provide cover at all heights. The heightfinder type 981 had a horn-fed paraboloid dish 14ft 6in high cut vertically to a 5ft width. It rotated continuously when in the search mode for low-flying aircraft, but for heightfinding it had to be put on to the target's bearing by 960 or 980 and the aerial stopped, whereafter the mirror 'nodded' in elevation until the target height had been read off. Both AQS and AQT aerials were fully stabilized and weighed some 7 tons each.[8]

The Tantallon trials lasted some eighteen months. On the whole, types 960 and 981 proved satisfactory, but type 980 and the UDU rather less so, the former because the hoped-for detection ranges were not achieved, the latter for environmental reasons due to the heat developed by the glass thermionic valves used before the days of semiconductors. The ability to predict radar performance was in its infancy, especially for sets as complicated as type 980. It is thus not surprising that the predicted performance was markedly in excess of that achieved in practice.

By 1949, the project had been modified for various reasons, political and technical, taking into account the results of the Tantallon trials, the shortage of skilled staff at ASE, and the fact that the new carrier *Eagle* (which needed to have the new FD system fitted) was due to complete in 1951. The following decisions were therefore taken:

- the original transmitter and receiver panels envisaged for types 980 and 981 were to be abandoned and replaced by the existing well-proven type 277P panels, but augmented by the new test equipments developed for 980/981 which greatly simplified the setting-up and tuning procedures, resulting in significantly improved range performance: the innovative aerials were to be retained, however;
- the resulting radar sets were to be renamed type 982 for the intercept set, 983 for the heightfinder;
- UDUs were to be fitted in radar offices for local monitoring and in the RDR as sector displays for IFF and height estimation, but were to be dropped *pro tem* in favour of earlier existing displays in remote positions such as the ADR;
- the type 960P with its bedstead aerial ANB was dropped, all future type 960 to have the old type 281 aerial ATE.

Meanwhile, development had started in 1947 of an even more ambitious FD project – no less than of a single giant 10cm '3–D' radar to perform all the functions of the previous 960/982/983 system – to be designated type 984. And this all-singing, all-dancing set was to go to sea with a new Comprehensive Display System (CDS). Being a postwar development, type 984 is outside the time frame of this book, so the reader should look elsewhere for a more detailed description of this important FD system.[9]

During 1950, while *Eagle*, building in Belfast, was being fitted with two type 982s and two 983s as well as type 960, single types 960, 982 and 983

54. Fighter direction radar (*left to right*) types 982 (intercept set with plan position), 983 (heightfinding) and 960 (long range) under trial at the RN Aircraft Direction Centre, Kete, Pembrokeshire, in 1949.

(NRT)

were doing trials ashore at the Royal Naval Aircraft Direction Centre at Kete in South Wales. *Eagle* commissioned in October 1951, followed in 1953 and 1954 by the light fleet carriers *Centaur*, *Albion* and *Bulwark*, fitted with similar radar outfits, except that each had only one type 983. The heavy carrier *Ark Royal* followed in 1955 with the same radar outfit as *Eagle*.

On the whole, type 983 proved very successful at sea, in both its heightfinding and low-air search roles, but type 982 was far from satisfactory, so from 1955 the enormously heavy triple-cheese aerial began to be replaced by a much lighter 26ft-wide near-parabolic cylinder, shaped to provide the necessary height cover while preserving a high gain for long-range detection. It was fed from a slotted waveguide and stabilized in azimuth only, resulting in a topweight saving of some 5 tons. This gave a beamwidth of only 1° and resulted in a significant improvement in the range performance. The reduction in weight meant that type 982 could be fitted in the *Salisbury* class aircraft direction frigates from 1957, with type 277Q as a heightfinder, type 960 for long range warning, and 293Q for surface search and target indication.

The first '3–D' type 984 was fitted in the reconstructed wartime carrier *Victorious* in 1957, together with the all-important comprehensive display system. The second 984 went in *Hermes*, completed in 1959, the third in *Eagle* (in place of 960/982/983) during her modernization in 1960–4, when her type 960 was replaced by type 965 (see page 259).

Though operationally most successful, the 11–ton weight of type 984's 15ft 'dustbin' radar lens aerial meant that it could only be fitted in the largest carriers and, furthermore, it was expensive to manufacture and maintain, so only three ever went to sea. The decision in 1966 to phase out the fixed-wing aircraft carrier finally put paid to plans for a successor to type 984 with enhanced performance against smaller targets and more sophisticated resistance to countermeasures.

* * *

To provide some link with the present day, the reader should turn to Admiral Black's Epilogue on p. 270 below.

12
Radar in the Wartime Navy: A Summing-up

And so we come to the end of our main story in which, as well as seeing how radar was used at sea in war, we have learnt:

- how development of radar in the Royal Navy started so slowly in 1936, starved of resources and at low priority largely owing to lack of interest and perhaps ignorance at the top, despite the efforts of C.S. Wright, in contrast to the Royal Air Force;
- how there was a dramatic improvement in 1938 when C.E. Horton had taken charge of radar development at HM Signal School, so much so that the Navy was able to enter the war with operational radar at sea: thereafter, it became a race between operational thought and technical achievement;
- how, with Admiral Somerville's initial encouragement during the first months of the war, so much was achieved in 1939, 1940 and 1941 in design and development by so few staff at Signal School working all hours God made;
- how those at sea learnt to use their new 'eyes' to the best advantage, and how those in authority ashore, stimulated by operational experience, began to give the highest priority to fitting ships with radar;
- how the Royal Navy owed a great debt to Canada – and to C.S. Wright, himself of Canadian origin, for arranging it – for the many Canadian physics graduates who kept so much of the British naval radar working at sea during the earlier years;
- how, by the end of 1941, such a high proportion of the Fleet was fitted with both warning and gunnery radar, and how the world's first operational centimetric radar got to sea in such a short time, to make a most significant impact upon the Battle of the Atlantic;
- how, by 1943, the lead time from conception to ship-fitting began to lengthen as increasing demands for new and more sophisticated systems competed with improvements to existing sets, and more Admiralty departments had to be associated in the design process,

55. ASE's Radar Division, senior civilian staff, 1945. *Left to right*: A. A. Symonds, A. W. Ross, J. D. S. Rawlinson, N. Shuttleworth, H. Noble, H. E. Hogben, C. E. Horton, S. E. A. Landale, D. Stewart Watson. J. F. Coales was absent when the photograph was taken.

(Times Newspapaers Ltd)

56. J. F. Coales.

(J. F. Coales)

particularly for weapon systems – except of course for preparations for the Normandy landings which continued to have the highest possible priority;

- how countermeasures and counter-countermeasures became more important as radar became more mature;
- how the radar set ceased to be thought of as a unit on its own but rather as part of a system – of air defence, of fighter direction, of gunnery;
- how Industry throughout the war collaborated closely with Signal School and ASE, seconding engineers and designers to work alongside the experimental staff so that manufacture could begin even before designs were finalized, with the result that new equipment was operational at sea less than a year after its development in Signal School;

- how the naval officers and men who used the radar exhibited great skill in coping with the vast increase in electronic equipment in a ship, and also in devising methods of using it effectively.

In 1939, the Fleet was not all that different from that which had fought the Battle of Jutland. Indeed, some of the ships were the same, as only 23 years had elapsed. Junior officers at Jutland were now senior and, for many, the battleship and the big gun still reigned supreme. To the credit of the Navy, there were now a few excellent carriers commissioned or about to be commissioned, but these were thought of mainly in an offensive role, for reconnaissance, and for gunnery spotting, not so much for the air defence of the Fleet. There were a few more AA guns, including the 'Chicago piano' 8-barrelled pom-pom, but these were less than effective because of the lack of accurate control, the long-range systems depending upon visual sighting and optical instruments, generally based on the assumption that the enemy aircraft would fly straight and level. (Because of financial stringency in the 1930s, the Navy had failed to invest in a tachymetric fire-control system.) There was a change for the better in the anti-submarine measures, however, as great efforts had been made in developing asdic, still very secret at the beginning of the war. Though the Navy was confident, the menace of aircraft attack had been underestimated and the use of U-boat pack tactics and their night attacks when surfaced had not been anticipated.

By 1945, six years later, larger ships could have an electronic plot (the PPI) of all aircraft above horizon range out to 120 miles and of all surface features out to horizon range. Aircraft height could be determined and electronic means showed whether they were friend or foe. Fighter aircraft of high performance could now be vectored onto attacking enemy aircraft. Wolf pack attacks against our convoys by U-boats were no longer productive because HF/DF and centimetric radar in escorts and maritime aircraft made it hazardous for hostile submarines to remain on the surface – though luckily for the Allies, the 'Schnorkel' arrived too late to influence the final result. Blind fire could be opened on aircraft and ships. Good predictors and proximity fuzes were just beginning to be introduced, ensuring effective AA defence at close ranges at least. Surface gunfire was vastly improved and the fall of shot could be spotted by radar. Many of the ships seen in 1939 were still in service, but now they were festooned with radio and radar aerials and dishes. No longer was the battleship the most important unit, the aircraft carrier having assumed that role, and, to fight the air war effectively, radar and the electronic facilities that went with it were vital.

In just six years, the whole system of command and control had moved from visual and optical methods to one that was heavily dependent upon electronics. The Jutland admiral Jellicoe would have found himself at home (more or less) in the 1939 Fleet, many of whose exercises still centred around 'Jutland' tactics, almost no thought being applied to the tactics of air defence; but could he have coped with the 1945 Fleet?

How did this transition occur? It was completely dependent upon the ability to generate higher and higher radio power at shorter and shorter wavelengths. The first technical achievement in the UK – though the Germans were probably ahead in Ship to Ship – was a radar to detect aircraft to beyond 50 miles so that ships could be at full readiness to repel air attack. Soon it became possible to detect ships out to 10 miles or so with reasonable bearing but (compared with optical methods) great range accuracy. Then came accurate ranging sets mounted upon gunnery directors, both high and low angle, the former proving far less effective than had been hoped because of poor predictors, but the latter exceeding expectations. From 1941, the 10cm magnetron made possible detection at horizon ranges against ships as well as precision bearing and elevation measurement. Shipborne and airborne 10cm radar profoundly influenced the Battle of the Atlantic in the Allied favour by making it hazardous for U-boats to be on the surface either by day or by night, either near a convoy or on passage.

The ability to detect an enemy to the limit of the range of the ship's armament in conditions of fog and darkness created an enormous tactical advantage. How different might have been the outcome of the Battle of Jutland if both sides had then had radar. Radar stripped away darkness. No longer was the bridge with its visual lookouts the tactical nerve centre of the ship. Tactical information now originated in the 'plot', which became the 'Operations Room'. Even so – up to 1945 anyway – captains were reluctant to leave their bridges, which meant that on occasions of night action it was not they but the Operations Room Officers who were fighting the ship.

From 1942 onwards, radar development was fast and furious, both in gunnery and warning radar. Fighter direction, whose success in the air defence of the Fleet made up for the poor showing of long-range AA gunnery, could not have happened without radar. And the development of the PPI, which could be fitted wherever needed in the ship, was an enormous boon, giving as it did a 'map' from which the radar situation could be interpreted without requiring specialist knowledge.

In the six years of World War 2, radar changed the whole aspect of naval warfare, just as the advent of steam had done a century earlier. Old concepts had to be discarded, old drills and fighting instructions rethought. In this, the Royal Navy played one of the most significant roles, to tell the story of which has been the aim of this book.

Epilogue: Forty Years On

Admiral Sir Jeremy Black,
GBE, KCB, DSO

My generation of naval officers was brought up to take the existence of radar for granted. Serving at sea from soon after the end of World War 2 until the 1980s, we spanned the era which began with simple warning and fire control radar, and ended with automatically controlled (hands-off) gun and missile systems in both ships and aircraft and radar satellites in orbit above us. My first destroyer, the War Emergency CO class HMS *Concord*, was fitted with type 275 fire control radar and types 291 and 293 for warning and target indication.

My last destroyer, the Guided Missile 'County' class HMS *Fife*, had a computer-controlled action information system with screens displaying both radar echoes and computer-controlled symbols recording additional information. It offered a choice between manual intervention and allowing the radar-fed computer to take decisions. HMS *Fife* had a long-range air warning radar, two surface warning sets, radar-controlled Seaslug and Seacat AA missile systems, Exocet active radar homing surface-to-surface missiles, fire control radar for automatic tracking gunnery control, and a radar-fitted Sea King anti-submarine helicopter. And *Fife* is now paid off and twenty years out of date.

HMS *Invincible*, the aircraft carrier which I took to the South Atlantic as part of the British Task Force to liberate the Falkland Islands, had the latest radars of her day together with radio links between all the action information systems of the ships in the Force which ensured that air, surface and sub-surface pictures between the ships were automatically aligned and kept up to date with the latest information such as numbers, friend or foe, mission, aircraft fuel and ammunition state. This was a dramatic advance over earlier systems and was of particular benefit in safeguarding our own aircraft. Notwithstanding the excellent picture, the speed of events under Super Etendard/Exocet attack told us that an automatic missile and gun response is necessary and that the well tried step-by-step acquisition processes are too slow in today's battle.

Much therefore has changed but the skills of the man remain as vital as they did in the battles of WW2. The Sea Harrier with its radar and homing missile has kept pace with the speeds and accuracy of modern weapons but the necessity for the skills of the pilot and his controller in the ship remain. On the first tense day inside the Total Exclusion Zone, 1 May 1982, when the Task Force faced over sixty attacking aircraft, it was the alertness of Lieutenant Davies, a direction officer in HMS *Invincible* responding to a single fleeting radar paint well north of the main battle area, which led to the shooting down of one of a formation of three Argentinian Canberras heading for the carriers, and the withdrawal of the remaining two of the flight.

On that first day, Sea Harriers made their mark and started to establish their reputation as the 'Black Death' when they shot down three Argentinian aircraft and damaged a fourth, who closed the Port Stanley airfield to make an emergency landing and was shot down by his own side, a case of mistaken identity no doubt resulting from an inferior action information system and associated radar.

Electronic surveillance equipment had made phenomenal strides between 1945 and 1982, such that radar was often switched off rather than risk its detection.

On 2 May there was intelligence of impending attack by the Argentinian aircraft carrier *Veintecinquo Mayo,* and two of *Invincible*'s Sea Harriers were launched on a surveillance mission. They flew out at low level, their radars switched off, relying on their passive warning receivers. When Flight Lieutenant Mortimer was at 150 miles, he climbed as briefed and switched on his Blue Fox radar to reveal a group of four or five ships less than ten miles ahead. The Argentinians, who must also have been radar silent, responded immediately by illuminating him with their fire control radar which was the GO9 as fitted in the British Type 42 destroyer but also in two Argentinian Type 42s. Mortimer detected the beams immediately and dived to break away. Following Mortimer's report, the British Task Force was deployed to meet the expected attack but the combination of all-up weights of the armed aircraft and light winds caused the Argentinian attack to be aborted.

Despite the technological advances both in radar and electronic surveillance, identification remains a difficult area, particularly in the highly charged politico/military atmosphere leading up to and perhaps continuing into period of warfare. In recent years, both the shooting down of a Korean airliner by the Soviets and the shooting down of an Iranian airliner by US forces in the Persian Gulf have sent serious ripples through diplomatic channels. In 1982, the Argentinians used a converted civilian Boeing 707 airliner to survey British Forces steaming south between Ascension and the Falklands. In such a case neither radar characteristics detected by Electronic Support Measures (ESM) nor our radar emissions could identify the aircraft in its military role. Even visual identification was far from positive. In the event we relied on flight patterns and the Boeing 707 radar search profile to identify them as hostile and even then they were protected by rules of engagement, those who set them obviously judging that the risk of error prior to the outbreak of hostilities outweighed the loss of intelligence to the enemy.

Likewise, later on, the Argentinians tracked the flight patterns of British aircraft in order to assess the position of the two aircraft carriers prior to Super Etendard/Exocet attack. Fortunately, the probability of their using this technique was foreseen and early in the campaign a low-level deceptive flight pattern for Sea Harriers and helicopters was devised to camouflage the carriers' position. The flight pattern was unpopular with aircrew as it used up valuable fuel for aircraft operating at extreme range, but it probably saved us from attack on the afternoon of 23 May when two Super Etendards returned to base without finding a target, and almost certainly did on the afternoon of 30 May when a combined force of Super Etendards and Skyhawks flew a cleverly planned sortie with the intention of passing the British Force to the south and then coming in behind the carriers where they were less well protected. But for our deception, I believe this might have worked and, at a time when the carriers had no Sea Wolf-fitted frigate in close proximity and when *Invincible*'s Sea Dart launcher was temporarily unserviceable, it could have had momentous consequences.

Active electronic countermeasures (ECM) or jamming have become a feature of the modern battle at sea. They have the capability both to deny an opponent information and to supply him with spoof information. There remains, however, the possibility of declaring one's own identity and position and therefore ECM must be used with caution.

Radar has made phenomenal advances when used in conjunction with computers to create automatic gun and missile systems which can shoot down a shell in flight without manual intervention. However, they do rely so heavily on electronics that they can easily become useless in a moment of crisis due to the unserviceability of a component or a confused radar signal, as occurred in the Battle of Clapp's Trap when HM ships *Coventry* and *Broadsword* were stationed north of San Carlos Water and to the west in order to intercept incoming raids. After five successful kills, a further raid approached the two ships. Although two Sea Harriers were in contact and closing, it was decided that their engagement would be too late and they were hauled off in favour of *Coventry*'s Sea Dart and *Broadsword*'s Sea Wolf. In the subsequent manoeuvring in this fast-moving scene, land echoes, a constant bugbear with radar, combined with low-level flying, prevented the Sea Dart's acquisition, and the Sea Wolf became confused at the moment it should have fired and slewed fore and aft.

Broadsword received one hit with a bomb which did not explode. The second pair of Argentinian Skyhawks made a similar run. Once again our CAP [combat air patrol] was hauled off, once again the Sea Dart failed to acquire, but this time Sea Wolf locked on, only to find *Coventry* in her line of fire. *Coventry* was hit by three bombs and sank.

As is so often the case in warfare, advances in one area are offset by countermeasures on the part of the enemy as the struggle for supremacy continues. As I have shown, the capabilities of radar assisted by computers have increased phenomenally since 1945 and much has been learned from the experience of WW2. Despite those enhanced capabilities, the constant search to obtain the battle-winning edge continues and is especially important in a world in which countries both large and small can field the most advanced conventional military systems.

Appendix A: The Skinner Report at Swanage (1940)

Transcribed from the original in RSRE TRE 4/4/457

Preliminary Reports on Tests on Ships using 10cm Waves – November, December 1940 by H. W. B. Skinner

The following tests were performed at Swanage by Members of TRE, and subsequently by Members of the Signal School, Portsmouth, who were temporarily attached here. The apparatus used was the GEC E.1198 Magnetron (9.1cm) as transmitter, and a crystal mixer with 45mc IF. The transmitter power was nominally 5kW, but the exact amount radiated is not exactly known. Transmitting and receiving aerials were placed at the foci of paraboloidal mirrors, 3ft in diameter, giving a gain of about 500. These were mounted so that their axes could be pointed towards the ship, manually or otherwise. In a few experiments, flat cylindrical-parabolic mirrors with an aperture of 6ft x 9in were used. The gain of these mirrors is about 250. We thought that the flat beam (narrow in a horizontal plane but about ±15° in a vertical plane) might be suitable for use on a ship, rolling in a rough sea. The alternative would be a complex system of gyro-stabilization.

Three sets of radio-apparatus were used – A and B, of which B was an almost exact copy of A. Apparatus C was also very similar, the main difference being that it was better engineered. The performances of the three sets of apparatus were not quite identical. This is unfortunately inevitable in the present early state of development of 10cm apparatus.

TEST 1. 7th November, 1940. Apparatus A – Site Leeson House (250ft)
Sea fairly calm.
The *Titlark,* a boat of 92 tons and steel construction was followed end-on to 7 nautical miles (signal approximately 3 times noise). The boat began to return towards the apparatus. Signals very clear, though fluctuating, the whole time.

TEST 2. 11th November, 1940. Apparatus B – Site Leeson House (250ft)
Sea fairly calm.
Submarine (U-class, the *Usk*) was seen up to about 7½ nautical miles. Strong signals up to 6 nautical miles. These figures apply to the stern-on view. The broadside-on view seemed to give signals about twice as strong. Horizontal polarization of the source seemed slightly preferable to vertical polarization. Between 4 and 4½ miles, the submarine dived and the signal disappeared. On coming to the surface, the signal re-appeared at a time when at most the periscope and conning-tower were visible. Signals at extreme range are always rather fluctuating, and at extreme range were liable to disappear for rather long intervals.

TEST 3. 8th December, 1940. Apparatus C – Site Leeson House (250ft)
Sea fairly smooth.
Titlark followed end-on to 12 nautical miles (signal equal to noise). It seems probable that in the interval between Tests 1 and 3, some slight improvement in the apparatus had been effected.

TEST 4. 15th December, 1940. Apparatus B – Site Leeson House (250ft)
Sea fairly calm.
Titlark followed end-on to 9 miles, but seen as strong echo at 10 miles when turning. From this it seems that Apparatus B was somewhat less sensitive than Apparatus C, but probably about the same as Apparatus A.

TEST 5. 16th December, 1940. Apparatus B – Site Peveril Point (60ft)
Sea calm.
Titlark followed end-on to 5 nautical miles; seen turning at 6.3 miles. Signals otherwise similar to those in the previous tests.

TEST 6. 17th December, 1940. Apparatus B – Site near Swanage Pier (20ft).
Sea fairly calm.
Titlark followed end-on 3.5 nautical miles.

TEST 7. 18th December, 1940. Apparatus B – Site near Swanage Pier (20ft).
Sea fairly calm.
The flat paraboloids were substituted for the circular mirrors. The *Titlark* was followed end-on to about 3 nautical miles. The decrease in range is due to the lower gain of the flat parabolic mirrors.

* * *

It is characteristic of ship-detection that the signal strength decreases very rapidly with range near the extreme range. This is much more marked than for an aeroplane signal, as would be expected theoretically. For a ship we may have a very clear echo at a certain range, but at a range of ½ or 1 mile greater, the ship may be practically invisible. This applies especially to observations from low altitudes. This fact makes it doubtful if a considerable increase of range can be obtained by increased transmitter power. On the other hand somewhat increased ranges would probably be obtained by the use of wavelengths even shorter than 9cm.

It would appear probable that the *Usk* gives somewhat weaker signals than the *Titlark*, though not very much weaker. The range obtained for the *Titlark* from different heights could very likely be reproduced for the *Usk*, with a maximum range difference of not more than about ½ mile.

H. W. B. SKINNER

TRE 4/4/457
20th December, 1940
HWBS/MT

This is the report referred to in History *ii, para.352, note (2). (See Chap.1, ref.4.)*

Appendix B: Radar in Convoy Protection (1942)

The paper which follows dated 4 October 1942 gives a good idea of the situation prevailing at that date – just eighteen months after type 271 first went to sea for the Orchis *trials – as well as quoting reports of some of the actions in which radar played a significant role. It was addressed to all 171 seagoing ships then in the Western Approaches Command – 57 destroyers, 23 sloops, 9 frigates, 9 coastguard cutters, 67 corvettes and 6 others, as well as many shore authorities. Drafted initially by Lieutenant A. J. B. Naish, RNVR, Command RDF Officer, it was signed by the Commander-in-Chief, Admiral Sir Percy Noble. There is a copy in ADM 220/78.*

Annotations in square brackets by the present author.

From . . . THE COMMANDER-IN-CHIEF, WESTERN APPROACHES
Date . . . 4th October, 1942 No.W.A.689/506

RDF IN CONVOY PROTECTION

POLICY

When the capabilities of Asdics became known to the enemy, U-boats adopted the policy of surface attack by night. This policy was immediately successful, and it soon became apparent that a means of detecting U-boats on the surface at night was a vital necessity.

2. RDF was introduced into the Western Approaches solely for this purpose.

3. In view of the varied uses to which RDF is now put, it is perhaps useful to stress that the detection of U-boats at night is the factor on which RDF policy in this Command must be based.

4. While experiments were in progress to adapt the then existing sets for A/S use, Signal School, with remarkable speed, produced type 271. It was immediately recognised that this was the answer to the surface U-boat, and the RDF policy of obtaining the maximum information of U-boats at night became identified with the policy of fitting the maximum number of RDF sets type 271.

5. The structural limitations of fitting type 271, involving compensation for top weight and in certain cases the removal of the director, are well known, but it is significant that these losses are accepted most willingly by those who have had most experience with type 271 in action against U-boats.

6. When type 271 was first used at sea, large numbers of breakdowns occurred, and even those sets which were in working order were frequently operating below the maximum efficiency. The electrical and structural weaknesses responsible for many of the early breakdowns have now been corrected, but the most important factor in the increased reliability of type 271 has been the appointment of experienced RDF officers to escort groups. It is now clear that type 271 can be made to work at sea, but it is also clear that occasional breakdown will continue to occur. The presence in the escort group of an experienced RDF

officer, equipped with an adequate supply of spare components, is an essential part of the present RDF policy.

7. It has seemed necessary to stress the technical side of the group RDF officer's work, as there are so far very few ratings capable of dealing with RDF defects, and this places the maintenance of RDF equipment in a rather different category from the maintenance of the rest of the ship's armament.

8. While a glance at the classified shipping losses is sufficient to show that RDF policy must be based on the U-boat, air warning cannot be neglected. The problem here is rather different. While it is most desirable that every escort should be fitted with type 271 to produce the most effective A/S screen, it is not essential for every ship to be fitted for air warning, and the remarkable success of H/F D/F has led, in certain ships, to the replacement of the air warning set by H/F D/F. While unable to compete with type 271 in the matter of U-boat detection, the air warning sets (types 286, 290 and 291) are designed to combine air and surface warning and in ships not fitted with type 271 they give valuable information of surface vessels and U-boats at close range. Until type 242 interrogators are fitted to type 271 the air warning sets provide the only means of identifying friendly ships and aircraft by IFF, and of making use of the shore RDF beacons.

[Radar beacons were set up ashore for navigational use principally for aircraft but responded to ships' 1.5m radar. *Harvester* with type 286M got a response from St John's, Newfoundland, out to the limit of the range scale, 60 miles, with a bearing accuracy of ±5°. (CAFO 2396/41, 11 Dec 1941.)]

9. The conflicting claims of RDF, H/F D/F, directors, searchlights, and secondary armament are constantly under review. Limitation in supply of materials and labour, the state of progress of research, the claims of new construction, future policy and the probable future policy of the enemy are some of the factors which have to be considered. It need hardly be said that the fullest consideration is given to reports from sea of the changing requirements of escort vessels.

USES OF RDF

10. The varied uses to which RDF is now put are perhaps best illustrated by the following extracts from reports:

A – Detection of U-boats

[*Convoy*] *OG 82* – 'At 2145/14th April [1942] *Vetch* detected by RDF (type 271) a U-boat on the port bow of the convoy. This was chased by *Vetch* and *Stork* and destroyed at 2309 . . . [*Vetch*'s] RDF obtained contact at 7,500 yards and held the target throughout a gunnery action until the U-boat submerged.'

SL 115 – HMS *Lulworth* had just engaged and rammed an Italian submarine.

'2107N [9.07pm in a time zone one hour behind Greenwich time] – They now surrendered, and as the U-boat had scraped across my forefoot, removing his starboard screw I imagine, I came alongside some fifty yards from his port side, that being the easier side to board owing to his port list, and sent my boarding party away. We ordered them not to destroy their papers which I am afraid they had already done, and lay off threatening them with our guns while the boarding party got aboard.

'2125N – I received an RDF report of the approach of the second U-boat, but was very reluctant to abandon my boarding party until I was satisfied they were in complete control of the Italian of whose condition I was not sure. As the new enemy was beyond the hull of the first which I was keeping illuminated by my forward and after searchlight, I felt I was safe from his torpedoes and justified in 'playing possum' with a view to giving him the impression he was undetected, and luring him into a perfect position for a counter attack. He obliged and when he was just clear of the Italian conning tower I raised the beam of the forward searchlight and opened fire at a thousand yards with my 0.5in machine guns. Going full speed ahead at the same time and cutting across the Italian's bows straight for him. I hoped to blind him with my searchlight and could see my tracer hitting his conning tower.'

OS 34 – 'RDF type 271 has once again proved its ability to co-operate with the Asdic set in preventing attack on convoy. It is interesting to note that the only attack which was able to be carried out on the convoy was pressed home when *Gorleston* was astern of station (2355N/18). As *Orissa* was not fitted with RDF, this left the starboard bow of the convoy uncovered.

'*Gorleston* obtained an RDF contact at 2350/19 but owing to an error in voice pipe communication the information did not reach the Commanding Officer. However, at 0002N/19 RDF contact was obtained at 3,200 yards [1.6 miles] which led to the chase of a U-boat. Contact was held during the chase and finally at 0035N/19, due probably to a defective engine, the U-boat dived at a range of 2,400 yards [1.2 miles]. The diving was correctly recorded on the RDF screen.

Subsequently at 0113N/*19 Folkestone* obtained an RDF contact at 5,000 yards [2.5 miles] and on illuminating observed a U-boat diving. Asdic contact and hydrophone effect were obtained immediately.'

Operation 'Pedestal' – the *Wolverine* (Cdr P.W. Gretton) sinks Italian U-boat *Dagabur* by ramming in western Mediterranean.

'At 0054B/12, dark night, no moon, bright stars, in position "O" for Orange in Screening Diagram No.5 on HMS *Furious* (Senior Officer of Screen HMS *Keppel*) Mean Line of Advance 262°, speed 21 knots. On port leg of zig zag No.12 (steering 232°), RDF type 271 contact was obtained, bearing 265°, range 5,000 yards [2.5 miles].

'At a range of 4,800 yards the Officer of the Watch altered course towards, I arrived on the bridge at range 4,500 yards. At a range of 3,500 yards, ordered the signal "JOHNNY 284" to be made by Broadcast Method to HMS *Keppel* on R/T. When range was about 1,000 yards [half a mile] I passed the same signal by V/S to HMS *Malcolm* who was in position "P" for Pudding.

'Depth charge pattern "A" for Apples was set and "B" gun was put on the bearing. Fire was not opened in order to make sure of ramming.

'At about 600 yards I identified contact as a submarine, increased to full speed, three boilers fortunately being connected, and altered course to ram. I also sounded "Crash Stations" on the alarm rattlers in order to clear the mess decks.

'The submarine was rammed amidships abreast the after end of the conning tower at an inclination of 90° to the right; my speed was 20 knots.

'The submarine rolled over and sank immediately in position 37°18′ North 1°55′ East. This was another triumph for RDF type 271, contact having been held down to a range of 600 yards. The set was in full working order after the ram, the shock of which was considerable.'

B – Escort joining convoy
AT 19 – '*Newark* made contact with AT 19 (SS *Queen Mary*) at 0706/7 by RDF type 271 at 18,200 yards [9.1 miles]. Three distinct echoes were obtained from this huge ship, which was not sighted until the range had closed at 4,000 yards [2 miles].'

C – Use in fog
SC 86 – 'The meet at WESTOMP. [The Western Ocean Meeting Point in about 49° West where the mid-ocean groups took over convoy escort from the Canadian groups based in Newfoundland who were responsible for escort from the Halifax Ocean Meeting Point (HOMP).]

'This was accomplished in a visibility of about 50 yards. Twelve months ago, before the days of H/F D/F and type 271 it would have been deemed miraculous. As it was, without any difficulty and in perfect safety, the Group was led in from astern at 12 knots and dispersed to their places on the screen as *Hesperus* continued through the convoy, greeted the Commodore by loud hailer in passing and arrived alongside *Walker*'s port quarter for the convoy documents by Coston gun line . . . '. Watson effect was used while moving among the convoy. [The first two to three hundred yards of the A-scan of type 271 were taken up by the 'ground wave' so that echoes within that range could not be observed in the normal way. However, a target as large as a merchant ship produced a chracteristic distortion of the ground wave itself, known as 'Watson effect', Although this effect did not give precise range and bearing, it enabled an experienced radar operator to provide the bridge with enough information to manoeuvre within the convoy without danger of collision, given fifty yards of visibility.]

ON(S) 106 – 'Dense fog was encountered at about 1630 on the 3rd July and did not lift until 1430 on the 4th.

'During this period, the position of all the five ships relative to *Ripley* was constantly before me by means of bearings and ranges reported by the type 271. During the hours of darkness, the *Hampton Lodge* was plotted drawing out from the convoy and she appeared to shape course direct for her destination, Wabana. By the time it got light enough to risk going within talking distance she had faded from the screen.'

D – Use of type 252 [ship IFF responding to type 286/290 and aircraft ASV.] for homing
HMS *Hesperus* (fitted 286P and 271)
'Type 252 has proved most helpful during continuously foggy weather. It was possible to plot the course and speed of *Vanessa*, who is fitted with this set, and to home her to the Group from a distance of twelve miles. The frequency of the "pip" is such that it is difficult to obtain an accurate bearing.'

E – Use of RDF in navigation
HMS *Rockingham* approached the Newfoundland coast on 20 April in heavy mist. By making use of soundings, a D/F bearing from St Johns W/T station and RDF ranges and bearings from land, Cape Race was rounded and when the fog cleared the DR position was only a quarter of a mile in error.

A speed of 16½ knots was maintained throughout and visibility was usually 1,000 yards.

HMS *Newark* reported ' . . . Thick weather was experienced all the way and very low cloud. RDF type 286P was used in conjunction with soundings to round Barra Head which was not seen, landfall being eventually made on Tiree and Rum Islands.'

F – Use of type 271 against aircraft
Results of a trial by HMS *Landguard* (fitted type 271 only)

'The capabilities of type 271 against aircraft have been worked out in some detail using a sextant to measure angles of sight and hence obtaining the height of the aircraft.

'They indicate that echoes can be obtained from aircraft at heights up to at least 1,500ft, even at 3,000 yards when the angle of sight is over 9°. In good visibility aircraft can be followed with glasses out to about 15,000 yards, but on several occasions with Hudsons and Sunderlands the RDF has obtained echoes out to 30,000 yards [15 miles].

'The general conclusion is that though aircraft are usually reported by lookouts before being picked up by type 271, once the bearing of the target has been indicated the set can be of considerable value in giving ranges and in following the movements of the aircraft, especially after it has passed out of sight, e.g. by giving warning when it turns round. Hence, although type 271 is not normally used when visibility is good, it is considered well worth operating the set in daytime when an enemy aircraft is sighted or known to be in the vicinity.'

G – Use of type 271 in spotting [gunnery]
Extract from a report by the Commanding Officer HMS *Dahlia*

'On the 18th May 1942 off Lough Foyle buoy, a practice shoot was carried out with the ship's 4in gun. Weather fine, visibility very good. Ten salvos were fired within the space of two minutes. Type of target, Battle Practice Pattern VI.

'Throughout the practice, RDF type 271 was in operation and the flight of the projectile, from soon after leaving the gun to the moment of its fall, was, on the whole, clearly followed.

'Although the echoes, both of the target and the projectile, were distinct, the base of the target echo occupied some 200–250 yards on the scale, and, occasionally, the smaller and swiftly moving projectile echo became incorporated with the former making it difficult to establish exactly the RDF distance of the fall of shot, particularly in the case of near misses.'

LIMITATIONS OF RDF

11. Briefly the present limitations of RDF are:
  - Sets must be reasonably compact, very robust, constructed as far as possible out of components and materials already in production, sufficiently simple for maintenance at sea to be possible, and capable of being produced quickly in large quantities.
  - Rough weather reduces the efficiency of type 271 by deflecting the narrow radiated beam away from the target [because of ship movement in bad weather]. The use of a wider beam has been tried but it causes too great a loss of range and too many side echoes to be acceptable. Even if the additional topweight involved in stabilizing the aerials could be accepted, the rough weather problem would not be solved entirely as small targets are frequently obscured by waves, and large waves give rise to unwanted echoes.
  - There is as yet no evidence that U-boats are able to D/F RDF transmissions, but if U-boats are eventually fitted for this purpose the greatest security will almost certainly be achieved by ships using type 271 in preference to other sets.

FUTURE DEVELOPMENTS IN RDF

12. Future developments in RDF will probably consist in providing:-
   - Plan Position Indicators to ships fitted with type 271. This gives a plan picture of the convoy and surroundings. The present straight scan will be retained in addition.
   - Type 242 interrogators [which] will enable type 271 to obtain IFF response from aircraft, ships fitted with type 252, and shore beacons.
   - Improved bearing indicators for type 271.
   - Mark IV 271 equipment, giving increased range and increased discrimination between targets at nearly the same range.
   - Improved gunnery facilities for type 271.
   - Increased ranges for air warning sets.
   - A simplified IFF situation giving greater security and reliability.

Appendix C: Height Determination by Radar

Basil Lythall

Originally intended for air warning in battleships and cruisers, the first naval radars achieved ranges large enough for fighter interception, always provided some information could be obtained about the height of the attackers. A rough indication of height was needed in time to despatch the fighters, with better estimates as they neared the interception zone. This appendix describes the pursuit of that goal.

During much of the war the term 'heightfinding' was applied indiscriminately to any method of deducing the height of an aircraft detected by a surface radar. It was later replaced by 'height determination' – while 'heightfinding' was reserved for special-purpose radars able to pinpoint targets in elevation and find height directly from the angle of sight and slant range. In fact no British radar wholly dedicated to heightfinding appeared before the end of the war and the Navy had to rely almost entirely on indirect methods, using types 79 and 281 for 'height estimation' – a much less precise process in which the height had to be deduced from the variation of echo strength with range.

To achieve and maintain a reasonable standard required constant training and practice, and depended heavily on the experience and skill of the operators. Ambiguities could arise, particularly with 281, and the method was also sensitive to the performance of the radar. In the absence of a reliable performance meter during much of the war, performance could only be checked by frequent flight trials with friendly aircraft – often difficult if not impossible in wartime for ships other than aircraft carriers.

Yet from rudimentary beginnings this rather crude process reached a surprising degree of refinement. With some initial help from the scientists, the techniques were largely developed through the initiative, enthusiasm and ingenuity of individual radar officers and ratings, especially in aircraft carriers, working closely with fighter direction officers and accumulating experience the hard way, often through enemy action against convoys and task groups in which the Navy suffered grievous losses. By 1945, the British Pacific Fleet was able to make a unique contribution to joint operations with the US Navy by providing long-range height estimation as a complement to the superior American capability in low cover and in numbers of true heightfinders.

PRINCIPLES OF HEIGHT ESTIMATION

Metric radars (Figure C.1) were designed with a broad vertical beam to provide simultaneous detection at all heights, but in practice unavoidable gaps in vertical cover were caused by reflection of radar energy at the sea surface, producing a

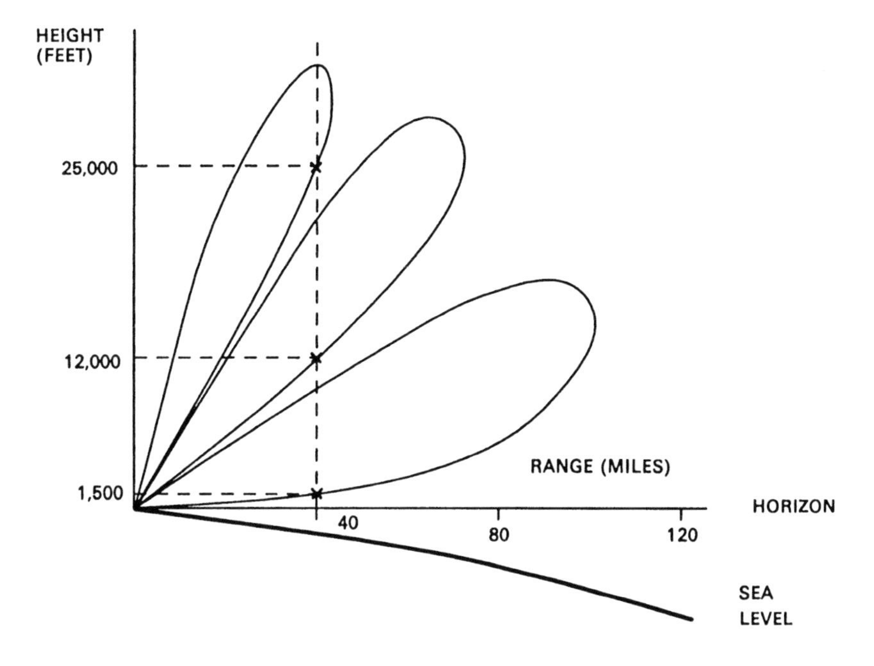

Figure C.1 The vertical coverage pattern of a metric-radar. There are many lobes between horizon and zenith, but our interest is confined to the lowest two or three, which are drawn with an expanded vertical scale. Ambiguities can occur if an aircraft is not detected in the first lobe, either because it passes over the top of it or because the set has been employed on sector sweeping and has not swept the target bearing for some period. A first detection at 40 miles could imply a height of 1,500, 12,000 or even 25,000 feet.

pattern of 'lobes' separated by successive 'minima'. Performance is poor at angles of sight near the horizon; that is, against low-flying aircraft, and again at certain higher angles. An aircraft approaching at constant height must fly through this pattern. As it enters the first lobe the echo strength will increase steadily, but will then decrease and fall below the threshold again as the aircraft reaches the first minimum; the cycle is then repeated through successive lobes, producing a characteristic variation of echo strength which will be different for each height.

Of particular interest is the range at first detection. Very small for low flyers, the range will increase with the height of the aircraft and so can provide an estimate of the height from the known coverage pattern. Above a certain height, however, it will decrease rapidly because the aircraft is flying above the top of the radar's first lobe and will not be detected until it reaches the second. Unless the top of this lobe is higher than the aircraft can fly there will be uncertainty about which lobe the aircraft is in when first detected, leading to two or more widely different estimates of height which may require considerable time to resolve. For type 79 the top of the first lobe was typically 35–40,000 feet; for the shorter wavelength type 281 it could be at least 10,000 feet lower.

Should the radar performance fall below standard – for example because of a drop in transmitter power or receiver sensitivity – the threshold contour will shrink, the estimate of height will be in error and the top of the first lobe will be lower. Similar uncertainties can be caused by differences in number, size or aspect of the attacking aircraft.

The 1½ metre air-warning sets on smaller ships (286 and 291) were virtually useless for height estimation because their aerials were much too high for their wavelength, producing a large number of lobes, with many ambiguities and very intermittent tracking. US warning sets also used this wavelength and had similar problems, although the more powerful SK, fitted in larger ships including a number of British escort carriers, had ranges similar to 79 and 281, allowing more time to eliminate the many ambiguities. A few British LSTs specially fitted as fighter direction ships had RAF GCI radars; these had a more useful lobe pattern, comparable in accuracy to 79 and 281, because the centre of the aerial was only about 30 ft high – the optimum for this wavelength.

GCI radars at RAF land sites used a more accurate method of determining height. The vertical aerial array was split into two equal parts at different heights, and signals from the two were compared. This was not practicable with 79 and 281 as the height difference between the two aerials would have had to be some tens of feet and in a ship there was no space available. An equivalent capability only became possible when carriers were fitted with both 79 and 281 and results from the two sets could be compared.

DEVELOPMENT OF THE TECHNIQUE

Almost from the start it was realized that a rough estimate of height should be possible from the range of first detection. In September 1939, *Curlew*, newly fitted with 79Z, was at Plymouth en route to working-up at Scapa. Lieutenant John Hodge, RNVR, the ship's radar officer, and Lieutenant Commander Hartley, an application officer from Signal School, organized trials with RAF aircraft, from which they constructed what was probably the first experimental graph of variation of detection range with height. Enemy air raids on Scapa soon had *Curlew* refining the data in real earnest, and she went on to develop the technique successfully during the Norwegian campaign, until she was regrettably sunk off Norway a few months later.[1]

Illustrious, the first aircraft carrier to be equipped with radar, reported in July 1940 after flight trials off Bermuda, stressing the 'supreme importance' of height information for directing interceptions, but also the weakness of having it only at first detection. Even this would be inaccurate if detection occurred at less than maximum range, for example because the radar had previously been operating in another sector before being trained all round.[2]

Meanwhile, at Signal School, A. W. Ross had derived formulae for variation of echo strength with range, but pointed out that estimates of height would be sensitive to the state of the radar, so that each ship would need to calibrate its own set using controlled aircraft, and then ensure that it remained as nearly as possible in the same condition. Failing a full calibration a simple formula would give a rough estimate of height from the range of first detection, and to supplement this Ross suggested a second method – recording the ranges at which the echo passed through successive *minima*. This method should not be sensitive to set performance, but might be too late for effective interception.[3]

The first method had already been demonstrated by *Curlew*; the second was found experimentally to be accurate provided the target could be continuously tracked. Ross also found that results from *Rodney* and *Valiant* indicated that both methods should be feasible, and in July 1940 all ships concerned were issued with a Signal School pamphlet describing both methods and advocating a judicious combination of the two. An addition to the type 79 handbook followed; by then the first method seems to have become limited to range of first detection rather than following the growth of the signal as Ross originally advocated. The formulae were in terms of heights above the radar than above the sea, and the ship's radar officer had to perform an additional calculation to allow for the curvature of the earth's surface.[4]

F. Hoyle joined Ross in December in 1940 and was asked to extend the study and to include type 281, which had recently had successful trials in *Dido*. To avoid computation by ship's officers, Hoyle devised a graphical presentation in which curves of constant height above the sea surface were superposed on the vertical polar diagram, so that true height could quickly be read off by eye. His first report did not allow for variations between individual radars and was not well received in the Fleet; it was followed in October 1941 by a more comprehensive paper illustrating the nominal performances of both 79 and 281 for a variety of installations in capital ships and carriers. This used an even simpler presentation, soon to be become universal, in which true heights were drawn as a series of horizontal lines.[5] Polar diagrams appear more distorted and the lobes appear thinner and turn upwards, as does the radar horizon. All these calculations embodied a correction to allow for the effect of water vapour in the atmosphere. This can be shown to be equivalent to increasing the earth's radius to 4/3 times its actual value.

This compendium gave a quick insight into the comparative performance of the two radars and different ship installations. The true height of the top of the lobes could be read off directly, showing that with 281 there was a much greater region of height ambiguity than with 79, but that the method of minima should be useful down to smaller heights. In carriers the transmitting and receiving aerials were usually at different heights; this reduced the maximum range but raised the ambiguity ceiling, so that the bottom lobe of 79 was rarely, if ever, overflown. The structure of the minima became confused, particularly in 281, and therefore less effective for height estimation. Each ship was recommended to construct its own individual diagram, either taken from the worked examples in the report or worked out theoretically, and then to calibrate it experimentally during flying trials. Hoyle followed this up by attending several such trials to confirm that the method really worked.[5]

THE CONTOUR METHOD

Practical techniques were soon taken a significant step further in *Victorious*, where it had occurred to the ship's radar officer, Lieutenant John Maynard, RCNVR, that a whole family of signal amplitude contours could provide a more continuous method of estimating height than relying on first detection alone (see Figure C.2). Using Hoyle's report as a basis, Maynard calculated a series of curves, each for a different level of signal amplitude in relation to noise.[6]

The scheme was tried out in flying exercises over two weeks in June 1942; signal amplitude was recorded at various ranges and the height estimated by noting where the observed range intersected the appropriate contour of signal amplitude. It was found that height estimates by this method were often correct to 1,000 feet, and that the radar operator could maintain plots on several aircraft at one time. There were limitations – for example ambiguities could occur if first

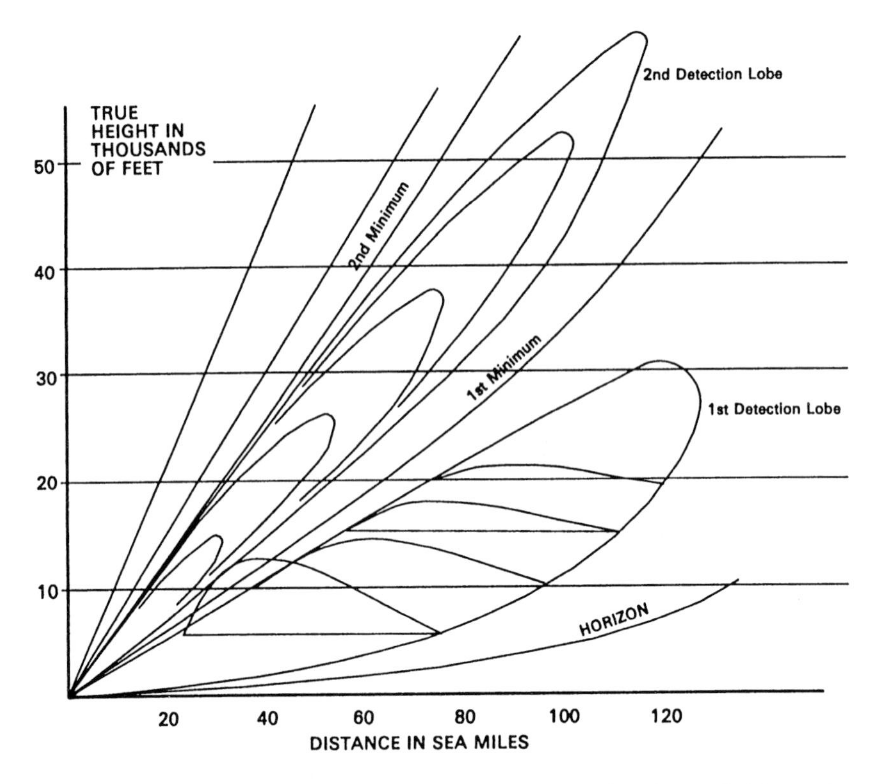

Figure C.2 A typical coverage pattern for type 281, plotted with true height as the *y*-coordinate. The lobes appear distorted and the horizon curves upwards. Examples of echo growth curves are plotted in the lowest lobe and contours in the lobe above it.

detection was at less than maximum range – nevertheless *Victorious* found the use of contours to be the most satisfactory method yet employed, although many more calibration tests were needed in order to assess the effects of number, size and aspect of aircraft, sea state and certain characteristics of the radar. Here perhaps was an answer to *Illustrious*'s earlier plea.[7]

ECHO GROWTH CURVES

Victorious's report of 22 June 1942 arrived at Signal School (by then ASE) just when the draft of an Admiralty Manual on 'Height Finding by RDF' was almost complete. CB 4224(42) had concentrated on the methods of First Detection and Minima and how to combine the two, but there had evidently been just enough time before publication in October to digest and include information from *Victorious.* An additional page at the end of the main text contained a brief résumé of the contour method, referred to experience gained to *Victorious* and recommended that the technique should be tried after familiarity with the more elementary methods had been acquired.

Also squeezed on to this last page was a short note on 'A practical method of combining the three heightfinding systems'. A series of height-lines is ruled on the vertical polar diagram at intervals of, say, 5000 ft. On these lines are then drawn reference curves of echo-amplitude versus range obtained during calibration. Ranges and echo-amplitudes reported by the operator are plotted on a suitably engraved sheet of perspex and superimposed on the height-line estimated from first detection. For each aircraft detected, the plot is then compared in shape with the reference curve and the estimated height revised if necessary to obtain a better fit.

By comparing the *shapes* of curves, even though their amplitudes might differ, this method exploited known relativities within the polar diagram without being sensitive to absolute quantities such as radar performance or size of raid, but the procedures required were considerably more difficult, with many graphs to be plotted and then compared with other graphs. And the method works best only when aircraft maintain a constant height; if they do not do so, recourse to the contour method may be necessary.[8]

First called 'Variation of Echo Amplitude' and then 'Echo Growth Curve', this elegant but rather laborious method became known throughout the Fleet as the 'Geddes method', after its originator Lieutenant-Commander Murray Geddes, RNZNVR. It remained the recommended practice throughout the war, but was not by any means universally adopted; indeed, there seems to have been a continuing controversy about the comparative merits and problems of contours and echo growth curves, until it eventually came to be accepted that the methods were complementary.

THE WAR IN THE MEDITERRANEAN

Before the 1942 manual could be distributed, ships were having to practice their existing height estimating skills in dire earnest. In August 1942, after exercising with other carriers en route for Gibraltar, *Victorious* became flagship of Operation 'Pedestal' – the first Malta convoy to include a carrier force equipped with metric radars and numerous fighters. Chapter 7 tells of the heavy shipping losses sustained, but there had also been conspicuous successes by fighters, and the surviving carriers, *Victorious* and *Indomitable,* had established close coordination in warning, height estimation and fighter direction. In a continuous series of joint conferences on the journey home the lessons were absorbed and ground rules established for improved layout of operational spaces, displays, communications and so on, which were quickly introduced into new carrier construction.

It was decided that, from mid-1943, the policy should be to fit all carriers and fighter direction ships with both 79 and 281, thereby considerably improving their potential for height estimation (see Figure C.3). Because the lobes and minima of the two sets did not coincide, many ambiguities which would have occurred with a single set were removed, or if not could quickly be resolved. Careful examination of the way the echo strength varied on each radar could give a much more confident estimate of height, and in principle at least it was possible to deduce when an aircraft was not flying level, or even flying down a minimum of particular radar in an attempt to escape detection.

The value of the combined installation was demonstrated during troop landing operations during 1943 which also led to two other innovations. 281 was given a sector selection display so that echo amplitudes on a chosen bearing could be measured without stopping aerial rotation and prejudicing the new skiatron plan position display. And because the new centimetric radars, particularly 273Q, had

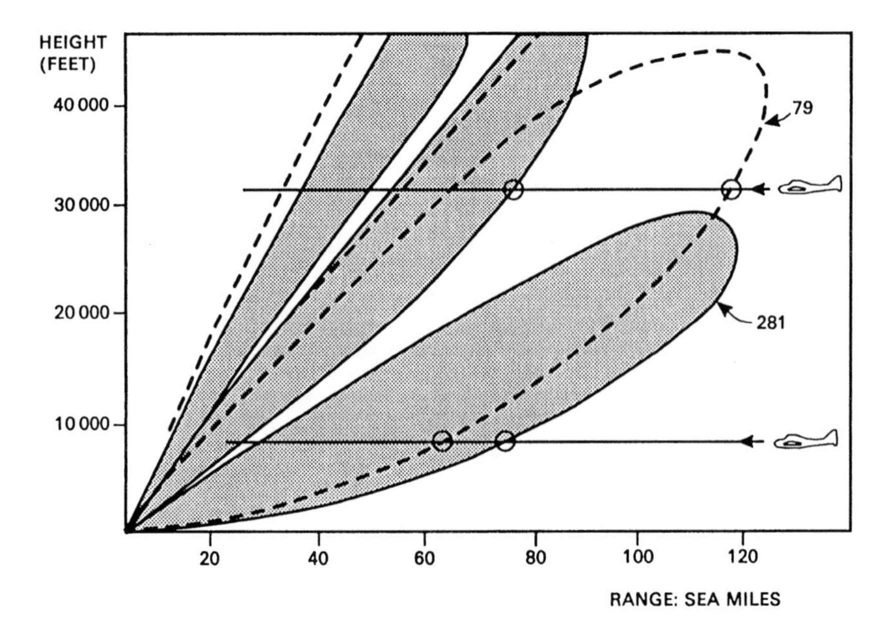

Figure C.3 Type 79 and type 281 in a single ship. The overlapping lobes remove many ambiguities and give some clue if the aircraft changes height.

shown their value in detecting low flyers as well as surface targets, type 277 was fitted with PPI's and a height position indicator, thereby being given a true heightfinding mode against low fliers, though only at much shorter ranges than high flyers could be detected by the metric sets.

All this led to the need for a Height Filtering Position (HFP), where height information from all sources could be assessed and combined before presentation to the FDO. Layouts and locations varied from ship to ship, but all contained the necessary plan, sector and height displays, together with the height plot – a large-scale representation of the vertical polar diagram, with contours and/or echo growth curves to taste, covered by a perspex sheet on which target data could be plotted. Height plots varied in design and layout according to the ideas of individual ships, and many different procedures were invented for using them.[9]

Even so there were marked differences in performance achieved, especially between aircraft carriers and other ships. Good height estimation was critically important to carriers for the direction of their own fighters. They were able to use routine flights for frequent radar calibrations and for regular training of operators, and particularly in Fleet carriers this led to the development of strongly motivated teams capable of consistent performance. The contour method seems to have been the most popular and *Victorious* continued to develop it in depth, gradually calibrating the system to take account of factors such as the number of aircraft in a group, whether they were approaching or receding, and the precise response law of the radar receiver.

Most other ships found it much more difficult to obtain flying trials for their own needs – indeed some radars were never calibrated at all! Height estimation was often unreliable and there were many complaints not only from Fleet commanders but from the ships themselves. The inability to monitor radar performance in a quick and simple way remained a serious weakness, and although a radar performance meter was developed by 1943, the war in Europe was over before it became widely fitted in the Fleet.

PREPARATION FOR OPERATIONS IN THE FAR EAST

By 1944, with the end of the war against Germany in sight, there was a much needed breathing space and it seems to have been time for stocktaking. A much more comprehensive edition of CB 4224 was issued in June, and an introductory paragraph set the scene as seen from ASE and Headquarters:

> There are numerous 'methods' of height-finding with WA sets in the Fleet, some peculiar to individual ships. Despite this, however, the standard of accuracy has been poor. Individual radar officers have been given considerable freedom to choose (or invent) whatever method they preferred, mainly because the subject was a new and unexplored one. Conditions have now changed. Sufficient experience is available to make it possible to choose one method and adopt it as standard practice. The adoption of such a standard method is made more necessary by the institution in larger ships of a Height Filtering Position or a Radar Display Room. To man and operate this, WA operators must be trained in height-finding, and it is impossible to do this with a variety of methods.[10]

Echo growth still remained the recommended standard. The contour method was not regarded as an alternative standard, although it was acknowledged that it might need to be used in certain cases. The use of echo growth curves with specimen results for a variety of situations was covered at length for types 79, 281, and for the two sets combined. Advice on calculating polar diagrams allowed a choice for the individual inclinations and abilities of radar officers, giving a choice between an approximate method and a more accurate one, and there were detailed instructions for carrying out calibration runs and applying the results.

For the first time there was mention of the human factors involved. For example, it was stressed that all height plots should be drawn with standard scale ratios to make it easier for operators transferring from one ship to another. The personnel needed at the Height Filtering Position were listed, with their functions, followed by a syllabus for their on-board training. The author evidently spoke from bitter experience in a previous seagoing appointment: 'Height-finding demands intelligence, and any training scheme is almost bound to fail if ratings are selected for the job merely because they are no good as WA operators.' And the first item in the training syllabus hints at practical problems, perhaps especially with the echo amplitude method: 'Simple graphs and how to read them. Experience has shown that the average rating finds graphs rather difficult as they involve unfamiliar thought processes.'[11]

WAR IN THE PACIFIC

CB 4224 (44) was intended as the heightfinder's 'bible' for British ships being prepared for Pacific operations, although in the event a number of them did not arrive before the war was over. Chapter 10 brings out the critical importance of height information in these operations; this still had to be obtained from lobe reading with the metric sets, apart from the one American SM-1 heightfinder in *Indomitable.*

For the first time since radar had been fitted, the majority of ships had carried out enough calibrations and radar/aircraft-direction exercises to ensure reasonable efficiency in height estimation, at least at first detection, and the many radars in the force could all contribute to the solution. By combining information from all ships, a significantly more reliable estimate could be obtained, and the Squadron FDO was generally able to produce an estimate within 1–2,000 feet using only the theoretical polar diagrams of the sets involved. In joint operations with the US Navy this long-range height information provided a valuable exchange for the improved low cover of the American 1½ metre and 10cm sets and their plentiful supply of heightfinders.

Height estimation after first detection was much less consistently successful. It worked well for targets at constant height, but could be rendered almost useless by intelligent tactics by the enemy, whereas these could usually be sorted out by the SM-1 heightfinder. It was also very slow, and once SM-1 had picked up a raid within 50 miles the results were so consistently accurate that it was no longer worthwhile to continue with height estimation. And in the presence of many friendly aircraft – frequently more than 1,500 – tracking of enemies could be so difficult with the broad-beamed metric sets that height estimation was again impracticable.[12]

POSTWAR

Lobe reading remained an essential technique for many years after the war. It was the only method of height determination between the outer limit of the warning radar, extended to 200 miles with 960, and the maximum range for heightfinding, increasing to 80 miles with the introduction of 983. Experience in the Pacific had shown that, if aircraft deliberately changed height or aspect, the use of echo growth curves could be difficult and at time dangerously misleading; the use of contours was therefore given full weight as a complementary method. A standardized height plot was recommended, containing both a full set of contours and echo growth curves at 5,000 feet intervals – making a truly formidable looking array of lines and curves even when drawn in several colours.

For a single ship using an intelligent combination of all methods it was reckoned to be possible to achieve an estimate within 2–3,000 feet on all aircraft above 5,000 feet by the time they had reached a range of 50–60 miles, with a rough estimate soon after first detection. For aircraft below 5,000 feet an estimate within 1,000 feet ought to be possible soon after detection. As regards initial detection, estimates from individual ships would be less accurate but, as had been found in the Pacific, combining all the information available would giver results within 1–2,000 feet.[13]

ANNEX: PRINCIPLES OF HEIGHTFINDING

A simple heightfinding radar employs a narrow vertical beam scanning up and down to pinpoint a target in elevation. Angle of sight and slant range, which together determine the height, can be presented on a height position indicator (HPI) in much the same way as bearing and range are shown on a PPI, and the absolute height read off directly from calibration lines. Moreover the *difference* in height between attacker and fighter, which is what matters during interception, will be independent of errors in absolute height.

With the advent of high-power centimetre-wave techniques in 1942, a shipborne radar of this type could be seen as feasible. At this wavelength the vertical aperture of the aerial need be no more than, say, 10–15 ft, but the need for triaxial stabilization will mean a heavy mounting suitable only for large ships. With its narrow vertical beam the radar can work down to much lower angles of sight, but to obtain full range performance it must stop rotating in azimuth while the height of a particular target is determined. To acquire each target the heightfinder beam must be put into the required bearing by the use of an 'Azicator Display' which presents the all-round picture from the warning radar together with a cursor indicating the azimuth of the heightfinder beam.

A radar capable of reliable heightfinding out to 80 miles did eventually emerge as type 981/983 (see Table C.1 on following page) but not until after the war. It was delayed in part by pressure for more ambitious schemes to provide height information on all targets in the fighter direction zone without interrupting the plan display picture. This would have involved rotating the heightfinder aerial at the same rate as the warning aerial and replacing its mechanical up-and-down nodding with movement with an electrical method; the idea was abandoned in 1944 in favour of developing 981 on a separate mounting.

Chapter 9 relates what followed. Type 277, soon to be fitted in increasing numbers, was provided with a heightfinding mode using the elevation stabilization mechanism to rock the aerial. Although it could detect low fliers well before the metric sets, its maximum reliable range of 25–30 miles for heightfinding was of little use for fighter direction, except in the crudest sense – 'if detected by 277 it must be low; if by a metric set it must be high'. And azimuth rotation had to be stopped to measure height, so that two sets were needed on the same ship unless the principle role of surface surveillance was to be prejudiced.

To fill the gap before 981 arrived, some ships were fitted with the USN SM-1, whose heightfinding range of 50 miles in normal conditions was a very notable improvement. One set, in *Indomitable,* saw action in the Pacific; this was the only case where a true heightfinding radar made a significant contribution to height determination in the RN during the war. Soon afterwards, 277 was provided with a larger aerial, to become 277Q, with a heightfinding performance similar to SM-1, although without conical scanning to assist in maintaining the target in view.

Table C.1 Official performance figures for heightfinding sets in 1949[14]

Radar set	Aerial outfit						Range of set for heightfinding		Height finding accuracy
	Letter	Polarization	Beam width		Stabilization	Minimum angle of elevation for Height-finding	Maximum reliable	Maximum	
			Vertical	Horizontal					
277P	AUK	H	4.5°	4.5°	Azimuth and elevation	4.5°		25 miles	½° to ¾° ±1,200 to ±2,000 ft. at 25 miles
277Q	ANU	V	2.5°	4.0°	Azimuth and elevation	2.5°	*40 miles	60 miles	½° to ¾°** ±3,000 ft at 60 miles
981/983	AQT	V	2.25°	6.0°	Triaxial to 6 arc-minutes	1.25°	80 miles	120 miles	±1,000 ft at 70 miles

* With a well-trained team.

** These represent maximum errors. The average over several 'cuts' gives ±1,500 ft at 60 miles in calm sea conditions.

Appendix D: Radar in Submarines

EARLY DAYS

The first British submarine to have radar was the large prewar *Proteus*, fitted experimentally with type 286P in the summer of 1941.[1] The aerial array was the same as that fitted in destroyers and sloops, on a telescopic mast just abaft the after periscope standard. The mast was raised so that the aerial was above the periscope standards for use on the surface and lowered to bring it below the jumping wire when submerged.

Submarine fitting posed many problems not met with in surface vessels, not the least of which was finding space below for the equipment needed, so that earlier fittings in existing boats tended to be in the larger ones.

The Naval Staff History describes *Proteus*'s first attack by radar in the Aegean in November 1941:

> *Proteus* then moved across to St. Georgio Island and on the evening of the 9th picked up a convoy by radar, one of the earliest reports of this form of 'sighting'. Lieutenant-Commander Francis decided to shadow the convoy (on the surface at night) and attack submerged after the moon was well up. Shadowing was carried out from 2050/9th until 0320/10th when the German (ex-Greek) steamer *Ithaka* (1,773 tons) was torpedoed and sunk.[2]

In December 1941, the Commander-in-Chief reported that in the Mediterranean, *Proteus* had achieved ranges of 6.5 miles on a battleship, 2.5 miles on a destroyer, 1 mile on a motor launch, 5 miles on an aircraft at 500ft, but 30 miles on one at 2,000ft.[3] Pointing out that these ranges were better than those achieved on trials, Admiral (Submarines) told the Admiralty that it was now clear that radar was very desirable in submarines. The primary object should be maximum warning of surface craft plus accurate bearing and range for weapon control, 'but that warning of approaching aircraft is rapidly assuming greater importance, especially at night, and may well eventually exceed the importance of surface craft warning. It follows that if aircraft are able to bomb submarines on the surface at night – and this form of anti-submarine warfare is rapidly developing – submarine operations are very severely handicapped.'[4] It was proposed that type 291 or 286P should be fitted in all the larger existing submarines, and in all new construction. By August 1943, over sixty operational submarines had been fitted with type 291W.[5] (Radar type numbers for submarines all had the suffix 'W'.)

In these early days, however, the 1.5m radar was far from popular with submarine commanding officers because it was feared it would give the submarine's position away. One CO recalls a report which circulated, telling how a submarine on passage in the Indian Ocean had switched on her type 291W and ten minutes later was attacked by Japanese bombers. This was taken, by an

57. Submarine *Scorcher*, 1945, with type 291W aerial in the raised position.

(IWM)

unreasoned but almost impossible logic, to have been cause and effect. But the word went round and, as with some justice it was felt that practically everyone's hand was against the submarine, most COs preferred not to invite what might be more trouble by using their radar.[6] There was also the reactionary emotion – a submarine CO to an ASE scientist: 'The Mark I eyeball is more effective than your radio stuff, and it takes up less room.'[7]

However, radar *was* used, if somewhat reluctantly, as shown by the following report from *Torbay* with type 291W in April 1943, returning to Algiers after a patrol in the Tyrrhenian Sea:

> At 2250, radar detected an echo to the south, range 1,500 yards. A dark shape was soon seen. Closed to attack, but even when range was down to 900 yards, target was only vaguely visible when on top of the heavy swell and its silhouette could not be made out or its course seen. Diesel HE [hydrophone effect] was heard on the asdic. Radar ranges then showed that it was going away, so gave chase. *Torbay* was just able to catch up again and in the meantime the enemy hauled gradually round to starboard to a course of 350°, the original course having been about 210°. It is thought that she had seen *Torbay* the first time the range was closed and had taken her for a patrol craft and decided to withdraw northward.
>
> When the range by radar had been closed to 750 yards, the U-boat dived in position 38°19'N 4°22'E. It was maddening to have had her in range for so long without being able to see her. The whole chase was carried out by radar . . .
>
> 0020 – Broke off the hunt. Passed enemy reports and steered for Algiers.[8]

The general feeling among British submarine officers at that time about radar – at 1.5m at any rate – is summed up in *Torbay*'s comment on that incident:

> While the radar is of enormous value on occasion, I have formed the opinion that it is easy to DF and it is only a question of time before the enemy get to know the frequencies used by HM Submarines, when it will be most dangerous, or at least compromising, to operate it in enemy waters.
>
> It should give warning of the approach of enemy forces, of course, but this is not positive as for ten minutes in every hour the set has to be switched off for cooling.[9]

However, seagoing opinion was not universally anti-radar. By operating type 291W continuously when on the surface during her patrol close inshore north of Corsica in May 1943, *Shakespeare* decided to make a test case for using radar near the enemy coast. No enemy anti-submarine measures ensued and there were considerable benefits, particularly with navigation.[10] (This patrol included entry into the port of Calvi and a bombardment of the aerodrome at a range of 800 yards.) *Trenchant* in the Far East always used radar on the surface by day for aircraft warning, and was never once surprised.[11]

Until 1943, Admiral (Submarines), the operational authority to whom the Admiralty turned for advice on submarine matters, was lukewarm on radar generally, and failed to press for a centimetric set (which could not easily be DFed) for surface warning and torpedo control, because, for space reasons, air warning would have had to be sacrificed. Then came a change of heart, largely brought about by news of the great successes of US submarines which since 1942 had been fitted with 10cm surface search radar, SJ, as well as SA, a fairly basic 1.5m air search radar (which gave warning and range of aircraft but no bearing). Their offensive achievements in night attacks – using radar to seek out the enemy and then using it for weapon control – as opposed to the defensive attitude, merely using it to warn when the enemy was near, fired Admiral (Submarines) with enthusiasm. 'The achievements of the US submarines,' he wrote in a letter to the whole British submarine service, 'show what can be done with the right equipment in the hands of determined men.'[12] Enclosed with that letter were extracts from reports by US and British submarines where radar had been successfully used (including that of *Shakespeare* mentioned above). Radar equipment, he said, was not 'just another gadget' but was an essential part of a submarine's ability to operate offensively, as well as being able to 'prepare herself against impending air attack and the enemy's A/S measures'.[13]

At ASE, considerable thought had already been given to the development of a suitable set for submarines but, as ASE told DSD, no firm action could be taken until the Admiral (Submarines) produced a Staff Requirement which spelt out priorities between surface and air capabilities.[14] This was eventually done on 19 June 1943, first priority being the detection of surface vessels; the second, provision of accurate data for torpedo and gunnery control; and last, warning against aircraft.[15]

ASE proposed to develop a hybrid set to be known as type 267W, with a PPI that could be switched to display either a 3cm 'seaguard' picture or the 1.5m 'airguard', the former not only giving a high-discrimination PPI picture, but also supplying accurate ranges and bearings of the target for plotting and fire control. The first version would not entirely meet the Staff Requirements and would suffer the disadvantage (overcome in postwar versions) that it could only be used when surfaced because the aerials would be on a fixed mast which protruded only a small amount above the periscope standards. However, compared with the old 291W without a PPI or accurate ranging, it would greatly enhance the submarine's capabilities. A contract had been placed with Metropolitan Vickers at Manchester and production was hoped to begin in March 1944 – but this was only a target. Not soon enough, said Admiral (Submarines),[16] though he fully realized that the delay in the start of the development at ASE was entirely due to the fact that he had not backed the surface warning set for submarines until 1943.[17] Indeed, it was to be March 1945 before type 267W was first used operationally, much of the later delay being due to the difficulty in obtaining 3cm magnetrons because, in the autumn of 1943, the RAF had the highest priority for supply for use in Bomber Command's Mark III H_2S.[18]

TYPE 291W IN AN ATTACK OFF NORWAY

In September 1944, HM submarine *Sceptre* (Lieutenant Ian McIntosh) on her tenth war patrol, was given a small patrol area off the SW coast of Norway near Egersund, containing a stretch of open coast protected in some degree by reported minefields. The weather was mostly fairly calm with some periods of fog and low visibility, improving from time to time. Within the area there was considerable activity by German anti-submarine trawlers, fishing vessels and some small coasters below the size that were allowed to be attacked, but no larger ships were sighted by day, probably because of a sustained period of day attacks on convoys by RAF Coastal Command.

Sceptre was fitted with type 291W airguard radar operating with an A-scan display and no PPI. In common with the then usual practice in submarines, it was seldom used as it was feared that it would give the submarine's position away. Against aircraft, reliance was placed on visual sightings and if on the surface the standard practice was to dive without trying to identify until safely under, experience having taught that 'friendly' aircraft were as liable to attack as enemy ones, often more damagingly.

In view of weather conditions and the lack of daytime targets, it was decided that continuous radar watch should be kept when surfaced each night, usually between 8.30pm and 5.15am, with the object of detecting any convoy movements on the inshore route, or any possible U-boat movements on the routes they were believed to take further to seaward. No ill effects resulted from this policy and it had its reward on 20 September, the day before *Sceptre* had been ordered to leave patrol.

Just before 9pm on the 20th, a possible target was detected by radar at a range of 5.5 miles, some 20 miles WSW of Egersund. *Sceptre* went to diving stations, altered course to close on the surface and, at 9.09, a plot was started. Range accuracy on that set was good but bearings were very unreliable. However, quite a good plot was generated using range only and by 9.25 the enemy was reckoned to be steering about 150° at 12 knots.

Sceptre altered course to run parallel 2 miles off the enemy track, and speed was increased to a maximum of 14½ knots to give chase and reach a good attacking position. By 10pm, bearing had been gained and course was altered 5° to port to reduce the distance-off-track. By 10.13, a suitable attacking position having been reached and the enemy course and speed from the plot confirmed, *Sceptre* altered to port to an attacking track, reducing speed to avoid closing too much. During this run-in, the target was sighted and, at just over half a mile, identified as an A/S trawler.

By now, three new echoes had been seen close together about 2 miles on *Sceptre*'s port bow and these were assumed to be a convoy, the final target. Course was altered to starboard to present a smaller visual target to the trawler escort and to pass under her stern to get inside the screen. Turning a full circle before steadying on a course of 090°, *Sceptre* closed the convoy on a 118° track, the enemy course being 152°, speed 12 knots. Three ships in line ahead were sighted – an A/S trawler followed by two merchant ships – silhouetted against Egero light which had just been switched on for the convoy's benefit.

At 10.31, six torpedoes were fired, aimed at the two merchant ships. Two hits were seen on each target and both were thought to have sunk because, after five minutes, the fires in each suddenly disappeared and the radar echoes were lost. *Sceptre* withdrew on the surface at top speed and was able to listen to the explosions of some 35 depth charges with equanimity.[19]

We have mentioned the apprehension felt by many submariners about giving away their position by using radar. *Sceptre*'s experiences in her next patrol in the same area off Norway is relevant here, when she tried, not unnaturally, to use her radar at night. Early in the morning of 18 October 1944, after five minutes of radar transmission, interference produced so much 'mush' on the A-scan as to obscure any possible echo information, the 'mush' extending 50° either side of the bearing 015°. On the evening of the 19th, just after surfacing, the same thing occurred after about five minutes. This was obviously deliberate jamming and, equally obviously, the enemy must know *Sceptre* was there.

On the 21st, Trafalgar Day, she used radar in short bursts and was not jammed but got no useful information. Later that night, she carried out a surface attack on a convoy using, *faute de mieux*, the 'Mark 1 Eyeball' with results very similar to the attack on the previous patrol. On surfacing the next night, she again had only four or five minutes transmitting before jamming obliged her to switch off and stand down the radar watch.

Probably lack of resources prevented the Germans from taking any offensive action against *Sceptre*. But although they seemed to know of her presence, their protection of the convoy on Trafalgar Day left much to be desired – which suited *Sceptre* very well.[20]

THE US TYPE SJ IN BRITISH SUBMARINES IN THE FAR EAST

In September 1943, Admiral (Submarines) asked the Admiralty to acquire a number of American 10cm type SJ sets to fill the gap until type 267W was ready.[21] The first two of these were fitted in *Tiptoe* and *Trump*, building at Barrow, completing in the summer of 1944. The aerial – or antenna in American terminology – was on a non-elevating mast abaft the periscope standards.

After working-up patrols in the North Sea, the submarines sailed for the Far East early in 1945, to be based in Fremantle. In their first war patrols, SJ's performance was very indifferent, but the situation was transformed in April 1945 when the maintenance staff in the British depot ship *Adamant* sought the help of the US Navy, which also had a submarine base at Fremantle. Tuning and a few modifications (to convert to SJa) soon improved matters. Both boats had complained of the lack of air cover, so *Adamant* fitted them with a modified 291W additional to the SJ, as well as a 1.5m search receiver SN4, using the same aerial.

In her next patrol, *Tiptoe* reported that SJ worked well, detecting 10 to 20 ton junks at 3 miles, a US submarine at 6 miles and land at 50 miles or more. During the first week of the patrol, at least twelve SJ contacts were investigated at night; in the event, they all turned out to be junks and the like, but would otherwise have passed without being detected. As a navigational aid, SJ was splendid, though rather less so in the Java and Flores Seas, not because of the radar but because of the inaccuracies of the charts. The 'lash-up' 291 was not satisfactory, however. *Trump* had much the same experience until her SJ broke down completely halfway through the patrol.

In their last war patrols off the south coast of Java from 16 July to 21 August 1945, *Trump* and *Tiptoe* worked together as a 'wolf pack'. In a highly successful patrol, SJ radar was used extensively not only for its prime purpose but also for communication between the two when surfaced at ranges of up to 8 miles, something which British radar equipment was not then capable of doing. The aerial was trained on the consort and the 10cm beam used with a morse key for wireless telegraphy in half-hourly routines. As this system was highly secure, no codes were needed, making possible very close understanding between commanding officers. 'This fluent freedom of thought,' wrote Lieutenant Anthony Catlow of the *Trump*, 'was only made possible by the SJ radar.' Lieutenant Richard Jay of *Tiptoe* reported that aircraft could be detected at about 7 miles with the new type of antenna reflector fitted by the US Navy in Fremantle. But sadly, he said, the 291 was useless . . . 'It is requested that it may be removed from the wardroom pantry at an early date.'[22]

TYPE 267W OPERATIONAL

The trials of the development model of the dual-wavelength type 267W were carried out in the submarine *Tuna* but the first two operational sets were fitted in *Turpin* (Lieutenant J.S. Stevens), building at Chatham, and *Tapir* (Lieutenant J.C.Y. Roxburgh), building at Barrow, at the end of 1944. In an account of *Tapir*'s first patrol, Admiral Roxburgh tells how wonderful it was to see the 3cm picture on the PPI showing the coast of Scotland as if it were a map. He goes on:

58. Submarine *Turpin*, 1944, with 3cm 'seaguard' aerial of type 267W forward (to the left) of the periscope standards, 1.5m 'airguard' aerial in the lowered position to the right.

(IWM)

Steve and I sailed in March/April 1945 for our first patrols [off the coast of Norway] in our brand new 'T' class submarines after work-up in the Clyde/Derry areas. We were allocated adjacent billets off a couple of lighthouses near Bergen. He was some five days ahead of me.

On the first evening in my billet we detected on our passive sonar what we took to be a dived submarine. We proceeded to track it and when we heard the contact blowing tanks we knew we had a submarine contact. I immediately came shallow to break surface and expose my type 267 aerial. This gave us an unmistakable surface contact which we plotted as going away. I promptly surfaced and gave chase, very soon obtaining splendid fire control information from the radar plot. It seemed the contact was taking me out of my area into Steve's, so I ordered a rushed enemy report to be prepared, informing *Turpin* an enemy submarine was approaching her area. At the same time, I continued to converge on the contact and had all my torpedoes ready for firing. Imagine my surprise when, just before I broke W/T silence to make my enemy report, an Aldis lamp flashed the challenge at me, immediately followed by, 'Don't shoot – Steve here!!'

Turpin had in fact spent their first five days on patrol in my area, having arrived in dense fog with some understandably shaky navigation. They had got an evening fix in clear weather the night that I arrived and discovered to their horror they were in my area, which they knew I was just entering. Steve had surfaced to get to hell out of it as quickly as possible, knowing I would not hesitate to attack if I came across him. (I actually sank a German U-boat there a week later!) Steve with his 267W radar had got an equally good plot of me after I had surfaced and was giving chase and could see our tracks converging. He also got interference on his 3cm set from another source on the same frequency. This convinced him it was me – hence his anxious signal. If we had any such interference we did not notice it – we were just convinced we were onto a good thing and were going for it.[23]

Type 267W had a fixed-height 'seaguard' aerial, usable only on the surface. Type 267PW – with the 3cm aerial array on a periscopic mast usable at periscope depth – did not come into service until after the war, the first trials being in *Amphion*, the first production model in *Alderney* in 1946.

TAILPIECES

• By Rowland T. Harris, sometime AB (Radar) in the submarines *Thrasher* and *Trenchant*, with type 291W, 1943–6.

. . . On our exercises out of Dunoon, I vividly recollect being asked – not ordered – to try to give the steering orders in order to negotiate a passage between two islands somewhere in the Minches. Needless to say, Skipper [Lieutenant A. R. Hezlet] and lookouts were on the bridge, but they were not needed. 291 was accurate within yards![24] [In fact, the navigating officer was standing alongside him in the Control Room with the captain checking for safety on the bridge.[25]]

• By Arthur Dilley, sometime AB (Radar) in *Thrasher* and *Trenchant*, 1943–6.

The most interesting bits came when, working as fall-of-shot spotter in gun action, occasionally some spirited Japanese characters would shoot back and the sight of echoes approaching the ground wave instead of the opposite concentrated the mind on giving accurate range changes . . .

The compensations for radar operators included diving stations in the Wireless Office where, once the shallow depth gauge had been shut off, we had no means of knowing how deep we were and were isolated from occasional slight panics caused by deep dives to escape the enemy's attention.

Interspersed with operating came the normal seaman's duties, i.e. lookout, helmsman, hydroplane operator, buoy jumper, harbour stations on the after crest, and cleaning the seamen's heads, this latter being part of the Admiralty's plan to keep Jack occupied when not at sea.[26]

Appendix E: Radar in Coastal Forces

THE FIRST COASTAL FORCES RADAR

After A. W. Ross and his Signal School team had designed the 'bedstead' aerial for destroyers in the summer of 1940 so that the RAF's ASV Mark II could be fitted in small ships as type 286M (p. 58), they turned their attention to designing a lightweight aerial suitable for Coastal Forces craft such as motor torpedo boats (MTBs), motor gun boats (MGBs), steam gunboats (SGBs) and motor launches (MLs). This aerial had three sets of fixed dipoles at the top of the craft's mast so that a transmitting array pointed right ahead and two receiving arrays pointed one on either bow, the relative strength of the echo signal between the two giving a measure of the bearing relative to the ship's head. It was similar to the arrays used in aircraft and was a lightweight version of the destroyer aerial. It had been tried out in an MTB by January 1941,[1] and by September, two MTBs, one MGB, and seventeen MLs had been fitted with type 286MU.[2] (For Coastal Forces radar, the suffix 'U' was added to the basic radar type number.) By November, general fitting had begun in all MTBs (except 60ft boats), MGBs, SGBs and MLs (except 72ft Harbour Defence MLs).[3] The boats in the 7th and 8th MTB Flotillas all had radar when they sailed for the Mediterranean in April 1942.

Meanwhile, a lightweight rotating aerial had been designed and was fitted in MTB 35 for trial towards the end of 1941.[4] General fitting began in mid-1942, all Coastal Forces fixed aerials eventually being replaced, so that the sets became type 286PU. The next development was the fitting of the more powerful all-naval type 291U with the same rotating aerial, the first development model being installed in MGB 320 in June 1942,[5] general fitting in new-construction boats starting in early 1943, though it was to be some time before all type 286s were replaced in existing craft.

Type 291U was to remain the main radar in Coastal Forces until the end of the war in Europe, used for aircraft warning and navigation in daytime and for surface warning and occasionally torpedo weapon control at night. But it had its limitations: being a 1.5m set, it was easy for the enemy to detect and had a wide beamwidth, so bearing accuracy, based on relative not compass bearings, was poor; the only display was an A-scan, the aerial was hand-rotated, and the plotting arrangements in the boats were rudimentary. British officers who saw the American 10cm type SO with PPI display and power-rotation – which came into service in the US PT Boats (Patrol Torpedo boats) early in 1943 – were green with envy. So much so that, in the Mediterranean in 1943, British MTB flotillas doing night attacks on Axis convoys would try to have a PT Boat in company so that they could make use of its superior radar and plotting arrangements.

59. An MTB, showing (1) WC type 291U; (2) fixed directional interrogator type 242.

(Radar Manual, 1945)

OPERATIONS OFF ELBA

A prime example of this cooperation occurred in two of the most successful operations by light forces in the Mediterranean in 1944, which were planned and executed by Commander Robert Allan, RNVR, who led a heterogeneous force of British LCGs (Landing Craft Gun), MTBs and MGBs and American PT boats based on Bastia in Corsica in attacks on German military supply convoys off the west coast of Italy. Allan had trained and exercised his sixteen vessels to work as a miniature battle fleet, in which the LCGs, armed with 4.7in guns, acted as capital ships; and he controlled them all by radar and radio from an American PT boat in which he sailed as 'admiral'.

In operation 'Gun' on the night of 27/28 March, Allan's force attacked a southbound convoy of armed barges escorted by two Italian destroyers off Capraia Island. All six barges were sunk and the two escorting destroyers were driven off by the PT boats. In a similar and even more successful operation, 'Newt' on 24/25 April, Allan in PT 218 had under him three LCGs, three MGBs, three MTBs, and seven PT boats. The first action took place just after midnight against a southbound convoy of three barges and a tug from Leghorn. All were destroyed, as were two armed barges patrolling nearby which came to their assistance, a third being driven ashore. Later in the night, the PT boats engaged three German torpedo boats minelaying west of Capraia, one of which struck a mine and was sunk by her own forces. There were no Allied casualties.

All of this was made possible, first by shore radar in Corsica tracking shipping in the channels between Corsica, Elba and the Italian mainland, and secondly by the PPIs of 10cm SO surface search radar fitted in the PT boats. (The British boats had the 1.5m type 291U air/surface warning radar – but no PPIs.)

Allan conducted the battle from the Combat Information Center of PT 218, where he had a clear picture of the situation and was able to vector craft as necessary, talking direct by voice radio. When necessary, he was able to detach an MTB unit under a Deputy Controller in another PT boat to deal with stragglers.[6,7]

As we saw in Chapter 9, vectoring of coastal forces from a ship was to be done again a few months later, during and after the assault on Normandy, this time from frigates. But once again, the British controllers had to use US radar.

US RADAR IN BRITISH BOATS

Such was the impact of the American 10cm type SO radar with its PPI display and power rotation, after the success of Operations 'Gun' and 'Newt', that the Senior Officers of the Fairmile 'D' flotillas in the Mediterranean entered negotiations for some unofficial Lend-Lease arrangement to secure it for their own boats.

60. MGB 658 in the Adriatic, October 1944, showing the radome of the American SO surface search radar.

(L.C. Reynolds)

Fortunately, with joint operations so firmly established over many months, relations with 'RON 15' (the USN's 15th PT Squadron) were so excellent that these negotiations were successful, and MGB 657 was the first Fairmile 'D' to be fitted with SO radar, in June 1944. It is said that the deal involved an unspecified quantity of Scotch whisky: but the detail is shrouded in mutual confidentiality.

By December 1944, at least six more 'D's were fitted with the set, enabling operations in the Adriatic to have the enormous search and plotting advantage it gave, especially as mixed units with PT boats were not normally possible.

Eight US-built Vosper MTBs equipped with the SO sets arrived in the Mediterranean in November 1944 and were formed into the 28th MTB Flotilla. They worked up diligently, specifically concentrating on deriving maximum operational value from a combination of an excellent search, detection and plotting radar with the availability of Mark VIII torpedoes with magnetic pistols. They achieved spectacular success in the first four months of 1945, operating in the northern Adriatic.

Once the enemy target had been picked up at extreme range and its course and speed plotted, the boats would set intercepting courses for an optimum firing position, and were able to move in at slow speed on silenced engines. This enabled surprise to be achieved with its benefits of uninterrupted approach and accuracy of torpedo firing angle.

In ten interceptions of enemy convoys by this flotilla, forty-three torpedoes were fired, twenty-four hits claimed, and nineteen ships sunk. None of this would have been possible without the superior PPI radar equipment.[8]

CENTIMETRIC RADAR FOR COASTAL FORCES

Let us now return to 1942. Our Coastal Forces badly needed centimetric radar, but how was it to be provided? British naval radar design and production facilities were stretched to the limit, and the 10cm type 271/2/3 radar then being fitted in corvettes and above could not be made compact enough or light enough for fitting in Coastal Forces craft. As mentioned on p. 234 above, the solution adopted was to ask the National Research Council of Canada to develop a high-resolution radar on the 3cm band, with power-rotation and PPI displays.

The first design work on this new radar – to be called type 268 – proceeded fast and a preliminary model was tested near Halifax in October 1942, with gratifying results. In April 1943, a set was sent by air to England, accompanied by three Canadian engineers led by K. C. Mann.[9] This was fitted in MGB 680.

And centimetric radar might do something else for Coastal Forces. The MGB was the equivalent afloat of the night fighter in the air, so, in the spring of 1942, Commander R. T. Young of the experimental department of the Navy's gunnery school, HMS *Excellent*, proposed that the lock-follow Air Interception radar (AIF or AISF) being developed at TRE for night fighters might be fitted to MGBs to make blind fire by their 2-pounder guns practicable against enemy vessels at night. That summer, trials were carried out tracking small vessels and low aircraft on the slipway at Lee-on-Solent, on the 'rolling platform' at Eastney, and at sea in MGB 614, where the AISF scanner was mounted on a wooden structure in the position of the forward gun. In the sea trial, in both calm and rough weather, the average lock errors in both azimuth and elevation were something less than ½°.[10]

In December 1942, however, TRE's AISF programme received a considerable setback from the loss in an aircraft crash of the prototype AISF. Luckily for the RAF this coincided with the arrival of the first American centimetric AI, the 3cm SCR 720 with a helical scanner rotating at 360 rpm, classified by the RAF (and the

Royal Navy) as AI Mark X. This was to remain the standard AI equipment in the RAF until 1957.[11]

In March 1943, AI Mark X – for naval purposes designated type 269 – was fitted to the 2-pounder gun mounting of MGB 680. Two months later, the Canadian tactical warning type 268 already mentioned was also fitted, with the scanner at the top of her mast. On 5 July, she gave a successful demonstration of these two 3cm radar sets in the Solent to senior naval officers and civilian scientists.

For type 268, the trial was most successful, resulting in an immediate Admiralty order for 1,000 sets to be produced in Canada, later increased to 1,500.[12]

However, type 269 was not so successful. Mr Robin Board, then commanding MTB 680, says that his memories of it are anecdotal:

- The dustbin-sized nacelle fixed to the pom-pom which so restricted the view from the bridge when coming alongside.
- The tendency to leave the target without warning and lock on to the Ryde ferry.
- The amazement of the ship's company watching sausages being cooked in the waveguide.
- The ship's company's certainty that the operator would become impotent, and their fascinated interest in him after each leave period.[13]

Type 269 never came into service and in October ASE reported that work on it had stopped.[14] Production of type 268 in Canada was plagued by difficulties – caused at least in part by suggested modifications from Britain to cope with the 'Schnorkel' threat – and the first batch of production sets (only eleven) did not reach England until January 1945,[15] too late to have any influence in the war in Europe – which was sad for Coastal Forces because it was to prove a most valuable piece of equipment. With hindsight, one wonders whether efforts could have been made to acquire the American SO search radar for British-built boats once the delays to type 268 became evident.

SURFACE FORCE DIRECTION

As mentioned on p. 109, a naval 10cm type 271 (army manned) was installed on the South Foreland north of Dover in July 1941, reporting to the combined services plot in Dover Castle. On the evening of 8 September, a German convoy was detected leaving Boulogne at a range of 24 miles. Dover plotted it tracking to the north and east and at 11.50pm directed three MTBs onto the attack off Calais. The target was held by radar until 1.45am as far as Gravelines (nearly 30 miles) and MTBs made a further attack at 2.01.[16]

Many such actions with Coastal Forces craft under W/T shore control, making use of shore radar, took place during the next year or so off Dover and elsewhere. True coastal force direction afloat, analogous to fighter direction, where the craft were actually given courses to steer – 'vectored' – by Coastal Forces control ships using R/T came into full operation in 1944 in Operation 'Neptune' as an anti-E-boat measure as described on p. 220 above. The idea was first suggested by Lieutenant M.G. Raleigh, RNVR,[17] Assistant Staff Officer (Operations) at Portsmouth and was enthusiastically welcomed by Captain Coastal Forces (Channel), who, with his staff (Lieutenant-Commander Peter Scott, RNVR, and Lieutenant Christopher Dreyer), followed it through and made it happen. As we saw above, something like it had already been used to some extent in Operations

'Gun' and 'Newt' against enemy convoys in the Mediterranean, but details of the tactics did not reach UK until July 1944.[18]

Coastal Forces control ships were destroyers or frigates with good radar and plotting facilities, particularly the US-built 'Captain' class frigates with SL search radar and automatic plotting tables. With specially trained control officers (ex-MTB COs) to supervise the plot and vector the attached craft, they were virtually secondary shore plots stationed so as to provide radar cover of the convoys routes wherever it could not be provided by shore radar.

The normal practice was to detail a frigate and two units of two or three MTBs as one force and to sail in company where practicable. Each control ship patrolled a six-mile line at 12 knots, coordinating its movements with adjacent patrolling frigates so that large gaps in the overall patrol line were avoided. The MTBs remained stopped at some convenient position (usually at each end of the frigate patrol line) so arranged that they were at all times visible on the radar scan of the directing ship. Once a detection had been plotted by the control ship, the attached craft were vectored by the control officer to intercept, subject to his orders until themselves in contact with the enemy, either by radar or visually. On occasions the control ship also provided illumination of the target. In the later stages of the evacuation of Le Havre, it became the practice to add one or two 'Hunt' class destroyers to the force. These acted as support for the MTB units, since some of the convoys were heavily escorted and were sailing close inshore, so that there was also trouble from shore batteries. The destroyers usually patrolled in company with the controlling frigate, keeping station on the frigate in line ahead.[19]

During the course of Operation 'Neptune', there were many examples of the value of the control ship technique, both in defensive actions to protect the cross-Channel convoy route against E-boat attack, and in offensive actions against German convoys evacuating Cherbourg and, later, Le Havre. It was most noticeable that contact with the enemy was almost always obtained.

Operating from Cherbourg with the US Navy in the later stages of the Channel War in August 1944, Peter Scott conducted one classic little operation – a splendid example of Allied cooperation. Embarked in a USN frigate, he controlled a mixed force of two British 'O' class destroyers, and three MTB units – one of US PT boats, one Canadian and one Free French. They had a brief and inconclusive action off Guernsey but sank one trawler and one 'M' class minesweeper.

After the Channel had been cleared by the end of August 1944, the MTB flotillas and control ships moved to the Nore Command and operated for the remainder of the war in Europe, mainly protecting the Thames–Scheldt convoy route against E-boat torpedo and minelaying attacks.

Appendix F: Naval Airborne Radar

David Brown, FRHistS

Two years, almost to the day, before the outbreak of the war, all the major future uses of airborne radar were demonstrated in a single bad-weather sortie by an Avro Anson aircraft attached to the Air Ministry's Radio Research Station. The experimental set then known as 'RDF 2', had been flown for the first time as recently as 17 August 1937, but, as mentioned on p. 18 above, on 4 September, in the hands of Dr E. G. Bowen, the leader of the airborne radar development team, it located HMS *Courageous* and her destroyer screen in low visibility, homed the aircraft to make a visual identification, detected *Courageous*'s fighters as they were scrambled. Finally, the radar permitted Bowen to navigate to an accurate landfall, in cloud that extended from 12,000 feet to near sea level. All that was lacking in the Anson was W/T, to report its sighting.

Despite this very early success, no operational aircraft was fitted with air to surface vessel radar (ASV) on 3 September 1939. Technical difficulties and rival demands on Dr Bowen's small team - particularly the priority given to the development of Air Interception (AI) radar – were largely responsible, but it was perhaps unfortunate that the Admiralty, which clearly had a major interest in ASV, was not given an official demonstration of its capabilities until May 1938. Six radar sets were requested for immediate delivery, but the niceties had to be observed and a joint Air Ministry–Admiralty conference had to be held to lay down guidelines for operational requirements, tactical development and trials. The conference took place in early August 1938, but the demonstration of the rival forward- and sideways-looking systems scheduled for September had to be cancelled owing to the Munich crisis. On 21 September, however, the AOC-in-C of Coastal Command, Air Marshal Sir Frederick Bowhill, recommended to the Air Ministry that ASV should be adopted immediately as standard equipment in maritime reconnaissance aircraft.

This recommendation, and persistent Admiralty demands, failed to produce any orders for pre-production ASV sets until July 1939, when 24 were ordered, of which six were to be fitted to Royal Navy aircraft. The design of the ASV Mark I had still not been finalized and the alternators could not be delivered for three months: not until November 1939 was the 1.5m ASV I flown in a Coastal Command Lockheed Hudson and the first sets were delivered to the Royal Navy in the following month, for trials installation in Supermarine Walrus and Fairey Swordfish aircraft.

At this stage, the Royal Navy's interest in ASV was for Fleet use, at night and in bad weather; the early trials were entirely devoted to the development of the appropriate 'night shadowing equipment' tactics to be used to gain and hold contact with enemy surface ships, as preliminaries to a carrier air strike or a capital ship duel. It was therefore left to Coastal Command to pioneer the anti-submarine development and use of ASV, in trials and on actual convoy escort missions.

The ASV I proved in service to be very fragile and its unreliability was not helped by a lack of instructional handbooks and maintenance manuals at squadron level. It also had a number of undesirable characteristics, not the least of which was interference with the parent aircraft's W/T reception. By February 1940, however, a full production model had been designed and engineered and two firms were asked to produce 4,000 ASV Mark II sets, for delivery from 1 August 1940. Unfortunately, the Air Ministry bureaucracy failed to recognise the importance of the programme and the formal order was not placed until April. A month later, AI production was given precedence over that of ASV, even to the extent of 80 ASV I transmitters being diverted. By the end of 1940, Coastal Command had only 49 aircraft equipped with ASV, while the Royal Navy had only its trials aircraft. The Admiralty itself contributed to the delay in availability of the ASV II, for it provided the basis of the type 286 radar, fitted in small surface ships from the late autumn of 1940. Production increased rapidly early in 1941 and during the first three months of that year components for 1,000 complete sets of ASV II were delivered to the Services.

ASV II remedied most of the pre-production Mark I's shortcomings and, with minimal development, the Royal Navy's variant, ASV IIN, remained in front-line service throughout the war. Operating in the ultra-high-frequency band, on 214MHz (the RAF's ASV II was tuned to 176MHz), with an initial peak power output of 7kW, it could give a range of up to 15 miles against a medium-sized ship and, under good conditions, up to 4 miles against a surfaced submarine. Fitted to the inter-wing struts of the Walrus and Swordfish aircraft, the distinctive splayed dipole antennae had little effect on the performance of biplanes which rejoiced in the Service nicknames of 'Shagbat' and 'Stringbag'.

The first naval front-line unit to be fitted with ASV IIN was the spotter–reconnaisance catapult flight embarked in HMS *King George* V, which received two radar-equipped Walrus amphibians in mid-March 1941. Other Home Fleet Walrus flights quickly followed and as individual Swordfish were fitted with radar they were delivered piecemeal to squadrons. In April 1941, 825 became the first Swordfish squadron to be entirely re-armed with ASV Swordfish and it was the nine aircraft of this unit which, flying from HMS *Victorious* to the south-west of Iceland, first used ASV IIN in anger, to locate and attack the *Bismarck*, just after midnight on 25 May 1941.

On the afternoon of 26 May, the *Ark Royal* launched her first torpedo striking force against the German battleship. Only the leader's aircraft was radar-equipped but the ASV picked up a target close to the briefed position and the Swordfish dived through thick cloud to deliver a surprise attack in low visibility. Eleven torpedoes were released and it was only a combination of a known defect in the firing pistol and *HMS Sheffield's* swift reactions which prevented the first radar-assisted 'own goal'. The subsequent attack was, of course, highly successful and led to the destruction by gunfire of the *Bismarck*. *King George V*'s Walrus flight played no part in the action, but remained in the battleship's hangars, their fuel drained down to reduce the fire hazard.

When the *Ark Royal* was torpedoed on 14 November 1941, most of her radar-equipped Swordfish were lost with her. 812 Squadron re-armed with aircraft from the pool held at Gibraltar, but there were only four ASV sets available when this unit was called upon to begin a little-known but highly successful short anti-submarine campaign. The German Navy was known, from 'Enigma' decrypts, to be passing batches of U-boats into the Mediterranean through the Straits of Gibraltar and 812 Squadron was ordered to reinforce the night surface patrols, the ASV II fitted to the RAF Catalinas based at Gibraltar being at this time so unreliable as to be useless. The patrols began on the night of 27/28 November

61. The tangled wreckage of two Swordfish, seen after an altercation aboard *Eagle* in the spring of 1942. The port reflecting-dipole receiving aerial of the victim is at centre bottom of the photograph, while the assailant's simple 'towel rail' transmitter extends along the leading edge of the upper wing centre-section, on three insulated supports. The shorter horizontal bar, mounted below the wing ahead of the pilot, is part of the torpedo sight.

(D. T. R. Martin)

and continued for two months. Of 22 submarines which attempted the passage, 1 was completely deterred by the Straits defences, five were damaged and forced to return by the Swordfish, whose efforts were crowned in the early hours of 21 December by the destruction of U.451, the first U-boat to be sunk at night by aircraft, or as a direct result of radar, or by any land-based aircraft, naval or RAF.

The intended replacement for the Swordfish, the Fairey Albacore, had been in service since the summer of 1940 but had won its laurels in the eastern Mediterranean without the benefit of ASV. Indeed, it was not until the end of 1941 that the problem of serious interference between the aircraft compass and the radar alternator was solved. Thereafter, the aircraft were equipped as quickly as possible and in March 1942 *Victorious*'s two squadrons used their radar to good effect in a torpedo attack on the *Tirpitz* which narrowly failed to score hits. Priority for Albacore re-equipment went to the Home Fleet squadrons and thus, when *Formidable* sailed for the Indian Ocean in February 1942, her squadron, like all but a handful of the Albacores aboard the *Indomitable*, which was already on station, lacked ASV and she had to 'borrow' a pair of ASV-fitted Swordfish. As Admiral Somerville's plan for dealing with the Japanese carrier fleet relied primarily on night search and strike, this was hardly a satisfactory state of equipment. In the event, one of the *Indomitable's* radar-equipped Albacores did

62. The Barracuda's ASV IIN antennae had to be mounted above the wings, owing to the peculiar geometry of the aircraft's wing-folding mechanism. They differed from previous fittings of this radar in that each antenna incorporated both the transmitter and receiver elements. This example is taking off from *Furious*, whose type 291 radar and type 72DM homing beacon are also well illustrated.
(IWM)

make the only contact between Somerville's carriers and ship of the Japanese fleet, as the latter was withdrawing beyond air torpedo striking range.

By mid-1942, sufficient ASV IIN sets were available to equip all embarked aircraft which could be fitted and work was well advanced on its replacement. The Admiralty had from the outset stated a requirement for all-round coverage but this was not practical until the development of centimetric radar, with its smaller transmitter–receiver scanner. In December 1941, a successful bid was made for a new centimetric radar specifically for its new torpedo-bomber, the Fairey Barracuda and design work began on ASV XI. In the meantime, the Barracuda appeared to be approaching acceptance for service and had to make do with ASV IIN: it could hardly be said that the overwing combined transmitter–receiver antennae marred the already 'over-loaded Christmas-tree' looks of the Navy's newest monoplane.

At the same time (late 1941), the Admiralty decided that embarked night-fighters were urgently required. The existing Fairey Fulmar two-seat fighter was chosen for trials with the RAF's AI Mark VI metric set, as an interim measure, but the intention was to fit its successor, the Firefly, with a dual-purpose centimetric radar, for air–surface and air–air use. In the event, this was not achieved and the centimetric AI XI which was installed in a small number of Fireflies proved to be unusable in any role.

63. Belt and Braces: the Swordfish IIIs operated by 860 Squadron (Royal Netherlands Navy) from the Dutch-flagged merchant aircraft carrier *Gadila* retained the ASV IIN fitted to the inter-plane struts even after the fitting of ASV XI. The positioning of the latter's 'bathtub' scanner dome below the engine meant that no weapons could be carried under the fuselage, while the weight of equipment was such that the crew had to be reduced to two and rocket assistance was necessary for take-off from the short decks of the MAC ships. (The tube of the port rocket can be seen projecting obliquely behind the radar dome.)
(RNethN)

ASV XI, on the other hand, was a success, with a range of up to 10 miles against a surfaced U-boat and 40 miles against a carrier; the same cannot be said of the aircraft for which it was intended. As the Barracuda was experiencing serious problems, priority was given to the fitting of the centimetric radar to the Swordfish, specifically for anti-submarine work. The first trials aircraft flew in February 1943, as the Battle of the Atlantic reached its climax, and the first squadron re-equipped with the ASV XI-fitted Swordfish III in November 1943. Afloat, these aircraft served only aboard the escort carriers, operating in the Atlantic and in the Arctic, where the squadrons scored some notable 'kills'. One successful Swordfish III squadron, 819, was given a quite different task in the summer of 1944, providing night anti-E-boat patrols before and during the Normandy invasion; so successful was it in this new role that it moved to Belgium in October 1944 and continued its coastal patrols until March 1945.

The ASV XI-fitted Barracuda III arrived in service as 819 Squadron began to demonstrate the usefulness of the centimetric radar in coastal waters. The equipment was limited by aircraft centre-of-gravity difficulties and it had been decided that only one new aircraft in four would be built as a Mark III and that these would not serve in Fleet carriers. In the event, only one Barracuda squadron armed with this Mark, the Canadian-crewed 821, went to sea at all, aboard the Canadian-manned *Puncher*, and it undertook only one offensive operation – a minelaying mission off the Norwegian coast. Several other squadrons formed, but all were employed on anti-E-boat and anti-midget submarine patrols in the southern North Sea.

It was obvious from the summer of 1944 that the ASV IIN Barracuda's days as the principal carrier strike aircraft were numbered. While it could cope with operations off the ill-defended coast of Norway, its performance in subtropical conditions was completely inadequate. Fortunately, the US Navy's Grumman Avenger was now being delivered in sufficient numbers to re-arm all the Fleet carrier squadrons. This aircraft was equipped with the standard American metric 'ASB' (air-to-surface, Type 'B') radar, a direct copy of ASV IIN, but towards the end of 1944 new aircraft began to arrive equipped with 'ASH' (officially designated AN/APS-4) radar.

This remarkable piece of kit had first been seen in Britain early in 1944 and large-scale orders had promptly been placed. A 3cm set, it had been specifically designed and miniaturised for external carriage on single-engined aircraft and the scanner, transmitter, receiver and alternator were all packed into a streamlined 350lb 'bomb'. It lacked the all-round scan of the ASV XI, but ASH's performance was as good and its light weight and lack of demand upon the parent aircraft's internal volume made it the obvious replacement, not just for ASV IIN but also for ASV XI. Even more telling was its air-to-air capability – ASH was the dual-purpose radar which the Admiralty had been seeking since late 1941.

The failure of the Firefly night-fighter programme had resulted in the Fulmar being retained, primarily as a training aircraft. Many of the crews which graduated from the RN Night Fighter School at Drem subsequently served with RAF squadrons, but in early 1944, with *Luftwaffe* guided missile squadrons beginning a night offensive against Gibraltar convoys, three-aircraft flights of night-fighter Fulmars, equipped with metric AI IV radar, embarked in escort carriers on that route. Of course, no convoy accompanied by one of these carriers was approached by enemy night bombers and it was not until a year later, in March 1945, on the Arctic convoy route, that the only 'scramble' of a Fulmar night-fighter was attempted: the fighter was too slow to catch the intruder and crashed on landing.

A single ASH-equipped Firefly night-fighter squadron (1790) arrived in the Pacific shortly before VJ day but too late to embark in HMS *Implacable*. Two more Firefly squadrons were then working up in the United Kingdom, as were two Grumman Hellcat squadrons, equipped with 'AIA' (AN/APS-6) radar, which resembled ASH but was designed specifically as a pilot-interpreted night-fighter radar, which had been used very successfully by the US Navy since the autumn of 1944.

At the end of the war, the metric ASV IIN (and the similar ASB) was still in use, having given excellent service for four and a half years. ASV XI had proved to be a useful radar during the last eighteen months of the war in Europe, but in the front line it did not survive the disbanding of the Swordfish and Barracuda squadrons. For the immediate future, ASH, with its virtues of light weight and acceptable performance, would fulfil all the Royal Navy's airborne needs; it had arrived too late for widespread service, but by the end of 1945 it had displaced every other radar in the aircraft of the British Pacific Fleet and it was to survive as the principal naval ASV radar for another twelve years.

The avenue opened by AIA was not explored by the Royal Navy after the return to the US Navy of the last Hellcat night-fighters in the spring of 1946. Night-fighting, it was reasoned – by the Observers' 'guild', which guarded jealously the black art of radar operation – was too complex for one-man operation and for the next thirty years, until the introduction of the Sea Harrier, this remained the province of the two-seat aircraft. In fifty years, however, no AI-assisted 'kill' has been scored by a naval fighter, in contrast with the conspicuous success of radar-equipped strike and anti-submarine aircraft.

Glossary of Technical Terms and Abbreviations

Abbreviations used in references are listed on p. 321.

AA	Anti-aircraft
A-arcs	The arcs on which all guns of a ship's main armament will bear, thus allowing them all to fire simultaneously at the enemy.
AB	Able Seaman
ACNS	Assistant Chief of Naval Staff
Acorn valve	A small high-frequency valve about 1 inch in diameter shaped like an acorn, with the leads arranged radially around the middle of the envelope.
ADEE	Air Defence Experimental Establishment (Army).
ADR	Aircraft direction room
ADRDE	Air Defence Research and Development Establishment (Army)
Admiralty organization	*Divisions of the Naval Staff* responsible for operational matters, such as the Operations, Plans, Training and Staff Duties, and Naval Intelligence Divisions, reported to the First Sea Lord through the Deputy and Assistant Chiefs of Naval Staff. *Admiralty Departments* generally responsible for *matériel* matters, such as the Naval Ordnance, Naval Construction, Scientific Research, and Signals Departments, reported to the Controller. From the outbreak of war, operational sections of the Signal Department and the Naval Ordnance Department were in London, *matériel* sections in Bath.
AEW	Airborne early warning
AFCC	Admiralty Fire Control Clock
AFCT	Admiralty Fire Control Table
AFO	Admiralty Fleet Order
AI	Air interception (airborne radar)
AIO	Action Information Organization
AMRE	Air Ministry Research Establishment
ANCXF	Allied Naval Commander Expeditionary Force
ARL	Admiralty Research Laboratory
A/S	Anti-submarine
'A'-scan	See *Display*

Asdic	Underwater detection equipment using acoustic waves. An acronym based on Allied Submarine Detection Investigational Committee (c.1920). US – *Sonar* (qv).
ASE	Admiralty Signal Establishment
ASV	Air to surface vessel (airborne radar)
Azimuth	Horizontal angle measured about a nominally vertical axis
Back echoes	The main beam of a directional aerial is accompanied by some radiation in other, unwanted, directions, known as side lobes or back lobes. Such lobes can give rise to echoes from large targets at short ranges in those directions, which can be wrongly attributed to the direction of the main beam. Such unwanted responses are known as side echoes or back echoes.
Beam riding	A method of controlling a missile in flight whereby a radar beam held on the target is used by the missile to derive its own guidance signals.
Beam-switching	A method of obtaining accurate bearing and/or elevation of a target by rapid switching of the radar beam between two directions, thus producing two echoes. When the echoes are equally matched, the bearing is correct.
Blind fire	Using radar bearing, range, and if necessary elevation to fire at an unseen target.
BPF	British Pacific Fleet
BPR	Bridge plotting room
BR	Book of Reference – an Admiralty publication with a security classification of Restricted or lower
BTH	The British Thomson-Houston Co Ltd
CA	Coast Artillery (army radar)
CAFO	Confidential Admiralty Fleet Order
CAM	Catapult Aircraft Merchantman
CAP	Combat air patrol
CB	Confidential Book – an Admiralty publication with a security classification above Restricted
CD	Coast Defence (army radar)
CDS	Comprehensive display system
CH	Chain Home (RAF radar)
CHL	Chain Home Low (RAF radar)
C-in-C	Commander-in-Chief
CO	Commanding Officer, or Chain Overseas (RAF radar).
Common aerial working	The use of a single aerial array for both transmitting and receiving.
CPO	Chief Petty Officer
CRBFD	Close range blind fire director
CRT	Cathode ray tube
CVD	Ultra short wave Communications Valve Development (committee)
CVE	Escort aircraft carrier (US)
CW	Continuous wave
DCT	Director control tower, on which gunnery radar aerials were mounted, placed high in the ship to get the best possible view of the target, visually and by radar. Its laying and training on the target was followed (via the *TS* (qv)) by

	individual turrets and guns when in 'director firing', the normal method. Often abbreviated to *Director*. See also *HA Director*.
DE	Destroyer escort (US)
DF	Direction finding
DGD	Director of the Gunnery and AA Warfare Division
Dipole	Aerial or element of an aerial consisting of two equal collinear rods, of total length equal to about one half the wavelength
Director	Common term for *DCT* (director control tower) (qv)
Discrimination	In radar, the ability to distinguish between (and if necessary range on) two targets close together. *Range discrimination* is achieved by very short pulse length, *bearing discrimination* by narrow beamwidth. Also used to define the ability of radio receivers to distinguish between signals of different frequency.
Display	The method of presenting radar echoes to the observer. The most common naval types were: *'A' Display* (formerly *A-scan*, the term generally used in this book), which shows the target's range (but no bearing) when the radar beam is trained on it. *PPI* (plan position indicator), which shows simultaneously both range and bearing of targets as bright spots on a CRT with a long afterglow. It thus gives a complete picture of the surroundings as detected by radar. PPI display requires the aerial to be kept spinning or sweeping. *Skiatron*, a display involving the optical projection of a form of PPI onto a ground glass screen, to facilitate plotting. Echoes appear as dark paints, as opposed to PPI where they are normally bright echoes on a dark background. *Sector display*, which shows on a type of 'A' Display all echoes within one or more selected small sectors of bearing. Used in height estimation and interrogation positions. *Meters*: are sometimes used to match echoes produced by beam-switching. *HPI* (height position indicator) which, for a given bearing, shows elevation, range and height on a display like that of a PPI.
DNI	Director of Naval Intelligence
DNO	Director of Naval Ordnance
DR	Dead reckoning
DRE	Director of Radio Equipment
DSC	Distinguished Service Cross
DSD	Director of the Signal Department/Division
DSM	Distinguished Service Medal
DSR	Director of Scientific Research and Experiment
DTM	Director of Torpedoes and Mining
ECM	Electronic countermeasures – anti-radio/radar techniques and tactics, particularly the jamming of transmissions. Formerly *RCM* (qv).
Elint	Electronic intelligence – the gathering of data by *ECM* (qv) techniques. Formerly known as *Y equipment* (qv).

EMI	Electric & Musical Industries Ltd.
ESM	Electronic support measures – passive measures of gathering electronic intelligence
FAA	Fleet Air Arm
FD	Fighter direction
FDO	Fighter direction officer
FDT	Fighter direction tender, a specially equipped vessel based on the hull of a landing ship tank (LST)
FKC	Fuze-keeping clock
GA	Gunnery Fire Control, Aircraft (RN radar from 1943)
Gain	Factor by which power is increased
GB	Gunnery Fire Control, Barrage (RN radar)
GC	Gunnery Fire Control, Close-range (RN radar)
GCI	Ground Control of Interception (RAF radar)
GDR	Gun direction room
GEC	The General Electric Company Limited (of England)
GL	Gun Laying (Army anti-aircraft radar)
GS	Gunnery Fire Control, Surface (RN radar)
HA	High angle (gunnery)
HACP	High Angle Control Position. Analogous to the *TS* (qv).
HACS	High angle control system
HA Director	A *DCT* (qv) when applied to long-range AA systems
HE	Hydrophone effect (asdic). See p. 146.
HF/DF	High frequency direction finding
HFP	Height Filtering Position, where all available radar height-finding information is received, filtered, and passed to where it is required
HMSS	His Majesty's Signal School, until 1941 housed in the Royal Naval Barracks, Portsmouth. HMSS's Experimental Department, responsible for radar development, became *ASE* in 1941.
HPI	Height position indicator. See *Display*.
IF	Intermediate frequency
IFF	Identification Friend or Foe – an ultra-high-frequency radio *interrogator* (qv) and *transponder* (qv) system used in association with warning radars to differentiate between friendly and hostile or unidentified contacts
Interrogator	A secondary radar which could activate an IFF *transponder* (qv). See *IFF*.
K-band	Electromagnetic wavelengths of about 1.5cm (20,000 MHz)
Klystron	Radio valve for amplifying or generating centimetric microwaves by forming electrons into bunches as they cross a gap.
LA	Low angle (gunnery)
LCG	Landing craft, gun
LCH	Landing craft, headquarters
LSF	Landing ship, fighter direction. *Boxer* (ex-LST) was the only one completed.
LST	Landing Ship, Tank
MADP	Main air display plot
Magnetron	Microwave generator employing an external magnetic field
MBE	Member of the Order of the British Empire
MGB	Motor gunboat

Mile	Where 'mile' is used in this book as a unit for distance at sea, the nautical or sea mile should be assumed – one minute of latitude, equivalent to 6,080 feet (usually rounded off to 2,000 yards in naval practice, including wartime British naval radar) and 1.8532 kilometres. The English land mile is 1,760 yards.
ML	Motor launch
MPI	Mean point of impact (gunnery)
MTB	Motor torpedo boat
NA	Naval attaché
n.m.	Nautical mile – See *Mile*
NOD	Naval Ordnance Department
NRC	National Research Council of Canada
PO	Petty Officer
PPI	Plan position indicator. See *Display.*
PT boat	Patrol torpedo boat (US)
Radar	An acronym from *RA*dio *D*etection *A*nd *R*anging
Radar beacon	A type of IFF transponder of which aircraft can obtain the range and bearing for homing purposes. Radar beacons are also used for distress purposes and ashore to assist ships and aircraft. Sometimes called a *Racon.*
Radome	Radar dome, protecting the aerial array from wind, weather and spray
RAE	Royal Aircraft Establishment (RAF)
Rate-aiding	A mechanical or sometimes electrical means of establishing the rate of change of a target's range, bearing or elevation by following its position as smoothly as possible. The fire-control director or radar aerial was then driven at these rates by remote power control (*RPC*)/*servo-mechanism* (qv) which helped the operators follow the target more smoothly, which in turn resulted in more accurate determination of the target rates.
RCM	Radio countermeasures (which include radar) – earlier name for *ECM* (qv)
RCNVR	Royal Canadian Naval Volunteer Reserve
RDF	Radar. See p. xix
RDR	Radar Display Room. The compartment of the Action Information Centre where displays from all warning sets, and the Height Filtering Position, were situated.
RF	Radio frequency
rms	Root-mean-square – square root of the arithmetic mean of the squares of a set of numbers (statistics)
RN	Royal Navy
RNR	Royal Naval Reserve
RNVR	Royal Naval Volunteer Reserve
RNZNVR	Royal New Zealand Naval Volunteer Reserve
RPC	Remote power control. See *Servo-mechanism.*
RRDE	Radar Research and Development Establishment (Army)
R/T	Radio telephony
SA	Ship-to-Air (RN radar up to 1943).
S-band	Electromagnetic wavelengths of about 10cm (3,000 MHz).

Scanning	The technical process whereby some radar sets automatically search in azimuth or in elevation, or in both simultaneously.
Servo-mechanism	A closed-cycle control system in which a small input power controls a large output power in a strictly proportionate manner. In this book, it is usually synonymous with remote power control.
SGB	Steam Gunboat
Side echoes	See *Back echoes.*
Skiatron	See *Display.*
Sonar	Underwater detection equipment. Acronym for **SO**und **N**avigation **A**nd **R**anging. The US version of British asdic (qv).
Sp.	The Special Branch of the RNVR, to which many wartime radar officers belonged, e.g. Lieutenant (Sp.) A.B. Jones, RNVR
Special Intelligence	Intelligence from a particularly sensitive and absolutely reliable source, available as a result of the solution of high-grade codes and cyphers, distributed to specially selected and severely restricted numbers of recipients by means of one-time pad cyphers. Code name: *Ultra* (qv).
SS	Ship-to-Ship (RN radar up to 1943)
STAAG	Stabilized tachymetric anti-aircraft gun
Staff Requirements	When the development of a new naval system or equipment was proposed, the operational requirements – in terms of range (maximum and minimum), accuracy, tactical deployment, and, for radar, operating frequency – would be prepared by the Naval Staff with help and discussions with Admiralty Departments. These would be formalized as 'Staff Requirements' and sanctioned or rejected by the Admiralty Board.
STC	Standard Telephones & Cables Ltd
Thyratron	A gas-filled valve with heated cathode, able to carry very high currents, which operated as a switch
TI	Target indication
TIR	Target indication room
TIU	Target indication unit
Transponder	A radio or radar device which, on receiving a signal, transmits a signal of its own, as with *IFF* (qv)
TRE	Telecommunications Research Establishment (RAF)
TS	Transmitting Station – A compartment between decks which housed the fire-control predictors. They were fed with target information (range, bearing, elevation) from optical instruments (sights and rangefinders) and/or radar equipment trained on the target; they calculated the future position of the target; and they transmitted to the gun mountings the predicted ranges, bearings and elevations for the guns to hit the target. In larger ships, the TS was concerned with LA (surface) fire control only, the HA equivalent being the *HACP* (qv).

TSR	Torpedo spotter reconnaissance (aircraft)
UDU	Universal display unit
UHF	Ultra high frequency, 300 to 3,000 MHz
Ultra	Code name and message prefix for *Special Intelligence* (qv), and for messages containing Special Intelligence
USN	United States Navy
VHF	Very high frequency, 30 to 300 MHz
V/S	Visual signalling
VT fuze	Variable time fuze – a proximity fuze which uses radar principles to initiate the detonation of a shell or bomb at a supposedly lethal distance from an air target or at a set height above a surface target.
WA	Warning of Aircraft (RN radar from 1943)
WC	Warning, Combined aircraft and surface (RN radar)
Window	Metal-foil strips dropped in quantity from aircraft which gave echoes on the enemy radar simulating our own forces.
Working up	A period spent in exercises working up the efficiency of the ship's company of ships newly commissioned after building or refit
WRNS	Women's Royal Naval Service
WS	Warning of Surface craft (RN radar)
W/T	Wireless telegraphy
WW1	World War 1, 1914–18
WW2	World War 2, 1939–45
X-band	Electromagnetic wavelengths of about 3cm (10,000 MHz).
Y equipment	Receiving and *DF* equipment employed for obtaining and analysing radio intelligence. Later known as *Elint* (qv).
Y service	The interception of enemy signals, including *DF*

References

For published works, full titles and sources are given only when they are not listed in the Bibliography.

References prefixed ADM, AIR, AVIA, PREM, etc., refer to documents at the Public Record Office, unless otherwise indicated.

The following abbreviations will be used:

ADR	Air Defence Research (Committee)
AFO	Admiralty Fleet Order
AL	Admiralty letter
ARE	Admiralty Research Establishment
ASE	Admiralty Signal Establishment
CAFO	Confidential Admiralty Fleet Order
CAC	Churchill Archive Centre, Churchill College, Cambridge, where the documents generated by the Naval Radar Trust project will be deposited
C-in-C	Commander-in-Chief
CSS	Captain, HM Signal School
CSSAD	Committee for the Scientific Study of Air Defence
DSD	Director, Signal Division, Admiralty
FO	Foreign Office
HF	Home Fleet
HMSO	HM Stationery Office
IWM	Imperial War Museum, London
Med	Mediterranean [Fleet]
NHB	Naval Historical Branch
NMM	National Maritime Museum, Greenwich
NRT	Naval Radar Trust
PRO	Public Record Office, London
RSRE	Royal Signals and Radar Establishment, Great Malvern

1 Setting the Scene (pages 1–10)

1. Printed in ASE Christmas card, 1943.
2. ADM 220/70. DSD to CSS, 13 Aug 1935.
3. For early radar in Britain, USA, Germany, France, Netherlands, USSR, Japan, and Italy, see Burns, *Radar Development*, and Swords, *Technical History*.
4. CAB 102/640, 641 & 642 are Parts I, II and III of typescript *History of the Development and Production of Radio and Radar*, hereafter cited as *History*. This particular source is in *History*, I, paras. 149–50. Two abridged versions of this – one official version with annotations (CAB 102/10), one on sale to the public without annotations – were published by HMSO in 1964 in the *History of the Second World War* series as Postan, Hay and Scott, *Design and Development of Weapons*.

5. CAC, NRT. Wright to Rawlinson, 20 July 1975. On the Alder proposals, see Burns and Coales & Rawlinson in Burns, *Radar Development*, 18–20 and 55.
6. On the Butement and Pollard proposals, see Burns in Burns, *Radar Development*, 20–2.
7. Rowe, *One Story*, 5.
8. CAC NRT. Freeman papers, Air Ministry minute 12 Nov 1934.
9. Ibid., 15 November 1934.
10. AIR 20/80. Reprinted in Swords, *Technical History*, 278–81.
11. Ibid., 281–5.
12. Montgomery Hyde, *British Air Policy*, 330.
13. *History* I, para. 24, n(2). For the early history of radio in the Royal Navy, see Hezlet, *The electron and sea power*.
14. Montgomery Hyde, *British Air Policy*, 322–3.
15. CAB 16/132, f.14: ADR minutes, 27 May 1935. Roskill, *Hankey*, ii, 144ff.
16. ADM 220/70. Shearing's notes on 'Meeting at the Air Ministry, Whitehall S.W.1, 18 March, 1935.' The official minutes are in AIR 20/181.
17. Ibid., Shearing's minute of 21 March 1935.
18. The minutes of the Tizard Committee are preserved in AIR 20/181, with the supporting papers in AIR 20/80 and 20/81. The minutes of the Swinton Committee are in CAB 16/132, with supporting papers in CAB 16/133 and 16/134. A more detailed and reasonably unbiased account of the Tizard and Swinton Committees will be found *inter alia* in Montgomery Hyde, *British Air Policy*, 322–33, which draws on published personal accounts by Churchill, Rowe and Watson-Watt, and on Clark's *Tizard*.
19. *History* II, para.134, n(3).
20. ADM 220/70. Signal School meeting, 1 Oct 1936, Enclosure C.
21. Watson–Watt, *Three Steps*, 123.
22. *History* II, para.170, n(3). For more detail on valve production and research, see Callick, *Metres to Microwaves*.
23. CAC, Wilkins papers. AWLK 1,27.
24. CAC, NRT ND 64. J.F. Coales, 'Naval RDF (radar), the formative years'.
25. ADM 220/70. Wimperis to Wright, 10 July 1935.
26. Ibid., DSD to CSS, 12 July 1935.
27. Ibid., CSS to DSD, 27 July 1935.
28. Ibid., DSD to CSS, 13 Aug 1935.
29. Ibid., Signal School meeting 1 Oct 1936, Encl. B & C.

2 1935–9: The Beginnings of the Navy's Radar (pages 11–29)

1. *History* II, paras.165, n.6. Yeo's notes can no longer be found.
2. CAC, Wilkins papers, AWLK 1,27.
3. ADM 220/70. Signal Dept meeting, 11 Oct 1935.
4. Ibid., Signal School meeting, 1 Oct 1936, Enclosure C.
5. Ibid., CSS to DSD, 28 Nov 1935.
6. CAB 16/132, f.31. ADR minutes, 9 Dec 1935.
7. Ratsey, 141. About 1972, O.L.Ratsey (one of the earliest Signal School scientists to be concerned with radar) compiled a history of the Signal School's Experimental Department and ASE, entitled *As We Were*. Never published, a copy at ARE Portsdown has been consulted and is hereafter cited as 'Ratsey'.
8. ADM 220/70. Signal School meeting 11 Oct 1935, Encl.B.
9. CAC NRT ND 64. Coales, 'Naval RDF . . . '.2215
10. ADM 220/70. Signal School meeting, 1 Oct 1936, with enclosures.

11. Ibid.
12. Ibid., CSS to DSD, 24 Dec 1936.
13. Ibid., 16 Feb 1937.
14. For the *Normandie* radar, see R.B. Molyneux-Berry in Burns, *Radar Development*, 46–8.
15. ADM 220/70. CSS to DSD, 16 Feb 1937.
16. ADM 220/70. Cecil minute, 27 July 1937, and Ratsey 141.
17. CAB 16/132, f.63. ADR minutes, 5 July 1937.
18. Bowen, *Radar Days*, 42–5, gives an eye-witness description of events. AIR 7/101 contains photographs of the radar scan during these trials.
19. ADM 220/70. CSS to DSD, 24 Aug 1937.
20. Ibid., Makeig-Jones to Wylie, 31 Aug 1937.
21. Ibid., CSS to DSD, 16 Sept 1937.
22. Ibid., DSD and DSR, No.SRE.1476/37 of 28 Oct 1937.
23. Coales, 'Naval RDF . . . '.
24. Ibid.
25. Ibid.
26. CAB 16/132, f.80, 15 May 1938.
27. ADM 220/70. Brundrett to Willett, 23 Oct 1937.
28. CAC NRT ND85. Coales, 'Preparing for War'.
29. Callick, *Metres to Microwaves*, 2–7.
30. ADM 220/70. Horton minute, May 1938.
31. Ibid., 'Detection of Aircraft and Ships by Radio Methods', enclosure to CSS to DSD, 22 Feb 1938.
32. Ibid., 2pp undated draft minutes of meeting, giving names of those attending; and DSD minute dated 31 May 1938 from docket SD.0345/38.
33. Ibid., DSD minute 'RDF development', 31 May 1938.
34. Ibid., CSS to Cdr Bowen of DSD, 25 Apr 1938.
35. Quoted in Rawlinson, J.D.S, *IEE Proceedings*, *132*, A (Oct 1985), 442. No copy of this important document (AL M/PD.06766/38 of 17 May 1938) has yet been found, but copies were almost certainly sent to other Commanders-in-Chief.
36. ADM 220/70, DSD paper 'RDF development', 31 May 1938, p.3.
37. CAB 16/132. ADR minutes, f.93, 14 Nov 1938.
38. ADM 220/70: AL SD.0345/38 of 16 Aug 1938 to C-in-C Portsmouth.
39. ADM 220/70, Head of CEI minute, 9 June 1938.
40. Rawlinson *Proc.IEE*, 132.
41. ADM 1/15193, CSS, 'RDF in HM Ships: a brief outline of progress', 26 July 1941.
42. ADM 116/3873. Combined Fleet Exercises 1939, serials II and XI.
43. ADM 220/70. CSS paper, 'Memorandum on expanded effort on RDF', 26 July 1939.
44. Ibid., Naval RDF Panel minutes, 27 July 1939.
45. Ibid., AL SD.0684/39 of 10 Aug 1939 to C-in-C Portsmouth.
46. CAC, NRT. Stainer to Howse, 9 March 1987.

3 1939–40: The Phoney War – But Not at Sea (pages 30–51)

1. CAC, NRT. Capt W.H. Watts to Howse, 23 Aug 1988.
2. ADM 220/71. *Rodney* to C-in-C HF, 2 Nov 1939.
3. For fleet bases, see Roskill, *War at sea*, i, 76–82.
4. ADM 277/1. Admiralty typescript, *Miscellaneous Weapon Development Department, Admiralty, 1940–45* (nd), p.5.
5. Watson-Watt, *Three Steps*, 304.

6. For an account of AMRE and TRE, see *History*, II, paras.95–121; for RRDE and ADRDE, see paras.176–205.
7. AIR 2/2942, minute 62a, note by Watson-Watt, 13 Sep 1939.
8. Ibid. minute 62b, note by Brundrett, 13 Sep 1939.
9. The Air Ministry side of the story is given in detail in AM File S.45641 in AIR 2/2942, and summarized in *History*, II, paras.92–4. The corresponding Admiralty files in ADM 1/10662 are very much thinner.
10. Roskill, *Hankey*, iii, 481ff.
11. CAC, SMVL 1/30. Somerville's pocket diary for 1939.
12. ADM 205/2. Churchill to Pound *et al.*, 22 Sep 1939.
13. AIR 2/2942, min.69b. Kingsley Wood to Churchill, 21 Sep 1939.
14. AIR 2/2942. Churchill to Kingsley Wood, 23 Sep 1939.
15. Cockcroft, J. D., 'Memories of radar research', *Proceedings IEE*, 132, A (Oct 1985), 327.
16. Somerville, 16 Sep 1939.
17. CAC, NRT, ND 86, J. F. Coales, 'The outbreak of war'.
18. CAC, NRT. Wright to Howse, 12 Feb 1989.
19. Ibid.
20. AVIA 46/36, No.11a. Note of Mr Jay's interview with Mr A.S.C.Phillips at ASE, 30 Oct 1946, f.4.
21. ADM 1/15193. 'RDF in HM Ships', 8.
22. ADM 220/71. Bawdsey's 24 May 1939 gives the full list of RDF code letters.
23. ADM 205/2. Somerville to Controller, 'Progress of RDF installation in HM Ships', 25 Sep 1939.
24. ADM 1/15193. CSS 'RDF in HM Ships', 5–6.
25. ADM 220/72. Minutes of RDF Policy Meeting, 15 Feb 1940.
26. ADM 220/74. Somerville to DNO and DNC, 'RDF Type SS (Type 281) – trials and first fitting' (SD.0143/40), 15 Feb 1940.
27. CAC, NRT. Rawlinson lecture text, 'The history of naval radar, 1935–1944', 12.
28. ADM 1/15193. CSS, 'RDF in HM Ships, 8.
29. CAC, NRT. Cdr F.C.Morgan, 1/87, vi–vii.
30. Roskill, letter to Editor, 'Radar for detecting surface ships', *Naval Review*, 67 (1979), 149.
31. Except where otherwise cited, most of the information in this section came from ADM 1/15193. CSS, 6–8, or from Coales op.cit ref.17.
32. Trenkle, *Die deutschen Funkmeßverfahren*, 116–7. I am grateful to Professor R. V. Jones for drawing my attention to this reference and to Herr Trenkle for his subsequent help.
33. Millington Drake, *The Drama of the Graf Spee*, 183.
34. ADM 1/9759. Naval Attaché, Buenos Aires, War Despatch No.4, 21 December 1939, Report by British prisoners from *Graf Spee*.
35. ADM 223/69. Decypher from Millington-Drake, 14 January 1940.
36. ADM 116/4472 provides most of the information in this account. We know that Bainbridge-Bell wrote at least five reports at the time (ADM 116/4472, NA Buenos Aires War Despatch No. 9, 15 Mar 1940 & DNI minute 13 Apr 1940), but a diligent search has failed to find any of them, presumably for security reasons. However a copy of his 'Report on visit to Montevideo to inspect the radio installation on the Control Tower in the "Admiral Graf Spee". January 1940', written two years after the event on 25 Jan 1942, giving the main technical details but none of the cloak-and-dagger aspects, is preserved in CAC NRT. IWM Misc.490 is a secondary account by Commander V. F. Smyth, an acquaintance of Bainbridge Bell's entitled 'Mr

Labouchère Bainbridge-Bell's Trip to Inspect the *Graf Spee*', dated 29 Apr 1988.
37. CAC, NRT. (SO(I) Montevideo to Admiralty for DNI 2235/7/2/40.
38. Ibid., P.C. Redgment to Howse, 4 Mar 1989.
39. ADM 116/472. Millington-Drake to FO No.84, 14 March 1940.
40. ADM 116/5475. Treasury to Admiralty, 9 October 1941.
41. Ibid., DSD minute, 14 December 1941.
42. Trenkle, *Funkmeßverfahren*
43. For the organization of British scientific intelligence, see Hinsley, *British Intelligence,* and Jones, *Most Secret War,* 618.
44. CAC, NRT. Searle, MS Recollections, 12 Aug 1987.
45. Ibid., Sir John Hodge, MS recollections, Feb 1989.

4 1940: Norway and the Fall of France (pages 52–71)

1. CAC, NRT. F.E. Butcher to Howse, 23 July 1987.
2. Sir John Hodge, personal communication.
3. Cdr F.C. Morgan, personal communication.
4. ADM 116/3873. Combined Fleet Exercises, Serial II, 1 Mar 1939.
5. Hezlet, *Electron,* 193.
6. CAC, NRT. Macdonald journal extracts, transcribed 16 Mar 1988.
7. HMS *Dryad* MSS, Coke to Schofield, 17 Aug 1973; private communications, Coke to Woolrych, 1987–9.
8. Ibid.
9. Bowen, *Radar Days,* 90.
10. CAC, NRT. Falloon, Fenwick, Ross, Shayler.
11. ADM 1/15193. CSS, 'RDF in HM ships – a brief outline of progress', 26 July 1941, 10.
12. Capt D.B.N. Mellis, personal communication.
13. ADM 220/86. CSS, trial results, 12 June 1940.
14. ADM 220/86. DSD to CSS, 18 July 1940.
15. ADM 220/77. DTSD memo 21 Sep 1940.
16. CAC, NRT. Admiral Sir Rae McKaig's recollections, 30 July 1987.
17. Ibid., Cdr F.C. Morgan No.1/87, xix–xx.
18. Hezlet, *Electron,* 194–5.
19. ADM 220/76. *Illustrious* to CSS, 27 July 1940.
20. Private communications, Coke, Pollock, Goodwin, Going, Sutton, Vincent-Jones to Woolrych, 1988–9.
21. Admiralty. BR 1736 (53) (1), *Naval Aviation,* I, 126 (ADM 234/383).
22. Ibid., 132.
23. As note 20.
24. D.L. P[ollock], 'Aircraft Direction during the War', *ND Bulletin 4* (June 1948), 10.
25. ADM 220/74. CSS to DSD, 'Trials of type 281 in HMS DIDO', 5 Nov 1940.
26. ADM 226/76. ASE to DNO, 24 Mar 1943.
27. Coales, Calpine & Watson, 'Naval Fire-control Radar', *Journal of the Institution of Electrical Engineers, 93* Part IIIA, No.2 (March–May 1946), 354.
28. Ibid.
29. Bowen, *Radar Days,* 143.
30. Coales's notebooks show that he visited Swanage between the end of August and the end of October, and J.R. Atkinson remembers that his visit was the first from Signal School.
31. CAC, NRT. Lovell to Howse, 9 Nov 1988.

32. Lovell, *Echoes of War*, 51.
33. RSRE MSS TRE 4/4/457. H. W. B. Skinner, 'Preliminary report on tests on ships using 10cm waves – November, December 1940', 20 Dec 1940. This paper is reproduced in full in Appendix A.
34. Much of this section has been drafted by C. A. Cochrane, a member of the Signal School team that went to Swanage in 1940. I am grateful also for the help of Sir Bernard Lovell and Sir Alan Hodgkin.
35. ADM 1/15193. CSS, RDF in HM ships', p. 13.
36. On the Tizard Mission, see Clark, *Tizard*, Chapter XI; Bowen, *Radar Days*, 150–95; Roskill, *Hankey*, ii, 475; and Roskill, *Churchill and the Admirals*, 109.

5 1941(1): Matapan and the Sinking of the *Bismarck* (pages 72–98)

1. Admiralty, BR 1736(48)(2), *Home Waters and Atlantic*, 276–8.
2. ADM 220/76. *Valiant* to RA 1st BS, 'Low angle trials of RDF short-range gunnery attachment', 10 Feb 1941.
3. Admiralty, *Home Waters*, 278–9.
4. ADM 220/86. May 1941.
5. Admiralty BR 1736(11), *Selected Mediterranean Convoys, 1941–42*, 11.
6. Pack, *Matapan*, 120.
7. ADM 1/11377, Battle of Cape Matapan, awards and decorations.
8. Ibid.
9. Quoted in Pack, *Matapan*, 77.
10. ADM 220/76. DSD to ASE, 6 Aug 1941.
11. Ibid.
12. Hezlet, *Electron*, 209.
13. Roskill, *War at Sea*, i, Map 36; and iii-2, Appendix T.
14. Churchill, *Second World War*, v, 6.
15. Ibid., iii. 106.
16. Roskill, *War at Sea*, i, 609.
17. ARE Archive Y. Admiralty CB 04110(41)/5, *Monthly report of progress in HM Signal School, May 1941.*
18. ADM 220/79. AL SDO.2053/41, 1 Oct 1941.
19. ADM 1/11065. *Walker* & *Vanoc*'s reports of proceedings under Captain (D) Liverpool's 22 March 1941.
20. Roskill, *War at Sea*, i, 476–7; Poolman, *Allied Escort Carriers*, 12; Brown, *Carrier Operations*, 42.
21. CAC, NRT. NRF1, 'XG2 Section of Admiralty Signals Establishment, November 1940 to November 1944', compiled by N. E. Davis.
22. *History*, para.353, written in 1946, cited the minutes of this important meeting as being in ASE File F1.271(I). In 1972–3, Ratsey reported them missing. Today, that file (now PRO ADM 220/78) starts in the middle of a paper of March 1941 and it is obvious that earlier papers have at some stage been removed.
23 ADM 220/78. DSD minute, 4 Jan 1941.
24. CAC NRT. Landale to Guy Hartcup, 23 March 1966. I am extremely grateful to Mr Hartcup for showing me this letter, written in connection with his book *The Challenge of War*.
25. ADM 220/78. CSS to DSD (personal), 29 March 1941.
26. CAC NRT. Orton to Howse, 7 Nov 1988.
27. Ibid., Raynsford to Howse, 9 March 1989.
28. ADM 1/11063. Landale & Hogben trial report, 5 May 1941.
29. ADM 220/78. *Orchis* to Captain (D) Greenock, 3 June 1941.

30. CAC NRT. Rawlinson text for lecture, 'The history of naval radar, 1935–1944'.
31. ADM 220/78. CinC Nore to Admiralty, 3 June & 21 Sept 1941.
32. Coales, 'The Outbreak of War.'
33. ADM 220/86. Drury to CSS, 17 May 1941.
34. Paddon, S. E., 'HMS Prince of Wales – radar officer', *Salty Dips*, i (NOAC, Ottawa, 1983).
35. ADM 220/76. H. J. W. Reeves manuscript report, 30 Apr 1941.
36. ADM 220/74. Horton, staff minute, 13 Feb 1941.
37. Ibid., Lockwood to Alexander, 5 April 1941.
38. Ibid., DSD to Controller, 8 Apr 1941.
39. CAC, ELLS 4/2. *Suffolk* to RA 1st CS, 11 June 1941, Encl.1.
40. CAC ELLS 3. R. M. Ellis unpublished autobiography, *When the rain's before the wind*, pp. 19.8 to 19.9.
41. Von Müllenheim-Rechberg, *Battleship Bismarck*, 1982 edition, 99–101.
42. Ellis, CAC ELLS 4/2.
43. Von Müllenheim-Rechberg, *Battleship Bismarck*, 101–2.
44. Ellis, CAC, ELLS 4/2.
45. Sieche, Erwin, 'German Naval Radar to 1945, Part 2', *Warship* 22 (1978), 146–8.
46. Von Müllenheim-Rechberg, *Battleship Bismarck*, 149.
47. Roskill, *War at Sea*, i, 404.
48. Paddon, HMS Prince of Wales,' 66.
49. NMM MS 88/100. R. Adm. A. D. Torlesse, 'History of a naval observer', Section 6(1), 4–5.
50. Von Müllenheim-Rechberg *Battleship Bismark*, 148.
51. Roskill, *War at Sea*, 409–18.
52. ADM 220/86. Wake-Walker to Warner personal, 12 June 1941.
53. ADM 1/11260. C-in-C HF's report, p. 17.
54. Ellis CAC ELLS 4/2 p. 15.7.
55. HMS *Dryad* MSS, Capt. D. G. Goodwin, 11 Apr 1983.

6 1941(2): The Battle of the Atlantic and Pearl Harbor (pages 99–127)

1. Admiralty, CAFO 186/41, 23 Jan 1941.
2. Admiralty, CAFO 1180/41, June 1941.
3. ADM 1/15193. CSS, 'RDF in HM Ships – a brief outline of progress', 26 July 1941, 14.
4. ADM 220/77. Minutes of meeting 22 May 1941.
5. ADM 220/81. Minutes of meeting 19 June 1941.
6. ADM 220/79. Gunnery/RDF staff requirements, 23 April 1943.
7. ADM 220/70. Minutes of Admiralty meeting, 11 Oct 1935.
8. Admiralty CB 04110/42(2), *ASE Monthly Report*, February 1942.
9. Lovell & Hurst, 'William Bennett Lewis', *Biographical Memoirs of Fellows of the Royal Society*, 34 (1988), 470–1.
10. ADM 220/76. Minute by H. M. Cecil ('B3'), 27 Mar 1941.
11. D. L. P[ollock], 'Aircraft direction during the war', *ND Bulletin*, 4 (June 1948), 13.
12. CAC, NRT. J. W. Sutherland reminiscences, 10 Aug 1987.
13. Admiralty, CB 04110/42(8), *ASE monthly report*, August 1942.
14. ADM 220/79. AL SDO.2053/41 of 1 Oct 1941.
15. Admiralty, CB 3090, *Instructions for installation . . .* (1943).
16. ADM 220/78. Chief Supt. ADRDE, report on Type 271X, no date but forwarded by CSS to DSD 15 Aug. 1941.
17. Ibid., Yates minute, 17 Oct 1941.

18. ADM 220/80. C-in-C HF to Admiralty, 23 Sept 1941.
19. Admiralty, *Selected Mediterranean Convoys*, 19.
20. Paddon, 'HMS *Prince of Wales*', 68.
21. ADM 220/80. C-in-C HF to Admiralty, 2 Nov 1941.
22. AFO 3368/40 of 5 Sept 1940.
23. Burton (ed.), *Canadian Naval Radar Officers*, 48.
24. ADM 220/78. Creasey to Willett, personal, 11 Sept 1941.
25. Beesly, *Very special intelligence*, 156. A good account of how Ultra affected naval operations is given in Winton, *Ultra at Sea*. See also Hinsley, *British Intelligence in the Second World War.*
26. ADM 220/78 Creasey to Willet.
27. Ibid.
28. Admiralty, CB 04050 series, *Monthly Anti-submarine Reports.*
29. CAC, NRT. Knight to Howse, 18 April 1989.
30. Dickens, Capt P., *Warship profile 20*, 6, 183–6.
31. ADM 1/12305. HMS *Marigold* – sinking of U.433.
32. Admiralty *Monthly Anti-Submarine Reports.*
33. Trenkle, *Funkmeßverfahren*, 130.
34. ADM 220/74. Various Naval Staff minutes in Docket NAD.998/41, Sept to Nov 1941.
35. For the wartime evolution of the Fighter Direction Branch, see Schofield, *HMS Dryad*, 93–108.
36. Roskill, *War at Sea*, i, 478–9; Poolman, *Allied escort Carriers*, 14–18.
37. Roskill, *War at Sea*, i, 479.
38. Paddon, 'HMS Prince of Wales', 72.
39. Ibid., 73.
40. ADM 1/12181. *Prince of Wales* and *Repulse* – reports from survivors, December 1941.
41. CAC, NRT. Armstrong to Howse, 3 Jan 1988.
42. Ibid., Fanning to Howse, 18 Jan 1988.
43. Ibid., Macdonald to Howse, 7 Feb 1988 & 6 Apr 1990.
44. Ibid., Butler to Howse, 9 July 1987.

7 1942: Malta Convoys and the Invasion of North Africa (pages 128–61)

1. Admiralty, BR 1736(7)(48), *Passage of the Scharnhorst, Gneisenau and Prinz Eugen.*
2. War Office (Sayer), *Army Radar*, 130, 147.
3. CAC, NRT. Fanning to Howse, 11 June 1989.
4. CAC, Willett papers. Memo, 22 April 1943, p. 2.
5. CAFO 861/43 of 29 April 1943.
6. Admiralty, CB 4497, *Simple Guide*, 23–25.
7. CAC, NRT. Morgan to Howse, 7 Sept 1989.
8. ADM 220/204. RDF Bulletin No.1, report 11.
9. CAC, NRT. Laws to Howse, 27 July 1989.
10. ADM 220/204. RDF Bulletin No. 1, report II.
11. Ibid.
12. Admiralty, CAFO 328/42, 19 Feb 1942, all ships except submarines, operational MTBs at home priority.
13. Bowen, *Radar Days*, 181.
14. Hezlet, *Electron*, 232.
15. Lovell, *Echoes of War*, 162.

16. I am grateful to Rear-Admiral P. W. Burnett for this description of HF/DF tactics.
17. Admiralty, *ASE Monthly Report*, January 1942.
18. Hezlet, *Electron*, 230.
19. For wartime HF/DF generally, see P. G. Redgment, 'HF/DF in the Royal Navy', *Journal of Naval Science*, 8 (1982), 32–43, 93–103.
20. ADM 220/204. RDF Bulletin No.1, report 3.
21. CAC, Willett papers. Creasey to Willett, 8 July 1942.
22. CAC, NRT. Yates memoirs for 1943.
23. Ibid., 1942.
24. C. A. Cochrane, private communication.
25. Ibid.
26. Pollock, 'Aircraft direction'.
27. Quoted in Pack, Sirte, 81–3.
28. Ibid., 134.
29. Admiralty BR 1736(11), *Selected Mediterranean Convoys*, 76.
30. AVIA 46/37, no.6. Interview with Young, 4 July 1945.
31. ADM 220/78. *Marigold* sea trials report, 22 May 1942.
32. ADM 220/80. ASE trials report, Landale and Yates, 18 July 1942.
33. Ibid., C-in-C HF to Admiralty, 5 Aug 1942.
34. Ibid., minute ASE to DSD, 16 July 1942.
35. CAC, Willett papers. Memo, 22 Apr 1943.
36. CAFO 67/43, 14 Jan 1943.
37. ADM 220/204. RDF Bulletin No.2, report No.2.
38. Paddon, 'HMS Prince of Wales', 75–6.
39. CAC, NRT. Hordern to Howse, 29 June 1989. Hamlet, III, iv.
40. NOAC, Ottawa, Salty Dips, i, 77.

8 Sicily, Salerno and the Sinking of the *Scharnhorst* (pages 162–90)

1. CAFO 2925/43, 1 July 1943.
2. CAFO 477/43, 11 March 1943.
3. ADM 220/79. Minutes of Admiralty meeting 'RDF priorities', 17 November 1942.
4. Ibid., ASE meeting 17 Dec 1942.
5. ADM 220/74. Minutes of Admiralty meeting, 8 Sept 1943.
6. Pollock, 'Aircraft Direction', 13.
7. ADM 220/204. Radar Bulletin No.5, Report No.3, August 1943.
8. Admiralty, BR 2435, *Technical Staff Monograph 1939–45*: Radar, 21.
9. ADM 220/204. Radar Bulletin No.5, Report No.2.
10. Pollock, 'Aircraft Direction', 14.
11. ADM 220/204 op.cit. Report No.1, July–Sept, 1943.
12. Ibid. Report No.3, August 1943.
13. Roskill, *The War at Sea*, iii(1), 30.
14. ADM 220/78. DSD minute SD.05577/42 of 14 Oct 1942, as amended 3 Nov.
15. Ibid.
16. ADM 220/204. Radar Bulletin No.2, Report No.9, Mar 1943.
17. CAFO 1590/43, 29 July 1943.
18. ADM 220/80 ASE Trials report, 18 July 1942.
19. ADM 220/81. Minutes of Haslemere meeting 'Requirements for future main armament RDF', 19 June 1941.
20. Ibid. Minutes of RDF/Gunnery staff requirements meeting VI, 16 October 1941.

21. Ibid.
22. ADM 220/228. Aust. & NZ Scientific Research Liaison, London, Report No.474, Sept 1943.
23. The document laying down the Staff Requirements has not been found. The figures quoted here for zone limits are from Admiralty, *Radar Manual*, and height accuracy from CAC NRT, Yates, Sydney Lecture No.1, para.41., 1945.
24. ADM 220/250. ASE paper 'Microwave fighter direction equipment', 13 March 1944.
25. Admiralty, *ASE Monthly Reports*, June to Dec 1943.
26. Winton, *Death of the Scharnhorst*, 80; Sieche, Erwin, 'German Naval Radar to 1945, Part 1', *Warship* 22 (1978), 8–10. There is, however, some doubt as to what radar was fitted at the time of *Scharnhorst*'s last sortie.
27. CAC NRT. Typescript, *Radar for the non-technical officer*, 115–6.
28. Ibid., 109.
29. In this account, I have used many sources including Admiralty, *Sinking of the Scharnhorst*, Roskill, War at Sea, iii(1), and Winton, *Death of the Scharnhorst.* But above all, I wish to thank Captain H. R. K. Bates, DSC, RN, for his help in criticizing and revising the draft.
30. CAC NRT. Bates to Howse, 17 June 1990.
31. ADM 1/16676. Sinking of the *Scharnhorst* – honours and awards.
32. Ibid.
33. Winton, *Death of the Scharnhorst.*
34. Quoted in Admiralty, *Sinking of the Scharnhorst.*
35. ADM 220/204. Radar Bulletin No.6, Report No.1, December 1943.
36. Ibid., Bulletin No.5, Report No.2, July 1943.
37. Ibid., Report No.1, July/September 1943.

9 1944: Normandy, Before and After (pages 191–231)

1. Roskill, *War at sea*, iii part 1, 308.
2. ADM 220/85. SO EG37 report, 26 Apr 1944.
3. Ibid., FO Levant & E. Med. to Admiralty, 31 Mar 1944.
4. Ibid., CRT to DTSD & DRE, 11 Sept 1944.
5. Ibid., Port Radar Officer Belfast to ASE, 14 Feb 1944.
6. Ibid., Cdre (D) HF to ASE, 23 Nov 1944.
7. HMS *Dryad* MSS. DRE to ASE, 13 July 1945, 'Reports by Commander Pollock', 18.
8. AFO 3208/45, 14 June 1945.
9. ADM 220/250. ASE paper 'Microwave fighter direction equipment', 13 March 1944.
10. Admiralty, AFO 1174/44, 9 Mar 1944.
11. Admiralty, BR 2435 – *Technical Staff History Monograph, Radar*, 39, 42.
12. Quoted in Roskill, *War at Sea*, iii, part 2, 13.
13. Ibid.
14. Roskill, *War at Sea*, chapters XIV and XV, 1–73.
15. CAC NRT. A. W. Ross – Reminiscences, Jan–Aug 1944 (March 1988).
16. Jones, *Most Secret War*, 405ff; Twinn, 'The use of Window (Chaff) to simulate the approach of a convoy of ships towards a coastline', in Burns, *Radar Development*, 416ff.
17. Jones, *Most Secret War*, 400ff.
18. Except where otherwise cited, the radar details in this section are based on Admiralty CB 004385 A, B & C (ADM 239/367, 368, 369), Report by ANCXF on operation 'Neptune' – from the signal and radar sections of the reports by

ANCXF (Ramsay), Commander Eastern Task Force (Vian), or the Naval Commanders of the British Forces 'S', 'G', and 'J' (Talbot, Douglas-Pennant, and Oliver).
19. Admiralty CB 004385A (ADM 239/367), Report by ANCXF, Appendix 10, para.104–120, pp. 115–7, gives further detail of RCM achievement.
20. CAC NRT. P. G. Redgment to Howse, 16 Feb 1990.
21. H. St. A. Malleson, 'The Decca Navigator system on D-day, 6 June 1944. An acid test', *Journal of Naval Science, 7, 4, (1981), 240–6.* C. Powell, 'Early history of the Decca Navigator system', *Journal of the Institute of Electrical and Radio Engineers*, 55, 6 (June 1985), 203–9.
22. E. M. B. Hoare, personal communication.
23. R. F. Hansford, 'The development of shipborne navigational radar', *Journal of the Institute of Navigation, 1*, 2 (April 1948), 121. On PPI prediction, see ADM 1/16013, ASE monographs M.666 and M.675.
24. ADM 1/16095. Minute by Staff FDO, Eastern Task Force, 14 June 1944.
25. Admiralty, CB 004385B (ADM 239/368), p.42, Enclosure No.9 to Report of Naval Commander Force 'J', para.14(b).
26. ADM 220/250. Aircraft Direction Bulletin No.3, Nov 1944, 10–12.
27. Hansford, 'Shipborne navigational radar', 123–4.
28. On the VT fuze, see Baldwin, *The Deadly Fuze.*
29. Knowles Middleton, *Radar development in Canada*, 63–5.
30. Admiralty BR 1736(42)(1), '*Neptune*', 8.
31. Roskill *War at Sea*, iii part 2, 114.
32. ADM 220/250 Aircraft Direction Bulletin No. 3 (November 1944), 5–6.
33. ADM 199/1094. *Bellona* to RA 1st CS, 17 November 1944. ADM 199/1094 contains the operation orders and reports of proceedings of operation 'Counterblast'.
34. Ibid.
35. CAC NRT. Lewin to Howse, 21 June 1989.
36. IWM MSS 86/11/1.

10 1945: The End of the War (pages 232–52)

1. ADM 199/337. Minute by DGAAWD, 9 March 1945.
2. Ibid. RA 1st CS to C-in-C HF, 26 Jan 1945.
3. Ibid. Capt (D) 17th DF to RA 1st CS, 17 Jan 1945.
4. Ibid. *Norfolk*'s gunnery report.
5. Ibid. *Apollo* to RA 1st CS, 12 Jan 1945.
6. Knowles Middleton, *Radar development in Canada*, 54.
7. ADM 220/216. 'Radar type 972'.
8. Hansford, 'Shipborne navigational radar', 124–6.
9. *Daily Sketch*, 15 Aug 1945, 5.
10. Yates, personal communication.
11. CAC, NRT. Yates, transcript of radar lecture no.1 given at Sydney on 11 July 1945, paras.9–12.
12. Ibid., para.33.
13. ADM 1/13247 – FD ships – draft staff requirements.
14. ADM 1/16373 – FD ships in assault operations.
15. ASE Monograph M.761 – Project Knobbly – Report of fitting out, operation, and sea trials in HMS *Boxer* (Jan 1946).
16. *ASE Bulletin*, Sept 1945 (RH.600(7)), 25.
17. Winton, *Forgotten Fleet*, 219–31, and Winton, *Haguro.*
18. ADM 199/116. Capt D.26 to VA 3rd BS, 18 May 1945.

19. ADM 116/5173.
20. HMS *Dryad* MSS. 'Reports by Commander Pollock', para.6(a).
21. Vian, *Action This Day*, 191.

11 Wartime Projects Postwar (pages 253–64)

1. For a good description of the development of British naval radar in the immediate postwar years, see Mitchell, Alastair, 'The Development of Radar in the Royal Navy 1945–60', *Warship, 17* (January 1981), 45–58. See also Friedman, *Naval Radar.*
2. Hansford, 'Shipborne navigational radar', 125–33.
3. ADM 220/21, 'ASE turn-over statement gunnery radar Cdr. A. F. P. Lewis to Lt.Cdr. F.C. Morgan, June 1946'. This file, together with ADM 220/222, 'Turn-over statement Cdr. F.C. Morgan to Cdr. T.W. Best, January 1949', gives an excellent summary of the development of gunnery radar between 1945 and 1949.
4. Part V Staff Requirements for LRS 1; detailed requirements for radar, dated 22 Feb 1945. This was issued by DGD under a cover note which expressed concern that the requirements might lead to excessive weight and complexity, but saying that this should be resisted.
5. AL GDO 2248/45 of 27 Aug 1945.
6. Mitchell 'Radar in the Royal Navy 1945–60', 47.
7. Admiralty, CB 4497, *Simple Guide to Naval Radar*, 8.
8. Ibid., 10; and ASE Monograph M.758 – C. A. Cochrane, *A description of types 960P, 980, and 981 and the radar display in the Action Information Centre*, January 1946 (not in PRO).
9. For general descriptions, see Mitchell, 'Radar in the Royal Navy 1945–60', 51, and Friedman, *Naval Radar*, 200. For more detailed information, see R. Benjamin, 'The post-war generation of tactical control systems', *Journal of Naval Science*, 15, 4 (1978), 263–76.

Appendix C Height Determination by Radar (pages 282–92)

1. CAC NRT. Hodge MS recollections, Feb 1989.
2. ADM 220/76. *Illustrious*, 'Interim Report on Directing Systems for Fleet Fighters', 24 July 1940.
3. Ibid. [A. W. Ross], 'Determination of the Height of an Aeroplane by R.D.F.', 30 July 1940.
4. Ibid., C-in-C HF to CSS, 29 June 1940; Ross minutes, 9 & 17 July 1940; CSS to C-in-C HF, 3 Aug 1940.
5. ASE Monograph M.382 (Oct 1941), [?F.Hoyle], 'Notes on the Vertical Polar Diagrams . . . '; also CAC NRT, Hoyle MS Recollections.
6. CAC NRT. Maynard to Howse, 12 Dec 1990 & 8 Feb 1991.
7. Ibid., *Victorious* to C-in-C HF, 22 July 1942, enclosed with Maynard, 12 Dec 1990.
8. Admiralty, CB 4224(42), *Heightfinding by RDF* (1942).
9. Admiralty, CB 4224(44), *Height-finding by Radar* (1944) refers to several examples. *Ajax* conceived a logarithmic plot, which once constructed allowed heights to be read from a series of standard curves. 15th Cruiser Squadron adopted the practice of adjusting the receiver gain control to maintain a constant echo level and drawing the corresponding contours on the height plot. *Pozarica* and *Malaya* each devised ingenious devices for determining

height at close range by exploiting fades through successive minima. Several other schemes had been proposed, for example by *Illustrious* (measuring rate of change of echo voltage) and *Palomares* (ratios of echo amplitudes from 79 and 281).
10. Ibid.
11. Ibid.
12. *Dryad* archive. D. Pollock, 'Aircraft Direction in the British Pacific Fleet', 13 June 1945; *Formidable* report on the operation of type 277, 2 Jan 1946; E.D.G. Lewin, 'Heightfinding by Radar', 15 Mar 1946.
13. Admiralty, CB 3180, *Heightfinding by Radar* (1949).
14. Ibid.

Appendix D Radar in Submarines (pages 293–300)

1. ADM 220/79. AL SDO 2053/41 of 1 October 1941, lists of ships fitted and to be fitted, p. 5.
2. Admiralty BR 1736(52), 2-*Mediterranean*, 60.
3. ADM 220/83. C-in-C Med signal to Admiralty, 2011B/21/12 1941.
4. Ibid. Admiral (Submarines) to Admiralty, 6 Jan 1942.
5. Admiralty, CB 3090, *Fitting of R.D.F. Equipment* (1943), 59.
6. CAC NRT. McIntosh to Howse, 1 Feb 1988.
7. CAC NRT. L.J. FitzGerald reminiscences, 12 Jan 1988.
8. ADM 220/204. Quoted in DSD, Radar Bulletin No. 4, 14 July 1943, p. 10.
9. Ibid., 11.
10. ADM 1/13476. Admiral (Submarines) to Captains (S), etc, 'The employment of radar in submarines', 3 July 1943.
11. CAC NRT. Hezlet to Howse, 24 Feb 1992.
12. Ibid.
13. Ibid.
14. ADM 220/83. ASE to DSD, 'Production dates for radar type 267W', 16 Aug 1943.
15. RN Submarine Museum A.1944/55. Admiral (Submarines) to Admiralty, 'Staff Requirements for Submarines', 19 June 1943.
16. ADM 220/83. Admiral (Submarines) to DSD, 7 July 1943.
17. Ibid. Admiral (Submarines) minute, 4 Sept 1943.
18. FitzGerald reminiscences, 12 Jan 1988.
19. CAC, NRT. McIntosh to Howse, 1 Feb 1988. See also ADM 199/1841.
20. Ibid., 22 Feb 1992.
21. ADM 1/13478. Admiral (Submarines) to Admiralty, 8 Sept 1943.
22. ADM 199/1867–8. *Tiptoe* and *Trump* patrol reports.
23. CAC NRT. Roxburgh to Howse, 12 Dec 1988. See also ADM 199/1845.
24. Ibid. Harris to Howse, Oct 1989.
25. Ibid. Hezlet to Howse, 24 Feb 1992.
26. Ibid. Dilley to Howse, 17 Oct 1989.

Appendix E Radar in Coastal Forces (pages 301–6)

1. Admiralty CB 04110(41)(1), *Monthly Report of Progress in HM Signal School*, Portsmouth, Jan 1941.
2. ADM 220/79. AL SDO.2053/41 of 1 Oct 1941.
3. Admiralty, CAFO 2295(F), 27 Nov 1941.
4. ADM 220/79, AL SDO. 2053/41 of 1 Oct 1941.

5. Admiralty, *Monthly Report*, June 1942.
6. ADM 199/268. Reports on Operations 'Gun' and 'Newt'.
7. J. Lennox Kerr & W.Granville, *The RNVR* (London, 1957), 199–202.
8. CAC NRT. L. C. Reynolds to Howse, 3 March 1992.
9. Knowles Middleton, *Radar Development in Canada*, 55.
10. Lovell, *Echoes of War*, 82–3.
11. Ibid.
12. Public Archives of Canada, NRC File 45.52.59, s.d. 27 August 1943, cited in Knowles Middleton, *Radar Development in Canada*, 55.
13. R. W. V. Board, private communication.
14. Admiralty, CB 04110(43)(10), *Admiralty Signal Establishment – Monthly Report of Progress in the Experimental Department*, Oct 1943.
15. Ibid., Jan 1945.
16. ADM 220/78. VA Dover's 0805A/9/9/41.
17. Admiralty BR 1736(42)(1), *'Neptune'*, 44.
18. CAC NRT. C. W. S. Dreyer to Howse, 3 March 1992.
19. Ibid. An appendix to Admiralty CB 03143, *Instructions for Coastal Forces Warfare*, describes the control ship technique in some detail.

Select Bibliography

General

By far the most important single primary source of information for this work has been the ADM 220 class of documents at the PRO, Kew, containing the surviving records of HM Signal School's Experimental Department and of ASE from 1934 to 1950. These, together with other Admiralty documents at the PRO, have been used to provide the frame of reference for the whole book.

To put flesh on this skeleton, the principal sources have been the communications from surviving Admiralty scientists and naval personnel – from Fellows of the Royal Society to laboratory assistants, from Admirals of the Fleet to Ordinary Seamen – those who actually designed, procured, used and maintained the wartime naval radar equipment. In 1986, the Naval Radar Trust began to collect monographs, reminiscences and recollections as well as primary material such as notebooks and diaries, and this was augmented after letters by the author to various journals and magazines in 1987. This archive is preserved in the Churchill Archive Centre, Churchill College, Cambridge, as Class 'NRT'.

Other repositories having significant quantities of primary material include the MoD's Naval Historical Branch in London, the Admiralty Research Establishment at Portsdown, and HMS *Collingwood* at Fareham.

Readers requiring more detailed contemporary information on user matters are recommended to consult the Admiralty's *Radar Manual (Use of Radar)* of 1945 (CB 4182/45, ADM 239/307). On technical matters, HMS *Collingwood* has a large collection of handbooks for wartime radar.

Secondary sources specific to British wartime naval radar are very sparse. Other than the Proceedings of the IEE's Radiolocation Convention of 1946 and a few monographs published in the Restricted MoD publication *Journal of Naval Science*, only the Admiralty's Technical Staff Monograph *Radar 1939–45* (1954, BR 2435, ADM 234/539), Hezlet's *The Electron and Sea Power* (1975), Friedman's *Naval Radar* (1981) and Callick's *Metres to Microwaves* (1990) go into any detail. However, the Naval Radar Trust intends to publish a volume containing a series of monographs on the more technical aspects to complement the present volume which is primarily addressed to the general reader.

Secondary sources for general naval history, many of which are listed below, are plentiful. Special mention must be made of Roskill's magisterial *War at Sea* (1954–61) and the excellent series of Naval Staff Histories and Battle Summaries in the Admiralty's BR 1736 series (in classes ADM 186 and ADM 234). However, as these were published before the lifting of security restrictions on the highly important wartime Special Intelligence ('Ultra'), works such as Beesly's *Very Special Intelligence* (1977), Winton's *Ultra at Sea* (1988) and Barnett's *Engage the Enemy more Closely* (1991) should be consulted.

Printed sources

ADMIRALTY, BR 1736 – *Naval Staff histories* (Select list)

(6) *Mediterranean, Selected Operations 1940* (1943) – ADM 234/325.

(7) *The Passage of the Scharnhorst, Gneisenau and Prinz Eugen Through the English Channel* (1948) – ADM 186/803.

(11) *Selected Convoys (Mediterranean), 1941–1942* (1957) – ADM 234/336.

(17) *Sinking of the Scharnhorst* (1950) – ADM 234/343.

(42) *Operation 'Neptune'* 2 vols (1947) – ADM 234/366–7.

(43) *Naval Operations in the Assault and Capture of Okinawa, March-June 1945 (Operation 'Iceberg')* (1950) – ADM 234/368.

(44) *Arctic Convoys, 1941–1945* (1954) – ADM 234/369.

(48/2) *Home Waters and Atlantic, April 1940* – December 1941 (1961) – ADM 234/372.

(50) *War with Japan – Vol. VI, The Advance on Japan* (1959) – ADM 234/379.

(51) *Defeat of the Enemy Attack on Shipping, 1939–1945* 2 vols. (1957) – ADM 234/578.

(52) *Submarines Vol. 1, Home & Atlantic; vol. 2, Mediterranean; vol. 3, Far East* (1953, 1955, 1956) – ADM 234/380–2.

(53) *The development of British Naval Aviation, 1919–1945.* 2 vols. (1954, 1956) – ADM 234/383–4.

—, Admiralty Fleet Orders (AFOs) and Confidential Admiralty Fleet Orders (CAFOs) – ADM 182 series.

—, CB 04050 series, *Monthly Anti-submarine Reports* – ADM 199/2057–62.

—, CB 04110 series, *HMSS/ASE Monthly Reports*, 1941–6 – copies ARE.

—, CB 04272 series, *Coastal Forces Periodic Review* – copies NHB.

—, CB 4224 (42), *Heightfinding by R.D.F.* (1942) – copy HMS *Collingwood.*

—, CB 04092/42, *Instructions for the Use of IFF Sets and RDF Beacons* (1942) – ADM 239/293.

—, CB 04092A/42, *Summary of RDF Identification (IFF, RDF Beacons and Interrogators)* (1942),- ADM239/294.

—, CB 04262, *Notes on the Direction of Fighters by HM Ships* (1942 & 1944–5) – ADM 239/352.

—, CB 3090, *Instructions for Installation and Fitting of R.D.F. Equipment and Associated Communications* (1943) – copy NHB & photocopy CAC; not found in PRO.

—, CB 04092/44, *Instructions for the Use of IFF Transponders and Radar Beacons by Allied Forces* (1944) – ADM 239/295.

—, CB 4224(44), *Height-finding by Radar* (1944) – copy HMS *Collingwood.*

—, CB 004385 A, B, C, *Report by the Allied Naval Commander-in-Chief Expeditionary Force on Operation 'Neptune'. (The Assault Phase of the Invasion of NW Europe, Operation 'Overlord'.)*, 3 vols. (Oct. 1944) – ADM 239/367.

—, CB 03143, *Instructions for Coastal Force Warfare* (1944), with appendix on Control Ship technique (1945) – ADM 239/220.

—, CB 4182/45, Radar Manual (Use of Radar) (1945) – ADM 239/307 & photocopy CAC.

—, CB 3180, *Height Determination by Radar* (1949) – copy HMS *Collingwood.*

—, CB 4497, *Simple Guide to Naval Radar* (1949) – copy NHB & photocopy CAC; not found in PRO.

—, BR 2435, ex-CB 3213, *Technical Staff Monograph: Radar 1939–45* (1954) – ADM 234/539 & photocopy CAC.

AIR MINISTRY (Air Historical Branch), Second World War, RAF Signals – Vol. IV (CD 1063), *Radar in Raid Reporting* (1950) – AIR 10/5519;

Vol. VI(SD736), *Radio in Maritime Warfare* (1954) – AIR 10/5555.

ALLISON, D.K., *New Eye for the Navy: the Origin of Radar in the Naval Research Laboratory* (Washington DC, NRL report 8466, 1981).
BALDWIN, Ralph B., *The Deadly Fuze. Secret Weapon of World War 2* (London, 1980).
BARNETT, Correlli, *Engage the Enemy More Closely: The Royal Navy in the Second World War* (London, 1991).
BASSETT, Ronald, *H.M.S. Sheffield: the Life and Times of 'Old Shiny'* (London, New York and Sydney, 1988).
BEESLY, Patrick, *Very Special Intelligence: the Story of the Admiralty's Operational Intelligence Centre* (London, 1977).
BLACKETT, P.M.S., see LOVELL.
BOWEN, E.G., *Radar Days* (Bristol, 1987).
BROWN, David, *Carrier Operations in World War II: Vol. 1, the Royal Navy* (London,1968, revised edition 1974).
—, *The Royal Navy and the Falklands War* (London, 1987).
BURNS, Russell (ed.), *Radar Development to 1945* (London, 1988).
BURRELL, Sir Henry, *Mermaids do Exist: the Autobiography of Vice-Admiral Sir Henry Burrell, Royal Australian Navy (Retired)* (Melbourne, 1986).
BURTON, E.F.(ed.), *Canadian Naval Radar Officers: the story of university graduates for whom preliminary training was given in the Department of Physics, University of Toronto* (Toronto, 1946).
CALLICK, E.B., *Metres to Microwaves: British development of active components of radar systems 1937 to 1944* (London, 1990).
CHURCHILL, Winston S., *The Second World War*, 6 vols (1948–54).
—, see also GILBERT.
CLARK, Ronald W., The *Rise of the Boffins* (London, 1962).
—, *Tizard* (London, 1965).
CLAYTON, Robert & ALGAR, Joan, *The GEC Research Laboratories, 1919–1984* (London, 1989).
CONNELL, G.G., *Valiant Quartet: His Majesty's Anti-aircraft Cruisers Curlew, Cairo, Calcutta and Coventry* (London, 1979).
CONWAY, *Conway's All the World's Fighting Ships, Part I, the Western Powers* (London, 1983).
COSTELLO, John & HUGHES, Terry, *The Battle of the Atlantic* (London, 1977).
CROWTHER, J.G. & WHIDDINGTON, R., *Science at War* (London, 1945).
CUNNINGHAM, Admiral of the Fleet Lord, *A Sailor's Odyssey* (London, 1951).
CUNNINGHAME GRAHAM, Angus, *Random Naval Recollections 1905–1951, Admiral Sir Angus Cunninghame Graham, KBE, CB, JP, of Gartmore* (Privately printed [1979], copy at CAC).
ELLIOTT, Peter, *Allied Escort Ships of World War II: a Complete Survey* (London, 1977).
FRIEDMAN, Norman, *Naval Radar* (Greenwich, 1981).
—, *The Postwar Naval Revolution* (London, 1986).
—, *British Carrier Aviation* (London, 1989).
GILBERT, Martin, Winston S. Churchill,
Vol. V – *1922–1939* (London, 1976),
Vol. VI – *Finest Hour, 1939–1941* (London, 1983),
Vol. VII – *Road to Victory, 1941–1945* (London, 1986).
GUERLAC, Henry E., *Radar in World War II*, 2 vols (New York, 1987).
HARTCUP, Guy, *The Challenge of War: Scientific and Engineering Contributions to World War Two* (Newton Abbot, 1970).
— & ALLIBONE, T.E., *Cockcroft and the Atom* (Bristol, 1981).
HEZLET, Sir Arthur, *The Submarine and Sea Power* (London, 1967).

—, *The Electron and Sea Power* (London, 1975).
HINSLEY, F. H. *et al.*, *British Intelligence in the Second World War*, 3 vols (1979, 1981, 1984).
HOLM, John, *No Place to Linger: Saga of a Wartime Atlantic Kiwi* (Wellington, NZ, 1985).
JONES, R. V., *Most Secret War* (London, 1978; citations from 1979 paperback edition).
KNOWLES MIDDLETON, W. E., *Radar Development in Canada: the Radio Branch of the National Research Council of Canada, 1939–1946* (Wilfred Laurier University, Ontario, 1981).
LOVELL, Bernard, *P. M. S. Blackett: a Biographical Memoir* (London, 1976).
—, *Echoes of War: The story of H_2S Radar* (Bristol, 1991).
—, see also SAWARD.
MACINTYRE, Donald, *U-boat Killer* (London, 1956).
—, *The Battle of the Atlantic* (London, 1961a).
—, *Fighting admiral: the Life of Admiral of the Fleet Sir James Somerville, GCB, GBE, DSO* (London, 1961b).
MILLINGTON DRAKE, Sir Eugene, *The Drama of the Graf Spee and the Battle of the Plate* (London, 1964).
MONTGOMERY HYDE, H., *British Air Policy Between the Wars, 1918–1939* (London, 1976).
OTTAWA, Naval Officers Association of Canada, *Salty Dips* Vol. 1 (Ottawa, 1983).
PACK, S. W. C., *Night Action Off Matapan* (London, 1972).
—, *The Battle of Sirte* (London, 1975).
PAGE, Robert Morris, *The Origin of Radar* (New York, 1962).
POOLMAN, Kenneth, *Escort Carriers*, 1941–1945 (London, 1972).
—, *Allied Escort Carriers in World War Two in Action* (London, 1988).
POPHAM, Hugh, *Into Wind* (London, 1969).
POSTAN, M. M., HAY, D. & SCOTT, J. D., *History of the Second World War – Design and Development of Weapons* (Chapter XV – The Development of Radar) (London, 1964).
POTTER, John Deane, *Fiasco: the Break-out of the German Battleships* (London, 1970).
PRICE, Alfred, *Instruments of Darkness: the History of Electronic Warfare*, 2nd edn. (London, 1977).
PRITCHARD, David, *The Radar War: Germany's Pioneering Achievements, 1905–1945* (Wellingborough, 1989).
RAVEN, A. & ROBERTS, J., *British Battleships in World War II* (London, 1976).
—, *British Cruisers in World War II* (London, 1980).
ROBERTSON, Terence, *Walker, R. N.* (London, 1958).
ROHWER, Jürgen, *The Critical Convoy Battles of March 1943: the Battle for HX.229/SC.122* (London, 1977).
ROSKILL, S. W., *The War at Sea 1939–1945*, 3 vols (London, 1954, 1956, 1960, 1961).
—, *Hankey*, 3 vols (London, 1970–4).
—, *Churchill and the Admirals* (London, 1977).
ROWE, A. P., *One Story of Radar* (Cambridge, 1948).
SAWARD, Dudley, *Bernard Lovell: a Biography* (London, 1984).
SCHOFIELD, Vice Admiral B. B., *The Loss of the Bismarck* (London, 1972).
—, *Navigation and Direction: the story of HMS Dryad* (Havant, 1977).
SCOTT, Peter, *The Battle of the Narrow Seas: a History of the Light Coastal Forces in the Channel and North Sea, 1939–1945* (London, 1945).
SMITH, Peter C., *Task Force 57: the British Pacific Fleet 1944–1945* (London, 1969).
SOMERVILLE, Sir James, see MACINTYRE.
SWORDS, S. S., *Technical History of the Beginnings of Radar* (London, 1986).

TAYLOR, Denis & WESTCOTT, C.H., *Principles of Radar* (Cambridge, 1948).
TIZARD, Sir Henry, see CLARK.
TRENKLE, Fritz, *Die deutschen Funkmeßverfahren bis 1945* (Heidelberg, 1986).
VIAN, Sir Philip, *Action This Day: a War Memoir* (London, 1960).
von MÜLLENHEIM-RECHBERG, Baron Burkard (SWEETMAN, J, tr.), *Battleship Bismarck: a Survivor's Story* (London, 1980: citations from Triad Grafton edition, 1982).
WAR OFFICE (Brig. A.P. Sayer), *Second World War, Army Radar* (1950).
WATSON-WATT, Sir Robert, *Three Steps to Victory* (London, 1957).
WHITLEY, M.J., *Destroyers of World War II: an International Encyclopedia* (London, Sydney, 1988).
WINTON, John, *The Forgotten Fleet* (London, 1969).
—, *Sink the Haguro* (London, 1978).
—, *Find, Fix and Strike* (London, 1980).
—, *The Death of the Scharnhorst* (Chichester & New York, 1983).
—, *Ultra at Sea* (London, 1988).
WOODWARD, Admiral Sandy, with ROBINSON, Patrick, *One Hundred Days: The Memoirs of the Falklands Battle Group Commander* (London, 1992).

Table of Type Numbers of Naval Radar Sets: Operational or Designed, 1935–45

AI	– Air interception	INT	– Interrogator
ASV	– Air to surface vessel	LA	– Low angle
BCN	– Beacon	Rg.	– In the ranging mode (types 279 and 281)
CCA	– Carrier Controlled Approach	Rx	– Receiver
FD	– Fighter direction	Sw.	– Sweeps
GA	– Gunnery fire control, aircraft, high angle (or combined low angle and high angle)	TI	– Target indication
GB	– Gunnery fire control, barrage	Tx	– Transmitter
GC	– Gunnery fire control, close range, high angle	WA	– Warning of aircraft
GS	– Gunnery fire control, surface (i.e. low angle)	WC	– Warning (combined aircraft and surface)
HA	– High angle	Wg.	– In the warning mode (types 279 and 281)
Ht-fndr.	– Height-finder	WS	– Warning of surface craft.
IFF	– Identification Friend or Foe	≈	– Variable around this frequency.

In the numbering of naval radar sets, the first, second and third major modifications to the basic set were indicated by the suffixes M, P and Q respectively, eg. type 286P. The suffix B indicated adaption to single-mast working.

Information not available is generally marked with a dash.

Type number	Classification	Wavelength nominal	Freq. MHz	Power kW	To sea (abandoned)	Description
79	WA	7.5m	39–42	70	1938	Long range air warning for large ships. See 279.
79B	WA	7.5m	39–42	70	1941	Single-masted version of 79 (originally 79M).
91	Jammer	50cm–3m	90–600	10–25W[1]	1941	Jamming of German metric and decimetric radar. Initially sine-wave modulation, ultimately noise.
241	INT	1.5m	214	–	1941	For use with 281 and IFF Mk.2N.
242	INT	1.5m	182 or 179	1	1943	For use with WS and WC sets and IFF Mk.3.
242M	INT	1.5m	182 or 179	2–10	1943	Ditto
243	INT	1.5m	179 or 171	1	1943	For use with 281 and IFF Mk.3.
244	INT	1.5m	–	–	1943	For use with US type SL and IFF Mk.3.
245	INT	1.5m	–	–	1944	For FD ships.
251/M/P	BCN	1.5m	176/177	7	1942	Modified RAF transponder coded to give ships' identity.
252	IFF	1.5m	38–52 & 195–220	–	1942	IFF Mk.2 in ships.
253/P	IFF	1.5m	Sw.157–187	10	1943	IFF in ships (Mk.3). Sweeps frequency.
253S	IFF	1.5m	Sw.157–187	10	1943	253 when fitted ashore.
255	BCN	1.5m	214	–	1944	Marker buoy for use only with 291.
256	BCN	1.5m	214	–	1944	Shore radar beacon for use only with 291.
257	CCA	3cm	–	–	1945?	Carrier controlled approach (BABS).
258	BCN	1.5m	179/182	–	1943	Shore radar beacon responding to 242.
259	BCN	1.5m	–	–	1944	Beacon for carriers responding to AI Mk.10.
261	WS	50cm	–	–	(1941)	Based on 282.
261W	WS	3cm	–	–	(1942)	Early 3cm development, leading to 267W.
262	GC	3cm	≈9,650	20	1945	STAAG, CRBFD.
263	GB	3cm	–	–	(1945)	Auto barrage for main or secondary armament, replacing 283.
267W/MW/PW	WS/WC	3cm & 1.5m	≈9,670 & 214	15–25 & 100	1945	Submarines. Hybrid WS/WC with common display.
268	WS	3cm	9,400	–	1945	For coastal forces, replacing 291U.

[1] Depending upon frequency.

cont. overleaf

Type number	Classification	Wavelength nominal	Freq. MHz	Power kW	To sea (abandoned)	Description
269	GS	3cm	10,000	–	(1943)	Modified 3cm AI set for coastal forces gunnery.
271/M/P	WS	10cm	≈3,000	5–10	1941	Small ships.
271Q	WS	10cm	≈3,000	70	1943	Small ships
272/M/P	WS	10cm	≈3,000	5–10	1941	Small cruisers, carriers, sloops, etc.
273/M/P	WS	10cm	≈3,000	5–10	1941	Large ships.
273Q	WS	10cm	≈3,000	70	1943	Large ships
274	GS	10cm	≈3,300	400	1944	Main armament directors, replacing 284.
275	GA	10cm	3,530	400	1945	HA directors (HA/LA directors in destroyers), replacing 285.
276	WS	10cm	3,000	500	1944	Small ships, replacing 271/2.
277/P/Q	WS	10cm	3,000	500	1944	Replaced 271/2/3. Could measure approximate elevation.
277S	WS/low air	10cm	3,000	500	1943	277 permanent shore installation for surface and low air.
277T	WS/low air	10cm	3,000	500	1943	Trailer-mounted 277. 'Monrads'.
279	WA	7.5m	39–42	Wg.70 Rg.60	1940	Long range air warning for large ships. Type 79 with gunnery ranging.
280	WA/GA	3.6m	82	25	1940	Based on Army GL1. In *Carlisle* and 'Bank' class ships only.
281	WA	3.5m	86–94	Wg. 600 Rg. 1,000	1940	Long range air warning for large ships.
281B	WA	3.5m	86–94	600	1943	Single-masted version of 281.
281BM/BP/BQ	WA	3.5m	86–94	350	1945	Continuous rotation.
282	GC	50cm	≈600	15	1941	Pom-pom directors, etc.
282M1/M2/M3	GC	50cm	≈600	60 or 80	1942	Increased power.
282M4	GC	50cm	≈600	60 or 80	1942	Beam switching.
282Q	GC	50cm	≈600	150	–	Beam switching and increased power.
283/M	GB	50cm	≈600	150	1943	Auto-barrage for main or secondary armaments.
284/M/P	GS	50cm	≈600	As 282	1940	Main armament directors.
285/M/P/Q	GA	50cm	≈600	As 282	1940	HA directors (HA/LA directors in destroyers).

Type number	*Classification*	*Wavelength nominal*	*Freq. MHz*	*Power kW*	*To sea (abandoned)*	*Description*
286M/P	WC	1.5m	214	7	1940	Small ships. 286M fixed aerial, 286P revolving aerial
286PQ	WC	1.5m	214	100	1943	Small ships. Higher power.
286U	WC	1.5m	214	7	1941	Coastal forces.
286W	WC	1.5m	214	7	1941	Submarines.
287	Minewatch	50cm	≈600	15	1941	284 adapted for minewatching ashore.
288(1)	GC	50cm	≈600	15	(1941)	284 adapted for Armed Merchant Cruisers.
288(2)	GC	50cm	≈600	15	–	284 adapted for training ashore.
289	GA	70cm	≈430	–	1940	Dutch. Fitted in *Isaac Sweers* and *Heemskerck* only.
290	WC	1.5m	214	100	1941	Small ships, replacing 286 but abandoned in favour of 291.
291/M	WC	1.5m	214	100	1941	Small ships, replacing 286/290.
291U	WC	1.5m	214	100	1943	Coastal forces.
291W	WC	1.5m	214	100	1943	Submarines.
293/M	WC/TI	10cm	3,000	500	1944	Destroyers and above. Replaced 271/2/3.
294	WC/FD	10cm	3,000	–	(1944)	Combined plan-display and heightfinding, replacing 277.
295	WC/FD	10cm	3,000	–	(1944)	Higher-powered 294.
650	Jammer	≈6m	≈50	10–20w	1944	Jamming of air-launched Fx.1400 and Hs.293 anti-ship guided weapons. Sine-wave modulation.
651	Jammer	≈6m	≈50	2–1kw	1944	As type 650 but capable of handling multi-missile attack. CW modulation.
930	GS/splash	10cm	3,000	7	1945	Splash-spotting; naval version of Army CA No.1, Mk.5 ('William')
931	GS/splash	1.25cm	≈24,000	–	1945	Splash-spotting, Canadian
940/1	INT	1.5m	209	1	1944	G-band interrogator with 281BP/BQ.
951	BCN	–	–	–	–	Marker beacon for use with 10cm WS and WC sets.
952	BCN	Rx 3cm Tx 1.5m	Rx≈9,400 Tx 182	–	1945	Portable combined ops. navigational. Triggered by X-band, response on type 242.
960	WA	3.4m	≈88	450	1946	Long-range air warning for large ships, replacing 281/79/279.

cont. overleaf

Type number	*Classification*	*Wavelength nominal*	*Freq. MHz*	*Power kW*	*To sea (abandoned)*	*Description*
961	CCA	3cm	≈9,320	–	–	Carrier controlled approach. Modified ASV II.
970	WS	10cm	≈3,300	–	1943	Combined operations. Modified RAF H_2S II.
971/M	WS	3cm	≈9,320	–	1945	As 970 but based on H_2S III.
972	WS	3cm	≈9,375	–	1946	Surveying.
980	WC/FD	10cm	3,000	500	(1949)	Fighter direction plan display, replacing 294/5.
981	Ht-fndr	10cm	3,000	500	(1949)	Fighter direction heightfinder, replacing 294/5.
990	WC	10cm	3,000	–	(1944)	Low cover, to go with 960, 294/5.
992	TI	10cm	3,000	–	1959	Target indication, replacing 293.
American sets fitted in British ships						
SA	WA	1.5m	–	–	1943	'Captain' and 'Colony' class frigates.
SG	WS	10cm	≈3,195	–	1943	*Indomitable, Victorious*, escort carriers.
SJ	WS	10cm	–	–	1945	Submarines Tiptoe and Trump.
SK	WA	1.5m	–	-	1943	Escort carriers.
SL	WS	10cm	–	–	1943	'Captain' and 'Colony' class frigates.
SM-1	FD	10.7cm	≈2,800	–	1944	Carriers *Indomitable, Ocean*; FD ships *Boxer, Palomares.*
SO	WS	10cm	–	–	1944	Coastal forces.
SQ	WS	12cm	–	–	1945	Big ships' portable 'after-action' set.
Naval Airborne Radar						
ASV Mk.II	ASV	1.5m	176	–	1940	Standard RAF version
ASV Mk.IIN	ASV	1.5m	214	7–22	1941	Naval version of ASV II. Swordfish, Walrus, Albacore, Barracuda
ASV Mk.XI	ASV	3cm	–	35–50	1943	Swordfish III, Barracuda III
ASB	ASV	1.5m	214	–	1944	US copy of ASV IIN. Avenger
ASH	ASV/AI	3cm	≈9,375	35	1944	US AN/APS-4. Avenger, Firefly. Barracuda V
AI Mk.IV	AI	1.5m	–	10	1944	Fulmar
AIA	AI	3cm	–	–	1945	US AN/APS-6. Hellcat

Principal source: Admiralty, CB 4497, *Simple Guide to Naval Radar* (1949).

General Index

Notes: Primary references to radio and radar equipments are gathered together in the 'Type Designation Index' on pages 378–83. Civilian staff of HM Signal School and Admiralty Signal Establishment are indicated by the terms HMSS, HMSS/ASE or ASE according to whether their period of service was before, through or after the change from HMSS to ASE (August 1941). Figure and diagram references are given by page numbers in **bold** type.

Type Designation Index (Radio and Radar Equipment)

Notes: brief descriptive notes are given in Table of Type Numbers, pp. 340–4; notes on nomenclature appear on pp. 13, 179–80, 293, 301.